Table of contents

Game Theory:
Mathematical Models of Conflict

Dedication:

To Tina

Talking of education, people have now a-days" (said he) "got a strange opinion that every thing should be taught by lectures. Now, I cannot see that lectures can do so much good as reading the books from which the lectures are taken. I know nothing that can be best taught by lectures, except where experiments are to be shewn. You may teach chymestry by lectures — You might teach making of shoes by lectures!"

James Boswell: *Life of Samuel Johnson, 1766* (1709-1784)

ABOUT OUR AUTHOR

Antonia J. Jones graduated from the University of Reading with a double first in mathematics and physics, and then gained a Ph.D. from Cambridge in number theory. During her distinguished career she has published numerous papers across a multi-faceted range of topics in number theory, genetic algorithms, neural networks, adaptive prediction, control and synchronisation of chaotic systems and other works on non-linear modelling and noise estimation. She has worked in mathematics and computer science at many renowned academic institutions in the UK and USA, including the Institute for Advanced Studies, Princeton N.J., University of Colorado, Boulder, Royal Holloway College, London, Imperial College, London, and the Computing Research Laboratory, NMSU Las Cruces. She is a Life Fellow of the Cambridge Philosphical Society.

She has helped to prepare artificial intelligence and neural network courses for Control Data Corporation and Bellcore, acted as consultant in advanced computing techniques for a number of major companies, has been an invited visitor to GMD, Cologne, Germany, the Oregon Graduate Institute, Weizmann Institute, Israel, Coppe Sistemas, UF Rio de Janeiro, and has a continuing research relationship with Laboratório de Inteligência Computacional, São Carlos-USP, Brazil.

Currently Professor of Evolutionary and Neural Computing at the University of Wales, Cardiff she has settled in a farm in the Brecon Beacons. The farm provides a peaceful environment for research and study and enables academic visitors from around the world to enjoy Welsh culture and the beauty of the mountains.

Game Theory:
Mathematical Models of Conflict

A.J. Jones
Professor of Computer Science
University of Cardiff

WP
WOODHEAD
PUBLISHING

Oxford   Cambridge   Philadelphia   New Delhi

Published by Woodhead Publishing Limited,
80 High Street, Sawston, Cambridge CB22 3HJ, UK
www.woodheadpublishing.com

Woodhead Publishing, 1518 Walnut Street, Suite 1100, Philadelphia,
PA 19102-3406, USA

Woodhead Publishing India Private Limited, G-2, Vardaan House, 7/28 Ansari Road,
Daryaganj, New Delhi – 110002, India
www.woodheadpublishingindia.com

First published in 2000 by Horwood Publishing Limited; reprinted 2004
Reprinted by Woodhead Publishing Limited, 2011

British Library Cataloguing in Publication Data
A catalogue record for this book is available from the British Library

ISBN 978-1-898563-14-3

Printed by Lightning Source.

LIST OF FIGURES

List of Figures

Author's Preface

This book originally evolved from a series of lectures in the theory of games which I gave some years ago at Royal Holloway College. I hope that it will serve to compliment the existing literature and perhaps encourage other teachers to experiment by offering a course in game theory. My experience has been that students are keen to learn the subject and, although it is not always what they expect, they enjoy it, provided one does not dwell on the technically more difficult proofs. One attractive aspect of the subject is that, unlike other quite pure mathematical disciplines, game theory offers room for interesting debates on, for example, what constitutes rational behaviour (the classical example being the Prisoner's Dilemma), how one might approach the general problem of social policy choice[1], and the nature of competitive markets in economic theory.

Before one is able to appreciate the virtue of proving things it is first necessary to have a firm intuitive grasp of the structure in question. With this objective in view I have written Chapter 1 with virtually no proofs at all. After working through this chapter the reader should have a firm grasp of the extensive form, the idea of a pure strategy, the normal form, the notion of a mixed strategy, and the minimax theorem, and be able to solve a simple two-person zero sum game starting from the rules. These ideas, particularly that of a pure strategy, are quite subtle, and the subtleties *can only* be grasped by working through several of the problems provided and making the usual mistakes.

Beyond Chapter 1 there are just two items which should be included in any course: the Nash[2] *equilibrium point* as a solution concept for a *n*-person non-cooperative game, which is introduced early in Chapter 2, and the brief discussion of utility, which is postponed until the beginning of Chapter 4. Granted these exceptions, Chapters 2-5 are logically independent and can be taken in any order or omitted entirely! Thus a variety of different courses can be

---

[1] Some very nice examples of such mathematical analyses can be found in *Game Theory and Social Choice - Selected papers of Kenjiro Nakamura*, Ed. Mitsuo Suzuki, Keiso Shobo Publishing Co. 2-23-15 Koraku, Bunkyo-ku, Tokyo, Japan, 1981. Unfortunately Kenjiro Nakamura, a graduate of the Tokyo Institute of Technology, died at the early age of 32. These selected papers are his gift to the subject to which he had devoted his life.

[2] An excellent account of the life of John Nash, who was awarded the Nobel Prize in Economics in 1994, is recently available in: *A Beautiful Mind*, Sylvia Nasar, Faber and Faber (Simon & Schuster in the USA), 1998, ISBN 0-571-17794-8).

assembled from the text, depending on the time available.

The mathematical background required to read most of the book is quite modest. Apart from the four items listed below, linear algebra, calculus and the idea of a continuous function from one Euclidean space to another will suffice. The exceptions are:

(i)   Section 2.4 where I have proved the results for compact topological spaces and the teacher is expected to rewrite the proofs for closed bounded subsets of Euclidean space if this is necessary.

(ii)   The proof of Theorem 2.2 which uses transfinite ordinals.

(iii)   The proof of Theorem 2.4 which uses the Brouwer fixed point theorem.

(iv)   The proof of Theorem 5.3 which uses the Kakutani fixed point theorem.

In the last three cases no loss of understanding results from omitting the proof in question.

Some teachers may be displeased with me for including fairly detailed solutions to the problems, but I remain unrepentant. Once the ideas have been grasped, numbers can be changed and slight variations introduced if this seems desirable. In my view any author of a mathematical text book should be *required* by the editor to produce detailed solutions for all the problems set, and these should be included where space permits. The advantage of such a practice will be obvious to anyone who has struggled for hours with an incorrectly worded or unrealistically difficult problem from a book without solutions. Moreover, even where the author has been conscientious in solving the problems, for students working alone detailed solutions are an essential check on their own understanding.

It is too much to hope that I have managed to produce an error-free text, but several successive years of critical students have been a great help in this respect.

ACKNOWLEDGEMENTS

The original acknowledgements stand, together with a sincere apology to Professor K. Binmore for the inadvertent inclusion, without acknowledgement, in the original version of some material from one of his papers.

The author duly acknowledges permission to reproduce the following material:

Two passages from R.D. Luce and H. Raiffa: *Games and Decisions*, J. Wiley and Sons (1957), New York.

The section on simplified Poker from H.W. Kuhn and A.W. Tucker: *Contributions to the Theory of Games*, Vol I. (Annals of Mathematical Studies, No. 24), Princeton University Press (1950), Princeton.

Problems 8, 10, 11, 12, 13 (Chapter 2) and their solutions from Robert R. Singleton and William F. Tyndall, *Games and Programs, Mathematics for modelling*, W.H. Freeman and Comp. (1974), San Francisco.

Exercise 3 (Chapter II) and Exercises 1, 7 (Chapter 4) from Kathleen Trustrum, *Linear Programming*, Routledge and Kegan Paul Ltd. (1971), London.

Thanks are due to many others. In particular: Dr. Alban P. M. Tsui who designed and produced such an attractive cover, Mary Carpenter who, some years ago, courageously produced the source files for the text in WordPerfect 5.1; and Christina Thomas who keeps the 'research farm' functioning; in fact nothing of any significance would get accomplished without her.

Antonia J. Jones
Brecon Beacons
Wales
1999

Glossary of Symbols

Free use is made of the logical quantifiers $\exists$ (there exists) and $\forall$ (for every) and of basic set notation. Thus $x \in S$ means x is an element of the set S (read: x belongs to S), $x \notin$ means x does not belong to S, and $S \subseteq T$ (S is a subset of T) means $x \in S$ implies $x \in T$. Classifier brackets $\{\ :\ \}$ are used throughout to specify a set. For example

$$S \cup T = \{x: x \in S \text{ or } x \in T\} \qquad \textbf{(union)},$$

is read as: S union T equals the set of all x such that $x \in S$ or $x \in T$. The commonly used set notations are

$$\phi = \text{the set with no elements} \qquad (\textbf{the empty set}),$$
$$S \cap T = \{x; x \in S \text{ and } x \in T\} \qquad (\textbf{intersect}),$$
$$S_1 \times ... \times S_n = \{(x_1, ..., x_n)\ ;\ x_i \in S_i\ (1 \leq i \leq n)\}, \qquad (\textbf{cartesian product}),$$
$$S \setminus T = \{x; x \in S \text{ and } x \notin T\} \qquad (\textbf{difference}),$$
$$\wp(S) = \{x\ ;\ x \subseteq S\} \qquad (\textbf{power set}),$$
$$|S| = \text{the number of elements in S} \qquad (\textbf{cardinal}).$$

For sets of numbers we use

$$N = \{1,\ 2,\ 3,\ ...\} \qquad (\textbf{the natural numbers}),$$
$$R = \text{the set of real numbers},$$
$$|N| = \aleph_0 \qquad (\text{aleph zero, the } \textbf{first transfinite cardinal}),$$

and for $x \in R$

$$|x| = \max\{x,\ -x\} \qquad (\textbf{absolute value}).$$

The set of ordered n-tuples $(x_1, ..., x_n)$, $x_i \in R$ $(1 \leq i \leq n)$, the cartesian product of n copies of R, is denoted by R^n. Elements of this set are n-dimensional vectors and are denoted by bold faced type

$$\mathbf{x} = (x_1, ..., x_n).$$

The **inner product** of two vectors $\mathbf{x} = (x_1, ..., x_n)$, $\mathbf{y} = (y_1, ..., y_n)$ in R^n is written as

$$\mathbf{x} \cdot \mathbf{y} = x_1 y_1 + \ldots + x_n y_n,$$

or as $\mathbf{x}\,\mathbf{y}^{\mathrm{T}}$ using the transpose (defined on page 101) and matrix multiplication.

Other vector notation includes the use of

$$\mathbf{J}_n = (1, 1, \ldots 1)$$

for the n-dimensional vector with all components equal to 1, and $\mathbf{x} \geq \mathbf{y}$, where $\mathbf{x}, \mathbf{y} \in \mathbf{R}^n$, for $x_i \geq y_i$ for all i, $1 \leq i \leq n$.

We also use the sum and product notation

$$\sum_{i=1}^{n} x_i = x_1 + x_2 + \ldots + x_n \,,$$

$$\prod_{i=1}^{n} x_i = x_1 x_2 \ldots x_n \,,$$

and

$$\binom{n}{r} = \frac{n!}{r!\,(n-r)!} \qquad (\text{read: } n \text{ choose } r)$$

for the number of ways of choosing r objects from n objects not counting different orderings of the r objects.

The following game theoretic notations are frequently employed:

2-person zero sum games

> σ, τ pure strategies,
> $\mathbf{x}$ a mixed strategy for player 1,
> X or $\Sigma^{(1)}$ set of all mixed strategies for player 1,
> $\mathbf{y}$ a mixed strategy for player 2,
> Y or $\Sigma^{(2)}$ set of all mixed strategies for player 2,
> $P(\mathbf{x}, \mathbf{y})$ payoff to player 1.

n-person non-cooperative games

> σ_i a pure strategy for player i,
> S_i set of all pure strategies for player i,
> x_i a mixed strategy for player i,
> X_i set of all mixed strategies for player i,
> $P_i(x_1, \ldots, x_n)$ payoff to player i,
> If $x = (x_1, \ldots, x_n)$, $x_i \in X_i$ ($1 \leq i \leq n$), then $x \parallel x_i' = (x_1, \ldots, x_{i-1}, x_i', x_{i+1}, x_n)$.

n-person cooperative games

$I = \{1, 2, ..., n\}$ the set of players,

$S, T \subseteq I$ coalitions S, T.

S_T set of all coordinated pure strategies for coalition T.

X_T set of all coordinated mixed strategies for coaltion T.

$\mathbf{x} \succ \mathbf{y}$ imputation $\mathbf{x}$ dominates imputation $\mathbf{y}$ (defined on page 182).

1

The name of the game

He thinks that I think that he thinks...

1.1 INTRODUCTION.

Our aim is to study various problems of conflict by abstracting common strategic features for study in theoretical 'models' that are termed 'games' because they are patterned on actual games such as bridge and poker. By stressing strategic aspects, that is aspects controlled by the participants, the theory of games goes beyond the classical theory of probability, in which the treatment of games is limited to aspects of pure chance. First broached in 1921 by E. Borel [1] the theory was established in 1928 by J. von Neumann who went on with O. Morgenstern [2] to lay the foundations for virtually the whole of what is now called the mathematical theory of games.

The origins of game *playing* are lost forever in the mists of time. Psychologists speculate that games, which enable children to practise roles and decision processes in relative security, are an essential feature of the learning process. It seems likely that many adult games were originally conceived as training aids for warfare. For games such as chess the military origins are clear, but could it be[3] that backgammon was designed to train tribesman in the art of filtering through a hostile tribe in ones and twos? Certainly a useful strategy in many common forms of backgammon is the 'buddy system' - never move alone. Whatever, the truth of the matter, there is a rich field for investigation in the historical origins of many common games, and we can probably learn a lot about a society by studying the games it invents.

There is no mystery, however, about the origins of game *theory*. Game theory is the mathematical study of conflict situations, and it was, to all intents and purposes, born in 1944 with the publication of a single book *Theory of Games and Economic Behaviour* by J. von Neumann and O. Morgenstern. Unfortunately that book is singularly difficult to read and, apart from the simplest case of the finite two person zero sum game, it was a long time before the contents became available to a wider audience. These days a comprehensive course in game theory is no more mathematically demanding than a similar course in classical mechanics. An exception is the theory of *differential games*, a topic founded by R. Isaacs and which is only lightly touched upon in this book.

---

[3] As suggested to me by W. Haythorn.

One does not expect to see a completely developed theory leap into existence with the publication of a single book, and with the hindsight of a further fifty odd years of development it is now clear that this did not in fact happen; von Neumann and Morgenstern made a monumental start, but only a start. Nevertheless to the economists and social scientists of the time it must have seemed that the answer to their prayers had magically appeared overnight, and game theory was hailed with great enthusiasm at its birth [3]. It opened up to mathematical attack a wide range of gamelike problems not only in economics but also in sociology, psychology, politics and war. A striking conceptual achievement of game theory has been to clarify the logical difficulties inherent in any rational analysis of many conflict situations.

In practical situations we might reasonably ask what individuals *usually* do when they find themselves in a gamelike situation. What to do in these situations is by no means transparent. See, for example, the highly enjoyable book *Beat the Dealer* by E. Thorp [4]. For practical decision making, it is clearly this approach which is important. The question asked by a game theorist is very different. It is:

- What would each player do, if all the players were doing as well for themselves as they possibly could?

The extent to which this question has been answered in practical applications is, we must frankly admit, sharply limited. Where it is not possible to give a full analysis, perhaps because of the amount of computation involved, analysing miniature or simplified forms of games can often throw light onto the original game. For example an exciting game of poker usually involves at least four or five players; but pot-limit, 2-person straight poker can be convincingly analysed in game-theoretic terms [5].[4] As another example, a convincing analysis of Monopoly would be of serious interest to economists.

The fact is, as we shall see in later chapters, in many contexts the very meaning of the question poses considerable philosophical difficulties and, even where its meaning can be clarified, the technical problems of seeking an answer are seldom easy. Thus the difficulties we can expect to encounter range from the practical impossibility of performing staggeringly vast computations to basic questions of philosophy. Nevertheless the question must be grappled with directly if progress is to be made.

1.2 EXTENSIVE FORMS AND PURE STRATEGIES

From the rules of any game one can proceed to several levels of abstraction. The first level, the extensive form of a game, eliminates all features which refer specifically to the means of playing it (for example, cards or dice). Thus the **extensive form** amounts to a literal translation of the rules into the technical terms of a formal system designed to describe all games. For the present we consider only 2-person games, but the ideas of this section apply quite naturally to any game.

Traditionally the game of *Nim* is played as follows. There are several piles of matches and two

---

[4] See Appendix 2 for poker terminology.

players who take alternate turns. At each turn the player selects a pile and removes at least one match from this pile. The players may take as many matches as they wish but only from the one pile. When all the matches have been removed there are no more turns; the player who removes the last match loses. This game is capable of complete analysis, and interested readers who wish to baffle their friends can find the details in Hardy and Wright [6]. Consider the following simplified version.

Example 1.1 (2-2 Nim). The game of 2-2 Nim is adequately described by the following rules:
Initial state: Four matches are set out in 2 piles of 2 matches each.
Admissible transformations: Two players take alternate turns. At each turn the player selects a pile that has at least one match and removes at least one match from this pile. He or she may take several matches, but only from one pile.
Terminal state: When both piles have no matches there are no more turns and the game is ended.
Outcome: The player who removes the last match loses.

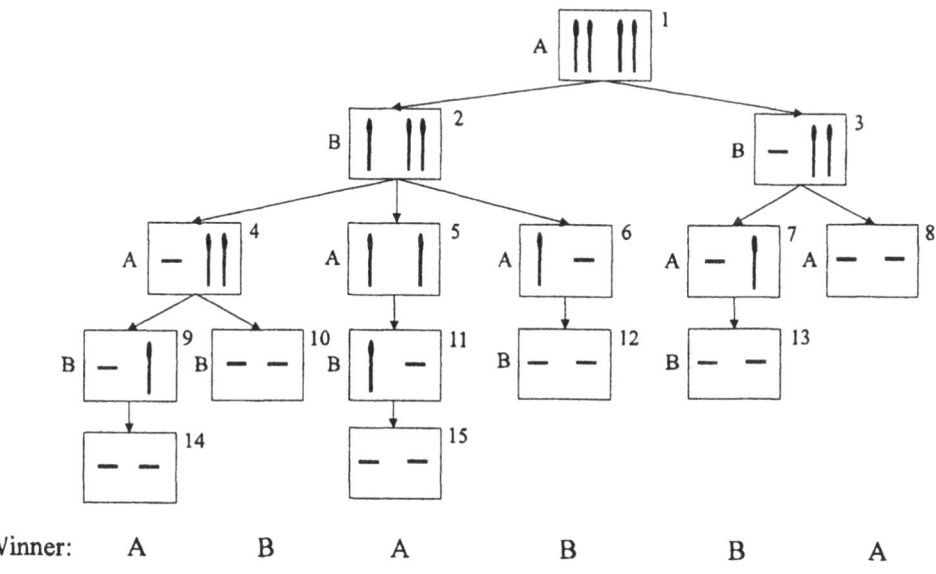

Figure 1.1 Game tree for 2-2 Nim.

The term **state** is used here in a precise technical sense, and a few words of explanation are in order. One is well aware that in a game such as chess the same position can be reached in a variety of different ways. In game theory we do not regard these same positions as identical states. That is to say, a state is determined not merely by the current position but also by the unique history of play which led to that position.

The conventional way of representing a game in extensive form is to use a 'game tree'. The states and their admissible transformations are represented in a decision tree. If the game tree faithfully reproduces each possible state, together with the possible decisions leading from it, and each possible outcome, it is said to represent the game in extensive form. Figure 1.1 represents 2-2 Nim in extensive form as a game tree.[5]

The boxes (that is, nodes or vertices) represent each state and occur at each point at which a player has to make a move or decision (that is, an admissible transformation). The interiors of the boxes in this example represent the piles of matches, the capital letter beside the box is the name of the player (A or B) whose turn it is to choose and the lines lead to the next decision nodes which can be reached from the current state. We have numbered the states for convenience.

Note at state 1 we have made use of symmetry to simplify the tree since A actually has four choices but the rules draw no distinction between the two piles; similarly at state 5.

It is important to realise that because of our convention about states *the situation illustrated in Figure 1.2 never occurs in a correctly drawn game tree.* In a tree the branches divide but never rejoin.

With the game tree before us it is now easy to do a 'bottom up' analysis. For example at state 14 A wins, so at state 9 B has already lost, so at state 4 A will choose to go to 9, so state 4 is already a winning position for A etc. In this way we discover that A has lost the game before it starts, for a 'winning strategy' for B is: if at state 2 go to state 6, if at state 3 go to state 7.

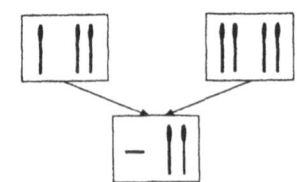

Figure 1.2 This does not happen.

2-2 Nim is a game in which each player knows at all times the precise state of the game and therefore the past history of the play; such a game is said to be of **perfect information**. Of course in many games a vital feature is that one or more of the players does not have perfect information, and in such cases it is essential that these aspects are precisely represented in the extensive form. There is a simple way of doing this as our next example illustrates.

> **Example 1.2** (Renée v Peter). This game is played with three cards: King, Ten and Deuce (two). Renée chooses one and lays it face down on the table. Peter calls 'High' or 'Low'. If he is right (King = High, Deuce = Low), he wins £3 from Renée, but if wrong he loses £2. If the card is the Ten he wins £2 if called Low, but Renée must choose between the King and the Deuce if he called High. This time after she lays her card face down on the table and he calls High or Low, he wins £1 if right but loses £3 if wrong.

Put R = Renée, P = Peter, K = King, D = Deuce, Hi = High, Lo = Low. The game tree is given in Figure 1.3.

---

[5]Following established tradition all trees are drawn with their roots at the top.

On his first move Peter is at one of three possible states, depending upon which card Renée chooses, but he doesn't know which. Similarly on his second move, if it occurs, he is at one of two possible states but again doesn't know which. In each case the shaded region (in complicated game trees such nodes can simply be joined by a dotted line) is called an **information set** (or more aptly an ignorance set). It indicates that when Peter makes his choice he is aware of the information set he is in but he does not know the precise node within this set. Irrespective of

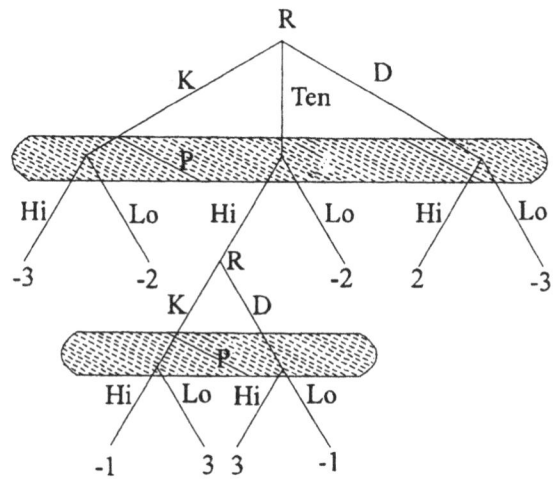

Figure 1.3 Game tree for Renee v Peter.

the node he is actually at in the information set his choices are the same.

The numbers at each terminal state in the game tree represent the payoffs to Renée which derive from the outcome. It is a tacit convention when drawing a game tree to indicate the payoffs to the first player (except if the game is not zero sum or involves more than two players: the conventions for these cases will be explained later).

Neither of the games we have considered involved any appeal to chance; no coin was tossed, no die was thrown and no card was dealt unseen by the dealer. In games with chance elements it is convenient to think of 'Nature' as a player and label the corresponding nodes of the game tree with an N as they represent states at which an appeal to chance is made. Here is a simple game of imperfect information and having chance elements.

> **Example 1.3** (Red-Black). A is dealt face down a red or black card. He or she looks at the card but does not reveal to the other player, B, whether it is red or black. If the card is red, A can either call for the toss of a coin or pass the move to B. If the card is black, A must pass the move to B. If and when B gets the move he or she then has to guess whether A's card is red or black. Otherwise the coin is tossed and the game is now ended. The payoffs for the various possible terminal states are given in the game tree of Figure 1.4.

In this figure 'red(½)' means Nature chooses a red card with probability ½ etc. Each decision line emerging from a node corresponding to Nature should always be labelled with the appropriate probability.

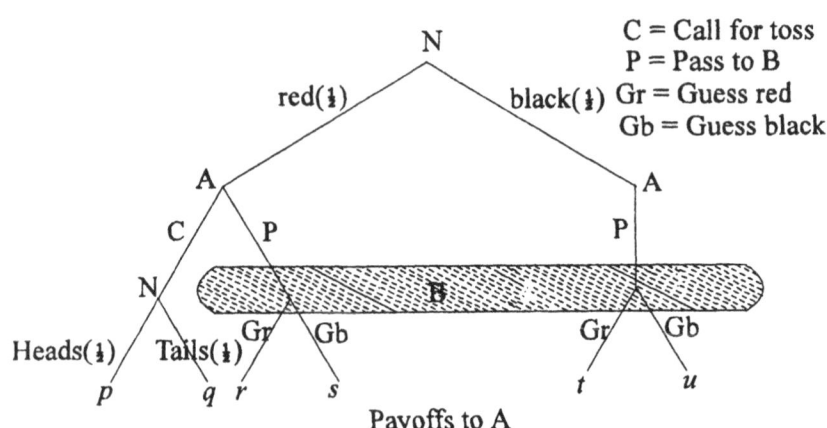

Figure 1.4 Game tree for Red-Black.

Table 1-1
Pure strategies for 2-2 Nim

	Pure strategies for A				*Pure strategies for B*		
Name of strategy	First turn	Second turn		Name of strategy	First turn		
		If at	Go to			If at	Go to
AI	$1 \to 2$	4	9	BI		2	4
						3	7
AII	$1 \to 2$	4	10	BII		2	5
						3	7
AIII	$1 \to 3$	-	-	BIII		2	6
						3	7
				BIV		2	4
						3	8
				BV		2	5
						3	8
				BVI		2	6
						3	8

Of the three games so far examined we have 'solved' only the first. Games of imperfect information are not susceptible to the simple analysis of Example 1.1. Our next step is to examine carefully the idea of a strategy. A **pure strategy** (the reason for the word 'pure' will become apparent in §1.4) is a prior, comprehensive statement of the choice to be made *at each decision point the decision maker might possibly meet.*

Using the game tree in Figure 1.1 we can draw up a table of all the pure strategies for both players in 2-2 Nim. This is given in Table 1-1.

The reader is urged to check Table 1-1 carefully. Since the pure strategies are abstracted from the game tree any mistake in drawing up the tree will result in the table of pure strategies itself being wrong and subsequent analysis based on the table will then be incorrect.[6]

Table 1.2
Pure strategies for Renée v Peter

Renée's pure strategies	Peter's pure strategies
I. Choose K.	I. Choose Hi; if again choose Hi.
II. Choose Ten; if again choose K.	II. Choose Hi; if again choose Lo.
III. Choose Ten; if again choose D.	III. Choose Lo.
IV. Choose D.	

In a similar way for Example 1.2 we obtain the pure strategy Table 1-2.

For games with chance elements, since Nature is nominally a player, it is often (although not invariably - see Example 1.5) helpful to list her 'pure strategies' in addition to those of the other players. Thus for Example 1.3 we can draw up Table 1-3.

Table 1.3
Pure strategies for Red-Black

Player A	Player B	Nature
AI If red then C	BI If P then Gr	NI Choose red; if C choose Heads(¼)
AII If red then P	BII If P then Gb	NII Choose red; if C choose Tails(¼)
		NIII Choose black. (½)

---

[6]In the author's experience the most common source of error is incorrect delineation of information sets in the game tree.

Notice we do not have to include in A's strategy statements the phrase 'If black then P' since this is forced by the rules; that is, in this case A has no decision to make. The numbers after each of Nature's strategies represent the overall probability of that strategy being chosen.

1.3 NORMAL FORMS AND SADDLE POINTS.

Given the rules of any game we are now able *in principle* to draw up the extensive form and list the pure strategies. Of course for any realistic game (such as chess) this is likely to be an impossibly arduous task, but this doesn't matter if we are trying to get to grips with the underlying ideas.

With a comprehensive list of pure strategies to hand, *any particular complete play of the game can be represented as a combination of one pure strategy from each player's list.*

Thus for 2-2 Nim we see from Table 1-1 that the strategy choices

AI and BI	yield the play $1 \to 2 \to 4 \to 9 \to 14$,
AII and BI	yield the play $1 \to 2 \to 4 \to 10$,
AIII and BII	yield the play $1 \to 3 \to 7 \to 13$, etc.

This property of pure strategies leads to another representation of the game called the *normal form*. This is an even more condensed version of the game, stripped of all features but the choice for each player of their pure strategies. We list the pure strategies for A (the first player) as rows and those for B as columns. Then since a specific pure strategy for A and one for B uniquely determine a play, and hence uniquely determine an outcome, we can list the outcome resulting from A's i^{th} pure strategy and B's j^{th} pure strategy in the (i,j) position; that is, put an A in the (i,j) position if A wins. For 2-2 Nim, in this way we obtain Table 1-4.

Table 1-4
Normal form of 2-2 Nim

		B					
		I	II	III	IV	V	VI
A	I	A(14)	A(15)	B(12)	A(14)	A(15)	B(12)
	II	B(10)	A(15)	B(12)	B(10)	A(15)	B(12)
	III	B(13)	B(13)	B(13)	A(8)	A(8)	A(8)

Here the numbers in brackets indicate the corresponding terminal state and have been inserted to facilitate checking (strictly speaking they are superfluous). Notice the column of B's under BIII which tells us this is a winning pure strategy for B (as we observed earlier).

Suppose now, for any game, we have found from the extensive form some rule such as Table 1-4 which, given each player's choice of pure strategy, determines the outcome. We can then define the normal form. The **normal form** of a game is another game, *equivalent* to the original game, played as follows. Aware of the rule (for example, knowing Table 1-4) each player selects a pure strategy not knowing how the others choose. Let us say they each inform an impartial umpire. Using the known rule the outcome is then determined.

Naturally many aspects of the original game are obscured by passing to the normal form. For example the normal form is by definition a game of imperfect information, whereas the original game may well have perfect information. However, the normal form is equivalent to the original game in the sense that the consequences of each strategic choice are preserved. This is the vital aspect if we want to know who wins and how.

Table 1-5
Matrix form of 2-2 Nim

		B					
		I	II	III	IV	V	VI
	I	1	1	-1	1	1	-1
A	II	-1	1	-1	-1	1	-1
	III	-1	-1	-1	1	1	1

payoffs to A

Let us now introduce a financial incentive into 2-2 Nim and specify the outcome by the rule 'loser pays the winner one pound'. If we enter the payoffs to the first player A (since the corresponding payoffs to B are just the negatives) the normal form of 2-2 Nim becomes the matrix form given in Table 1-5 and is called a **matrix game**.

In a matrix game like this it is conventional that the payoffs are to the row player.

Table 1-6

State	14	10	15	12	13	8
A's payoff	1	1	1	-1	-1	1
B's payoff	-1	-1	-1	1	1	-1

Reading from left to right in Figure 1.1 the states 14, 10, 15, 12, 13, 8 represent terminal states of the game. With each terminal state of 2-2 Nim is associated one of two possible outcomes: A wins or B wins. In general such a game is called a **win-lose** game. With the rule 'loser pays winner one pound' we can summarise the payoffs to each player by Table 1-6.

Note that the sum of the payoffs for the players at each terminal state is zero. Such a game is called a **zero sum** game. This definition applies equally to games with several players, but in the case of two players if the game is zero sum it means that what one player wins the other loses and *vice versa*. Thus in this special case their interests are diametrically opposed, and such games are often called games of **pure conflict** or **antagonistic** games.

Table 1-7

State	14	10	15	12	13	8
A's payoff	1	-4	1	-4	-4	1
B's payoff	-5	1	-5	1	1	-3

If instead we were to specify that 'the loser must pay the winner a pound and burn the number of pounds equal to the number of turns in that play' the summary of payoffs becomes Table 1-7.

Table 1-8
Bimatrix form of Anti-inflationary 2-2 Nim

		B					
		I	II	III	IV	V	VI
	I	(1,-5)	(1,-5)	(-4,1)	(1,-5)	(1,-5)	(-4,1)
A	II	(-4,1)	(1,-5)	(-4,1)	(-4,1)	(1,-5)	(-4,1)
	III	(-4,1)	(-4,1)	(-4,1)	(1,-3)	(1,-3)	(1,-3)

In this situation the game is no longer zero sum. Actually it is not really the same game and so deserves a new name. Let us call it 'Anti-inflationary 2-2 Nim'. In the case where a game is not zero sum, or involves more than two players, the payoffs to each player must be given at the terminal states of the extensive form and in the normal form. In this example the normal form is given in Table 1-8.

The payoffs to A appear first in each entry, and for obvious reasons this is called a **bimatrix game**.

From Figure 1.3 and Table 1-2 the reader should now be able to construct the normal form for Renée v Peter. Since this is a 2-person zero sum game the normal form will actually be the matrix game which is given in Table 1-9.

Table 1-9

Matrix form of Renée v Peter

		P		
		I	II	III
	I	-3	-3	2
R	II	-1	3	-2
	III	3	-1	-2
	IV	2	2	-3

Table 1-10

Intermediate table for Red-Black

	NI(¼)		NII(¼)		NIII(½)	
	BI	BII	BI	BII	BI	BII
AI	p	p	q	q	t	u
AII	r	s	r	s	t	u

We next consider the game of Red-Black. With chance elements present a pair of pure strategies, one for A and one for B, no longer uniquely determines the outcome of a play. We must take account of Nature's choices. Obviously we should replace an actual payoff resulting from a pair of pure strategies with an expected payoff (that is, a probabilistic average). It is arguable that this involves a subtle change of philosophy; perhaps we should now imagine the game played many times, but it is an entirely reasonable line to adopt. To facilitate working out the expected payoff from each pair of pure strategies we construct from Figure 1.4 and Table 1-3 an intermediate table, Table 1-10.

If we now consider the pair AII, BII, for example, we see from this table that

NI   gives payoff s with probability ¼,
NII   gives payoff s with probability ¼,
NIII   gives payoff u with probability ½.

Thus the expected payoff from this pair of pure strategies is

$$E(AII, \ BII) \ = \ \frac{1}{4}s \ + \ \frac{1}{4}s \ + \ \frac{1}{2}u \ = \ \frac{s + u}{2}$$

In this way we obtain the normal form given in Table 1-11.

A game *in extensive form* is said to be
finite if the rules ensure the existence of
some number N such that *every possible*
complete play of the game has at most N
moves. This means that there is an
effective stop rule. For example in chess
it is easy to construct sequences of moves,
particularly in the end-game, which can
go on *ad infinitum*. The simplest are
periodic, that is infinite repetition of the

Table 1-11

Matrix form of Red-Black

	BI	BII
AI	$(p+q+2t)/4$	$(p+q+2u)/4$
AII	$(r+t)/2$	$(s+u)/2$

same cycle of moves, but there also exist some which are non-periodic. All of these offer a
player in danger of losing, a very real possibility of prolonging the end indefinitely. For this
reason various stop rules are in use to prevent the phenomenon. One well known rule is: Any
cycle of moves, when three times repeated, terminates the play by a tie. This rule excludes most
but not all of the infinite sequences, and hence is not really effective. Another stop rule is: If
no pawn has been moved and no officer[7] taken (both steps being irreversible) for 40 moves,
then the play is terminated by a tie. It is easy to see this rule is effective, although the
corresponding N is enormous. Such a finite game will necessarily have a finite number of pure
strategies for each player, and so it is consistent to define a game *in normal form* to be **finite**
if each player possesses a finite number of pure strategies. (If at this point you are speculating
doubtfully about the implied concept of an infinite game, several examples are given in §1.7).
All the games so far considered have been finite.

There is one other assumption which we have made, and will continue to make until Chapter
4, which is that the games considered are non-cooperative. A **non-cooperative** game is one
where the players receive at each terminal state the payoff assigned to them by the rules of the
game, thus they are each seeking to maximise their own individual payoff. This concept will
be discussed at length in Chapter 4, but until then we shall continue to be concerned exclusively
with non-cooperative games.

With the techniques illustrated above we are now able in principle to construct the normal form
of a finite non-cooperative game. For this reason we begin §2.2 with a formal definition of a
non-cooperative game which amounts to starting off with all such games in normal form.

For the remainder of the present chapter we shall restrict our attention to 2-person zero sum
games with numerical payoffs. Such games are automatically non-cooperative, and the normal
form is then a matrix game. It is therefore sufficient to study matrix games, and the question
to be answered is roughly: given the matrix how do we 'solve' the game?

Consider the following game.

> **Example 1.4** A chooses a number (s) from the set $\{-1,0,1\}$, and then B, not
> knowing A's choice, does the same (t). After the choices are revealed B pays

---

[7] An *officer* is any piece other than a pawn.

A $s(t - s) + t(t + s)$.

The function $P(s,t) = s(t - s) + t(t + s)$, which determines the payoff to A, is an example of a payoff kernel. We can readily construct the payoff matrix given in Table 1-12.

Table 1-12

	B				
	I $t = -1$	II $t = 0$	III $t = 1$	ROW MIN	
I $s = -1$	2	-1	-2	-2	
A II $s = 0$	1	0	1	0	← MAX of ROW MIN
III $s = 1$	-2	-1	2	-2	(pure MAXIMIN)
COLUMN MAX	2	0	2		
		↑ MIN of COLUMN MAX (pure MINIMAX)			

We have also introduced an extra row and column, as indicated, for reasons which will become clear.

A wishes to obtain the highest possible of the payoff entries. If A chooses $s = 1$, B may choose $t = -1$, and then A will lose 2; similarly if A chooses $s = -1$. If A chooses $s=0$ the worst that happens is that he or she gains nothing and loses nothing. Similarly B decides to choose $t = 0$ because then he or she cannot have a worse payoff than 0; for any other choice of t, B may lose 2.

If both players play in this way, *neither has cause for regret once the opponent's choice is known* because both notice that, given the opponent's choice, they would have done worse by choosing differently.

Plainly this can happen because there is an entry in the payoff matrix which is the smallest in its row and at the same time the largest in its column. For a general matrix game let g_{ij} denote the entry in the ith row and j^{th} column ($1 \leq i \leq m$, $1 \leq j \leq n$). We define

$$pure\ MAXIMIN = \max_{1 \le i \le m}\ \min_{1 \le j \le n}\ g_{ij}$$

$$pure\ MINIMAX = \min_{1 \le j \le n}\ \max_{1 \le i \le m}\ g_{ij}$$

Then our previous observation amounts to saying that for Example 1.4

$$pure\ MAXIMIN = pure\ MINIMAX$$

We call the position of this entry (the particular pair i,j for which equality occurs) a **saddle point**, the entry itself the **value** of the game, and the pair of pure strategies leading to it **optimal** pure strategies. Obviously a game is **fair** if its value is zero. A game with a saddle point is called **strictly determined** (which means it is soluble in pure strategies). An easy exercise for the reader is to show that if more than one saddle point exists in a matrix game the corresponding entries are equal, which justifies calling this number *the* value. It may help to visualise MAXIMIN as the floor beneath A's gains, and MINIMAX as the ceiling above B's losses. If they touch at one or more points determined by a pair of pure strategies, this point is a saddle point. In this case a **solution** to the game consists of three things: an optimal pure strategy for A, an optimal pure strategy for B and the value of the game.

From Table 1-5 we see that in the matrix form of 2-2 Nim there are three saddle points at (I,III), (II,III), (III,III) and the value of the game is -1.

We have thus far considered games of perfect or imperfect information and with or without chance elements. Table 1-13 classifies some of the more or less innocent family games according to this scheme.

Table 1-13

	Perfect information	*Imperfect information*
No chance moves	Chess, Draughts, Tic-tac-toe (Noughts and crosses)	Paper-stone-scissors
Chance moves	Ludo, Backgammon Snap	Poker, Bridge, Scrabble

Historically the first result on games in extensive form was stated by Zermelo and given a complete proof by von Neumann. The theorem asserts that any finite 2-person game of perfect information is strictly determined. This means that such a game can always be solved in pure strategies since the normal form has a saddle point. For example in chess this means that one, and only one, of the following three alternatives is valid.

(i) White has a pure strategy which wins, no matter what Black does.

(ii) Both players have pure strategies that ensure at least a draw, no matter what the other does.

(iii) Black has a pure strategy which wins, no matter what White does.

Thus if the theory of chess were really fully known there would be nothing left to play. The theorem guarantees the existence of a saddle point, and in principle we know how to find it, but the stupendous amount of computation required prevents us from ever carrying out this program. From the viewpoint of a fusty old game theorist it might be true to say that: chess is a rather trivial game which displays no interesting features because it is in principle soluble once and for all! This just demonstrates yet again that Life and Mathematics are not quite synonymous.

In the next example we put to work many of the ideas developed so far. At the same time we discover that in games with chance elements the concept of 'a pure strategy for Nature' is not always helpful; sometimes it can be downright misleading.

Example 1.5 (Russian roulette)[8]. Two Muscovite guards buy a black market American sports kit. It consists of a carton of Larks, a Colt 45 six shooter, one bullet and rules for Russian roulette. Each player antes[9] a pack of cigarettes. As senior officer A plays first. He may add two packs to the pot and pass the gun to B, or he may add one pack, spin the cylinder, test fire at his own head and then (Supreme Soviet willing) hand the gun to B. If and when B gets the gun he has the same options; he may hand it back, duly adding two packs, or he may add one pack and try the gun. The game is now ended. Each picks up half the pot, if feasible; else the bereaved takes it all.

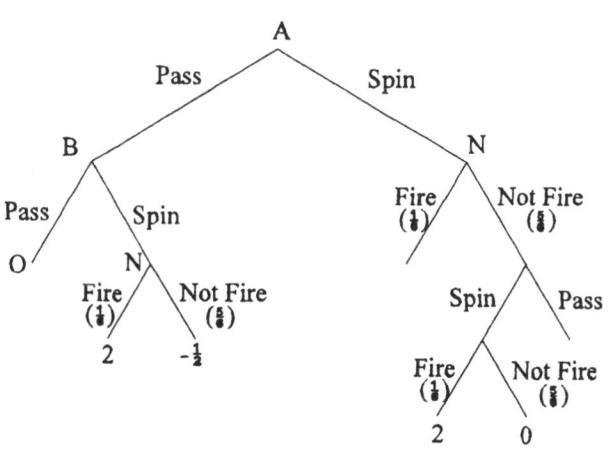

Figure 1.5 Extensive form of Russian roulette.

First we draw up the game tree in Figure 1.5. Using the game tree we list each player's strategies as in Table 1-14.

[8]This example is due to J. D. Williams [7] who reports that in Russia it is called French roulette!

[9]**Ante:** a bet made before any action.

Table 1-14

Pure strategies for A	Pure strategies for B
I Pass	I If A passes, pass; If A survives, spin.
II Spin	II If A passes, pass; If A survives, pass.
	III If A passes, pass; If A survives, spin.
	IV If A passes, spin; If A survives, pass.

It is plainly nonsense, for example, to speak of 'If A spins, fire; if B spins, fire', as being a pure strategy for Nature; for what probability are we to attach to this statement? The simplest approach is to write down the game matrix directly from the tree without an intervening initial table. Thus

$$E(AI, BI) = E(AI, BII) = 0$$

$$E(AI, BIII) = E(AI, BIV) = \frac{1}{6}(2) + \frac{5}{6}(-\frac{1}{2}) = -\frac{1}{12}$$

$$E(AII, BI) = E(AII, BIII) = \frac{1}{6}(-2) + \frac{5}{6}\left(\frac{1}{6}.2 + \frac{5}{6}.0\right) = -\frac{1}{18}$$

$$E(AII, BII) = E(AII, BIV) = \frac{1}{6}(-2) + \frac{5}{6}(\frac{1}{2}) = \frac{1}{12}$$

Hence the game matrix is given in Table 1-15

Table 1-15 Expected payoffs in Russian roulette.

		B			
		I	II	III	IV
A	I	0	0	$-\frac{1}{12}$	$-\frac{1}{12}$
	II	$\frac{1}{18}$	$\frac{1}{12}$	$-\frac{1}{18}$	$\frac{1}{12}$

which has a saddle point at (II,III). We could have predicted this by the Zermelo-von Neumann theorem since Russian roulette is a finite 2-person zero sum game of perfect information. Thus both players should spin at every opportunity (it was after all an American sports kit) and A, as the senior officer, should expect a loss of 1/18 a pack of Larks per game!

1.4 MIXED STRATEGIES AND THE MINIMAX THEOREM.

It is hardly surprising, however regrettable it may seem, that not all matrix games have saddle points. Indeed you may already have noticed that the payoff matrix for Renée v Peter, given in Table 1-9, has pure MAXIMIN=-2 and pure MINIMAX=2, and therfore does not possess a saddle point. In terms of pure strategies this game has no solution and no value. This is sometimes expressed by saying that, in terms of pure strategies, the game has a minorant value -2 (A's gain floor) and a majorant value 2 (B's loss ceiling).

The most we can say is that we always have

pure MAXIMIN ≤ pure MINIMAX.

You should check this now although it is in fact a special case of Lemma 2.2.

We have reached a point in our study of games analogous to a situation which occurs in the theory of equations. The equation $x^2 + 1 = 0$ has no solution in real numbers, and the classical resolution of this difficulty was the invention of complex numbers. The real numbers are then regarded as a particular subset of the set of all complex numbers. Plainly we need to generalise the concept of 'a solution of a game', and this means essentially to generalise the concept of a pure strategy.

Consider the matrix game

		B	
		I	II
A	I	1	-1
	II	-1	1

This game has no saddle point and so no solution in pure strategies. Here premature disclosure of a choice of pure strategy would be to a player's disadvantage. A way out of the difficulty consists of deciding *not on a single pure strategy but on a choice between pure strategies regulated by chance.* Such a probability combination of the original pure strategies is called a **mixed strategy.** For this particular game if both players choose each of their pure strategies with probability ½, then each will have expected payoff zero. We have an intuitively acceptable solution.

In general if A has m pure strategies a mixed strategy for A consists of an ordered m-tuple

$$(x_1, \ldots, x_m), \ 0 \leq x_i \leq 1, \ \sum_{i=1}^{m} x_i = 1$$

where x_i denotes the probability that A will select the ith pure strategy. Similarly if B has n pure strategies then

$$(y_1, \dots, y_n), \ 0 \le y_j \le 1, \quad \sum_{j=1}^{n} y_j = 1$$

is a typical mixed strategy for B.

We can imagine a player to be equipped with a device for randomly selecting each pure strategy with preassigned probability. The player chooses a mixed strategy and thereafter, in one or many plays of the game, implements the pure strategy indicated by the device. (For the game above such a device might be a fair coin. Where more complicated mixed strategies are required the second hand of a watch can often be pressed into service).

If A uses mixed strategy $x = (x_1, \dots, x_m)$ and B uses $y = (y_1, \dots, y_n)$ in a game with payoff matrix (g_{ij}) $(1 \le i \le m, \ 1 \le j \le n)$ the *expected* payoff to A is easily shown to be

$$P(x, y) = \sum_{i=1}^{m} \sum_{j=1}^{n} x_i g_{ij} y_j.$$

A pure strategy is now a special kind of mixed strategy. For example the ith pure strategy of A is $(0, \dots, 0, 1, 0, \dots, 0)$, where the 1 occurs in the ith position.

Suppose B is made to announce a mixed strategy, y_0 say, in advance. Then A, seeking to maximise the payoff, will obviously choose x_0 so that

$$P(x_0, y_0) = \max_{x \in X} P(x, y_0),$$

where X denotes the set of all A's mixed strategies. Under these circumstances the best that B can do is to announce y_0 so that

$$\max_{x \in X} P(x, y_0) = \min_{y \in Y} \max_{x \in X} P(x, y) = \bar{v} \quad (= \text{MINIMAX}) ,$$

where Y denotes the set of all B's mixed strategies. Thus $\bar{v}$ is the most that B can expect to lose under these conditions.

Now reverse the situation and make A announce a mixed strategy, x_0 say, in advance; since B will obviously choose y_0 so that

$$P(x_0, y_0) = \min_{y \in Y} P(x_0, y) ,$$

the best A can do is to announce x_0 so that

$$\min_{y \in Y} P(x_0, y) = \max_{x \in X} \min_{y \in Y} P(x, y) = \underline{v} \quad (= \text{MAXIMIN}).$$

Here $\underline{v}$ is the least that A can expect to win under these conditions.

It is true, as before with pure strategies, that $\underline{v} \le \bar{v}$ (Lemma 2.2). However, if this were all we could say we should be no further advanced. The reason the introduction of mixed strategies is successful as a new solution concept is that in 1928 von Neumann was able to show that, for any matrix game,

$$\text{MAXIMIN} = \text{MINIMAX} \ (= v, \text{ say}).$$

(This is Theorem 2.6.) This particular theorem, first conjectured by Zermelo, turned out to be the keystone of the theory of finite 2-person zero sum games. It shows that any such game has a solution consisting of:

  (i) An optimal mixed strategy which ensures A an expected gain of at least v.
  (ii) An optimal mixed strategy which ensures B an expected loss of at most v.
  (iii) The value v itself.

Given the optimal mixed strategies (i) and (ii), premature disclosure of one's strategy is no longer disadvantageous. Neither player could have obtained a better expectation by mixing their pure strategies in any other proportions, hence neither has cause for regret.

From now on it should cause no confusion if we speak of 'payoff' rather than 'expected payoff'.

If we think we know a solution to a matrix game it is very easy to check if in fact it is a solution. For example a solution for Renée v Peter is:

  Renée's optimal strategy: (½, 0, 0, ½),
  Peter's optimal strategy: (¼, ¼, ½),
  Value = -½.

Let us check this. We first compute Renée's expectation against each of Peter's pure strategies when she uses the mixed strategy (½, 0, 0, ½). If the value of the game is -½ we should find that in each case her payoff is *at least* -½, and indeed this is the case. The point here is that if Renée wins at least -½ against each of Peter's *pure* strategies no *mixed* strategy employed by Peter can improve his situation. Secondly we must compute Peter's expectation against each of Renée's pure strategies when he uses the mixed strategy (¼, ¼, ½). Again we find that in each case Peter's losses are *at most* -½. Thus no mixed strategy employed by Renée can improve her situation either, and so the value of the game is indeed -½.

This calculation can best be set out by adding an auxiliary row and column to the game matrix, Table 1-9. The result is displayed below.

			Peter			
		¼	¼	½		
		I	II	III	Peter's losses	
½ I		-3	-3	2	-½	
Renée 0 II		-1	3	-2	-½	Payoffs to Renée
0 III		3	-1	-2	-½	
½ IV		2	2	-3	-½	
Renée's gains		-½	-½	-½	$v = -½$	

Thus the first entry in the row representing Renée's gains is computed as

$$½(-3) + 0(-1) + 0(3) + ½(2) = -½$$

and is her expected payoff against PI. Similarly for Peter's losses

$$¼(-3) + ¼(-3) + ½(2) = -½,$$

the first entry in the column, is his expected loss aginst RI. That each entry in the auxiliary row and column is equal to -½ is fortuitous; the important point is that entries in the row are *at least* -½, and entries in the column are *at most* -½. Every solution of a matrix game should be checked in this way.

Although von Neumann's minimax theorem assures us that every matrix game has a solution in mixed strategies, it is not necessarily the case that the solution is unique. Indeed it frequently happens that there are many mixed strategies which are optimal. For example in Renée v Peter you should check, as above, that a more general solution is

Renée's optimal strategy: $(\frac{3}{8} + \frac{1}{8}\lambda, \frac{5}{16} - \frac{5}{16}\lambda, \frac{5}{16} - \frac{5}{16}\lambda, \frac{1}{2}\lambda), \quad (0 \leq \lambda \leq 1)$

Peter's optimal strategy: $(\frac{1}{4}, \frac{1}{4}, \frac{1}{2})$

Value = $-\frac{1}{2}$

Here λ can take any value in the range $0 \leq \lambda \leq 1$, and the previous solution corresponds to $\lambda = 1$. Thus there are infinitely many solutions (in fact uncountably many). This situation occurs quite generally, namely if more than one optimal strategy exists there are continuum many.

We have now reduced the problem of matrix games to that of finding optimal mixed strategies. In the next two sections we examine some techniques which often suffice for small matrix games. A general method which will always find a solution is explained in Chapter 3.

1.5 DOMINANCE OF STRATEGIES.

Consider the following matrix game.

		B		
		I	II	III
	I	1	-1	2
A	II	-1	1	3
	III	-3	-2	4

Player B, wishing to minimise the payoff, observes that BIII is an undesirable strategy compared with BI, since $1 < 2$, $-1 < 3$, $-3 < 4$. Whatever A does, BI is preferred by B over BIII. We say BI *dominates* BIII.

In general, for a 2-person zero sum game in normal form, let X and Y denote the sets of strategies (pure or mixed) for A and B respectively. Suppose that if A chooses $x \in X$ and B $y \in Y$, the payoff to A is $P(x, y)$.

For player A we say $x_1 \in X$ **dominates** $x_2 \in X$ if $P(x_1, y) \geq P(x_2, y)$ for all $y \in Y$.
For player B we say $y_1 \in Y$ **dominates** $y_2 \in Y$ if $P(x, y_1) \leq P(x, y_2)$ for all $x \in X$.
The dominance is said to be **strict** if the corresponding inequality is strict for all choices of the opponent's strategy.

We concentrate first on the simple case where one pure strategy dominates another. In the example above, BI strictly dominates BIII, which suggests that BIII will not appear as a component of an optimal strategy for B. (This conclusion is correct and justified by Theorem 2.11). Thus from B's viewpoint the game is effectively reduced to

		B	
		I	II
	I	1	-1
A	II	-1	1
	III	-3	-2

If A is quick enough to notice this he or she will see that since $1 > -3$, $-1 > -2$, AI (or AII) now dominates AIII. This effectively reduces the game to

However, we have seen this game before and know that both players should play each pure strategy with probability ½, and the value of the game is zero.

At this point we conclude that a solution to the original 3 × 3 game is probably

A's optimal strategy: (½, ½, 0),
B's optimal strategy: (½, ½, 0),
Value = 0.

Using the procedure explained in §1.4 we can quickly check that this is indeed a solution. That this process of reduction works quite generally if it works at all, that is, that a solution to the small game 'lifts back' to a solution of the large game, is proved in Theorem 2.10. Moreover one can show that if *every* dominance used is a *strict* dominance no optimal strategies will be 'lost', that is, every solution of the large game will derive from a solution of the small game. In this particular example, since the small game has a unique solution and every dominance used was strict, it follows that the solution we found for the 3 × 3 game is the only solution. The deletion of rows and columns by non-strict dominance can proceed just as above except it should be borne in mind that if a non-strict dominance is used at any stage a complete set of solutions for the small game will not necessarily lead to a complete set for the original game. Of course this does not matter if one merely seeks *some* solution.

It has been remarked that every finite 2-person game of perfect information will possess a saddle point in its matrix form. However, by no means all matrices with saddle points correspond to perfect information games. The matrices of perfect information games have a further property, namely that the matrix can be reduced to a single element by successive deletion of rows and columns by dominance. (The proof, which is rather tiresome to write out, can be obtained by modifying the argument used in Theorem 2.1.). Thus the matrix

$$\begin{pmatrix} 2 & 3 & 4 \\ 1 & 4 & 0 \\ 1 & 0 & 6 \end{pmatrix}$$

which has a saddle point in the upper left corner does *not* correspond to a game of perfect information since no reduction by dominance is possible.

It is helpful to introduce some notation for dominance. If A's k^{th} pure strategy dominates his or her l^{th} pure strategy $(k{\neq}l)$ we write $A(k) \geq A(l)$ and $A(k) > A(l)$ if the dominance is strict.

Similarly if $B(k)$ dominates $B(l)$ we write $B(k) \geq B(l)$ or $B(k) > B(l)$if the dominance is strict. According to the definition of dominance given at the beginning of the section, this means, if the payoff matrix is (g_{ij}) $(1 \leq i \leq m, 1 \leq j \leq n)$, that

$$A(k) \geq A(l) \text{ if } g_{kj} \geq g_{lj} \text{ for } j = 1, 2, ..., n,$$

$$A(k) > A(l) \text{ if } g_{kj} > g_{lj} \text{ for } j = 1, 2, ..., n,$$

$$B(k) \geq B(l) \text{ if } g_{ik} \leq g_{il} \text{ for } i = 1, 2, ..., m,$$

$$B(k) > B(l) \text{ if } g_{ik} < g_{il} \text{ for } i = 1, 2, ..., m.$$

Note the inequalities for B are reversed since B is trying to minimise the payoff (to A).

A pure strategy can also be dominated by a mixed strategy (and *vice-versa*, but this is not helpful since we are trying to reduce the size of the matrix, which means eliminating *pure* strategies as these correspond to rows or columns). Provided that the mixed strategy does not involve this pure strategy as a component we can then eliminate the pure strategy. For example in the game

		B		
		I	II	III
A	I	1	1	1
	II	1	2	0
	III	1	0	2

AI is dominated (non-strictly) by the mixed strategy $(0, ½, ½) = ½AII + ½AIII$, say because

$$\frac{1}{2}.1 + \frac{1}{2}.1 \geq 1, \quad \frac{1}{2}.2 + \frac{1}{2}.0 \geq 1, \quad \frac{1}{2}.0 + \frac{1}{2}.2 \geq 1.$$

(In fact ½AII + ½AIII is equivalent to AI). This enables us to reduce the game to

		B		
		I	II	III
A	I	1	2	0
	II	1	0	2

Similarly for B, $(0, ½, ½) = ½BII + ½BIII$ is equivalent to BI. Thus it is legitimate to say that ½BII + ½BIII dominates BI. The game is now reduced to

		B	
		II	III
A	II	2	0
	III	0	2

which, pretty obviously, leads to the solution

$$\text{A's optimal strategy: } (0, \tfrac{1}{2}, \tfrac{1}{2}),$$
$$\text{B's optimal strategy: } (0, \tfrac{1}{2}, \tfrac{1}{2}),$$
$$\text{Value} = 1,$$

for the original game.

However, because a non-strict dominance has been used (in fact two) we have not actually found all solutions. Firstly the game has a saddle point at (I, I) so another solution, indeed a more sensible one, is

$$\text{A's optimal strategy: } (1, 0, 0),$$
$$\text{B's optimal strategy: } (1, 0, 0),$$
$$\text{Value} = 1.$$

(Incidentally this example illustrates that a game with even a single saddle point can still possess a variety of mixed strategy solutions.) Secondly, one can readily check that if λ, μ are *any* numbers with $0 \le \lambda, \mu \le 1$ a solution is

$$\text{A's optimal strategy: } (\lambda, \tfrac{1}{2}(1 - \lambda), \tfrac{1}{2}(1 - \lambda)),$$
$$\text{B's optimal strategy: } (\mu, \tfrac{1}{2}(1 - \mu), \tfrac{1}{2}(1 - \mu)),$$
$$\text{Value} = 1.$$

Thus the original game was positively bursting with mixed strategy solutions.

In general, finding a mixed strategy which dominates a pure strategy is not particularly easy. Nevertheless, faced with a relatively small game matrix having no pure strategy dominances (and no saddle point!) it is often worth spending a little time looking for a mixed strategy dominance.

1.6 2 × n AND SYMMETRIC GAMES.

Our primary concern with a matrix game is the set of optimal strategies for each player. The von Neumann minimax theorem tells us that these sets are non-empty. A detailed analysis for $m \times n$ matrix games (Theorem 3.15) discloses that *each player possesses at least one optimal strategy which has at most min{m, n} non-zero components*. This fact can be quite useful. In particular for a 2 × n (or m × 2) game it tells us that if we solve every 2 × 2 subgame, that is, study each 2 × 2 submatrix of the original matrix, then at least one of these solutions will correspond to a solution of the original game. Fortunately, for such games it is rarely necessary to look at *every* 2 × 2 subgame. We can proceed by a rather neat graphical method illustrated by the next example.

Example 1.6 Consider the 2 × 3 matrix game

$$\begin{pmatrix} 1 & -1 & 3 \\ 3 & 5 & -3 \end{pmatrix}$$

From now on it will be understood that in such a game A chooses rows, B chooses columns and the payoffs are to A. This game has, in terms of pure strategies, minorant value -1, majorant value 3, and so does not possess a saddle point.

Suppose A employs pure strategy AI with probability p (and hence AII with probability $1-p$). To solve the game for A we have to find two numbers p, v such that

(a)   A's expectation from using AI with probability p and AII with probability $1-p$ is at least v.

(b) v is as large as possible.

This will be accomplished if the mixed strategy $(p, 1-p)$ guarantees A at least v against all of B's pure strategies. For then A is guaranteed at least v no matter what B does, that is, against every strategy pure or mixed which B cares to employ. Looking at the payoff matrix we see that the problem amounts to finding p, v so that:

$$p + 3(1 - p) \geq v \quad \text{(against BI),}$$

$$-p + 5(1 - p) \geq v \quad \text{(against BII),}$$

$$3p - 3(1 - p) \geq v \quad \text{(against BIII),}$$

where $0 \leq p \leq 1$ and we require v maximal.

This is an example of a *linear programming* problem. Any matrix game can be reduced to a problem of this type, and Chapter 3 is devoted to methods of solution. For 2 × n games we proceed as follows. First we draw a graph, which summarises A's situation, as in Figure 1.6.

Here the line BI,

$$y = p + 3(1-p),$$

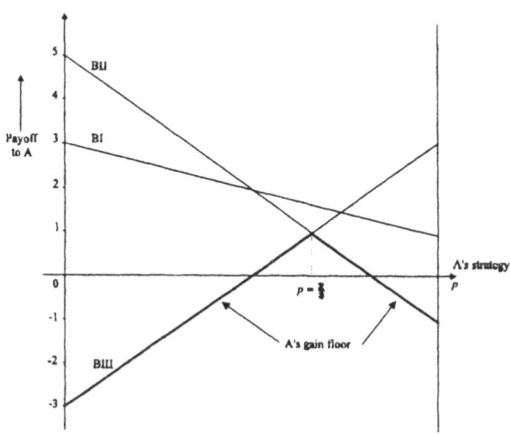

Figure 1.6 Graphical representation of Example 1.6.

represents the payoff to A from the strategy $(p, 1-p)$, as p varies from zero to one, assuming B uses only BI. Similarly for BII and BIII.

From the graph it is apparent that the strategy $(2/3, 1/3)$ maximises A's gain floor, that is, is optimal, and gives $v = 1$.

To find B's optimal strategy we observe from the graph that if A uses $(2/3, 1/3)$ the payoffs from BII and BIII are both $v = 1$, but the payoff from BI is $5/3$. Thus B stands to lose more than 1 if he employs BI. In this circumstance, where a pure strategy played against an opponent's optimal strategy yields less than the value of the game, we may ignore BI since it will appear with zero probability in any optimal mixed strategy for B. (This is proved in Theorem 2.9). The alert reader may notice that in this particular case we have the weak domination $\frac{1}{2}$BII + $\frac{1}{2}$BIII ≥ BI which would again justify eliminating BI; however, we are not making use of this.

We therefore select from the original 2 × 3 game the 2 × 2 subgame in which B employs BII with probability q and BIII with probability $1-q$.

To solve this game for B we have to find q, u so that:

$-q + 3(1 - q) \leq u$ (against AI)
$5q - 3(1 - q) \leq u$ (against AII)
where $0 \leq q$ and we require u minimal

		B	
		q	$1-q$
		II	III
A	I	-1	3
	II	5	-3

Drawing a new graph for this game we obtain Figure 1.7.

In this 2 × 2 subgame B's optimal strategy is $(\frac{1}{2}, \frac{1}{2})$ which gives $u = 1$ (a good sign). Is $(0, \frac{1}{2}, \frac{1}{2})$ an optimal strategy for B in the original 2 × 3 game? Writing out our solution check scheme we get the table below.

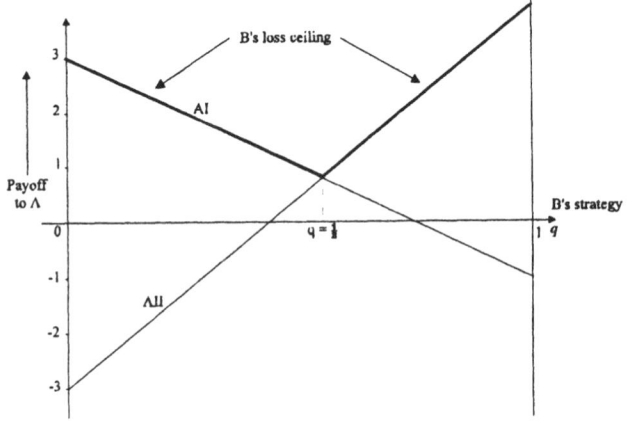

Figure 1.7 Graphical representation of Example 1.6.

		B					
		0	½	½			
		I	II	III		B's losses	
A 2/3	I	1	-1	3		1	ceiling = 1
1/3	II	3	5	-3		1	
A's gains		5/3	1	1		Value = 1	
			floor = 1				

Thus we have indeed found a solution.

Example 1.7 The game

$$\begin{pmatrix} 1 & 2 & 4 & 0 \\ 0 & -2 & -3 & 2 \end{pmatrix}$$

has the graph given in Figure 1.8.

In this game an optimal strategy for A is (2/3, 1/3) and $v = 2/3$. This information is not very helpful to B. To find optimal strategies for B we proceed as follows.

Note that the point $p = 2/3$, $v = 2/3$ lies on the intersection of three lines BI, BII, BIV. We can therefore ignore BIII because the payoff against A's optimal strategy is 5/3, that is, a worse yield for B than the value of the game. (It is worth observing here that BIII is not dominated by *any* mixed strategy combination of the remaining pure strategies and so could not be eliminated on grounds of domination). Hence three possible 2 × 2 subgames arise, one from each pair of BI, BII, BIV.

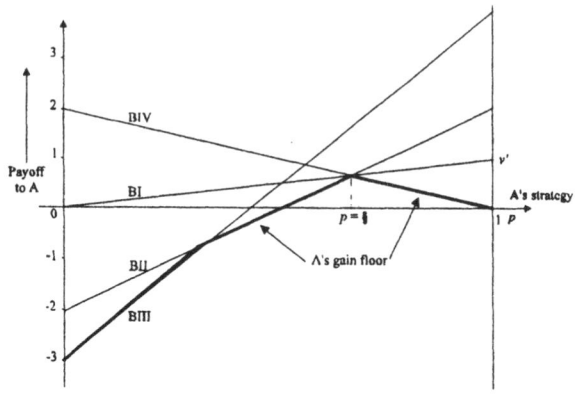

Figure 1.8 A's Gain floor in Example 1.7.

From the remarks made at the beginning of this section we know that at least one of these subgames will lead to a solution of the original game, that is, it is not the case that every optimal strategy for B involves each of BI, BII, BIV with positive probability.

Table 1-26 2 X 2 Sub-games.

			B					B	
			2/3	1/3				1/3	2/3
			I	IV				II	IV
A	2/3	I	1	0		2/3	I	2	0
	1/3	II	0	2		1/3	II	-2	2
			$v = 2/3$					$v = 2/3$	

Examining the graph we see that a further reduction of the number of 2×2 games to be considered is possible. The subgame involving BI and BII would have its solution at $v' = 1$ not at $v = 2/3$, and v' is not the value of the original 2×4 game. Hence it is only necessary to examine two 2×2 subgames. These are easily solved graphically to get Table 1-26.

For the original 2×4 game, A should use the strategy $(2/3, 1/3)$ whereas we can quickly check that for B either of the strategies

$$y_1 = (2/3, 0, 0, 1/3) \text{ or } y_2 = (0, 1/3, 0, 2/3)$$

is optimal.

Can B use any other mixed strategies in optimal play? The answer is yes. For, as we now proceed to verify, any combination $\lambda y_1 + (1 - \lambda)y_2$ with $0 \le \lambda \le 1$, that is, a **convex combination**, is also an optimal strategy for B. (This has a simple geometrical interpretation. If we regard y_1 and y_2 as points in Euclidean four-dimensional space then the assertion is that every point on the line segment joining y_1 and y_2 is an optimal strategy, cf Theorem 2.8.) Firstly

$$\lambda y_1 + (1 - \lambda)y_2 = (\tfrac{2}{3}\lambda, \tfrac{1}{3}(1 - \lambda), 0, \tfrac{2}{3} - \tfrac{1}{3}\lambda).$$

Since $0 \le \lambda \le 1$ every component is positive. Moreover

$$\tfrac{2}{3}\lambda + \tfrac{1}{3}(1 - \lambda) + \tfrac{2}{3} - \tfrac{1}{3}\lambda = 1$$

Thus $\lambda y_1 + (1-\lambda)y_2$ *is* a mixed strategy. To check that it is also optimal we draw up the usual scheme.

		B				
		$^2/_3\lambda$	$^1/_3(1-\lambda)$	0	$^2/_3-^1/_3\lambda$	
		I	II	III	IV	B's losses
A	2/3 I	1	2	4	0	$^2/_3\lambda+^2/_3(1-\lambda) = {}^2/_3$
	1/3 II	0	-2	-3	2	$-^2/_3(1-\lambda)+2(^2/_3-^1/_3\lambda) = {}^2/_3$
	A's gains	2/3	2/3	5/3	2/3	$v = 2/3$

We find that A's gain floor equals B's loss ceiling, hence the mixed strategies indicated are indeed optimal and the value of the game, as we already knew, is 2/3.

We next turn our attention to a rather different kind of matrix game and pose the question: what property must a matrix possess if the two players (Clare and Doug) can exchange roles as row and column players and thereby obtain a new payoff matrix which is identical to the original one? This means each player is faced with the same set of strategic choices, with corresponding choices leading to identical consequences from each viewpoint. In particular an individual has no reason to prefer to be player A or player B. Plainly the matrix must be square. We shall consider first 3 × 3 games.

Suppose the initial choice of rules gives

		B		
		DI	DII	DIII
A	CI	d	a	b
	CII	g	e	c
	CIII	h	i	f

Payoffs to Clare

If now Clare and Doug interchange roles (in this modern age happily they feel quite free to do so) but the payoffs for a given strategy in each case remain the same, we get

		B		
		CI	CII	CIII
	DI	d	g	h
A	DII	a	e	i
	DIII	b	c	f

Payoffs still to Clare

Converting this matrix to standard form (that is, payoffs to A) we get

		B		
		CI	CII	CIII
	DI	$-d$	$-g$	$-h$
A	DII	$-a$	$-e$	$-i$
	DIII	$-b$	$-c$	$-f$

Payoffs to Doug.

If the first and last matrices are to be identical we get

$$d = e = f = 0, \quad g = -a, \quad h = -b, \quad i = -c.$$

Thus the original payoff matrix should look like

		B		
		I	II	III
	I	0	a	b
A	II	$-a$	0	c
	III	$-b$	$-c$	0

A similar analysis obviously works in general for an arbitrary square matrix, and so we make the following definition. A matrix game (g_{ij}) is **symmetric** if (g_{ij}) is square and $g_{ji} = -g_{ij}$ for all pairs i, j. In particular note that $g_{ii} = 0$ for all i. Thus a matrix game is symmetric *if its matrix is skew-symmetric.*

It is intuitively apparent (the sceptical reader is referred to Theorem 2.12) that the value of any symmetric matrix game is zero and that any strategy which is optimal for one player is also optimal for the other.

Many games, as actually played, are symmetrised by asssigning to Nature a prior move to decide which player plays first, often by tossing a coin. Chess, as customarily played, is preceded by a chance move in which one player guesses the colour of a piece concealed by the other.

Assume, for a moment, that a player, A say, knows how to solve every conceivable symmetric matrix game. Suppose A is then asked to play a non-symmetric game G. How can A replace G by a symmetric game in order to apply this special knowledge? In fact there are many ways, but it could be done like this. Imagine A to be playing *simultaneously* two games G; say G′ and G″. In G′ A takes the role of the first player, in G″ the role of the second player. The payoff to A is then the combined payoff of the first player in G′ and the second player in G″. For the opponent the positions are reversed. The total game H, consisting of G′ and G″ together, is clearly symmetric. Hence A will know how to play H and consequently also its parts, say G′, that is, G.

In a practical sense this 'reduction' of G to H is rather futile since if G is $m \times n$ then H will be $p \times p$ where $p = mn$. However, in a theoretical sense it is quite interesting since it shows that the study of matrix games reduces in principle to that of symmetric matrix games. Moreover, there are other more efficient 'reduction' procedures (which are also more complicated).

Our last general observation on symmetric matrix games is a rather pretty theorem due to Kaplansky [8]. A necessary condition for an $m \times m$ symmetric game to be completely mixed (that is for all pure strategies to appear with positive probability in *every* optimal strategy) is that m is odd.

For the 3×3 symmetric game any one of the following three conditions leads to a saddle point as shown:

$$a \geq 0 \text{ and } b \geq 0; 0 \text{ in the } (1, 1) \text{ position,}$$
$$a \leq 0 \text{ and } c \geq 0; 0 \text{ in the } (2, 2) \text{ position,}$$
$$b \leq 0 \text{ and } c \leq 0; 0 \text{ in the } (3, 3) \text{ position.}$$

Thus in order to have a mixed strategy solution a, b, c, must be such as to contradict *all* these conditions.

Suppose $a < 0$. Then we must have $c < 0$ (second condition) and so $b > 0$ (third condition).

Thus the case $a < 0$, $b > 0$ and $c < 0$,

$$\begin{pmatrix} 0 & - & + \\ + & 0 & - \\ - & + & 0 \end{pmatrix}$$

will lead to a mixed strategy solution. The other case is $a > 0$, $b < 0$ and $c > 0$,

$$\begin{pmatrix} 0 & + & - \\ - & 0 & + \\ + & - & 0 \end{pmatrix}$$

In fact these two situations are strategically equivalent since we may interchange the last two rows and the last two columns (that is, relabel the strategies) of the first to get the second. Skew-symmetry of the matrix is preserved by this operation. It therefore suffices to solve the case $a > 0$, $b < 0$ and $c > 0$.

Let A's optimal strategy be (x_1, x_2, x_3), $x_i \geq 0$, $\Sigma x_i = 1$, then A's expected gains against each of B's pure strategies are

$$g_1 = -ax_2 - bx_3 \geq 0 \qquad \text{(against BI)},$$

$$g_2 = ax_1 - cx_3 \geq 0 \qquad \text{(against BII)},$$

$$g_3 = bx_1 + cx_2 \geq 0 \qquad \text{(against BIII)}.$$

Each g_i is non-negative, for since the game is symmetric its value is zero, hence since (x_1, x_2, x_3) is optimal it must produce a payoff of at least zero against any strategy that B cares to adopt, in particular against B's pure strategies.

As $a > 0$, $b < 0$ and $c > 0$ we obtain from these inequalities

$$\frac{x_3}{a} \geq \frac{x_2}{-b}, \quad \frac{x_1}{c} \geq \frac{x_3}{a}, \quad \frac{x_2}{-b} \geq \frac{x_1}{c},$$

so that

$$\frac{x_3}{a} \geq \frac{x_2}{-b} \geq \frac{x_1}{c} \geq \frac{x_3}{a}$$

and equality must hold throughout. Solving, by setting each ratio equal to t, gives

$$x_3 = at, \quad x_2 = -bt, \quad x_1 = ct,$$

where, since $x_1 + x_2 + x_3 = 1$, we have

$$t = \frac{1}{a-b+c}.$$

Hence the optimal strategy (for B as well as for A) is

$$x_1 = \frac{c}{a-b+c}, \quad x_2 = \frac{-b}{a-b+c}, \quad x_3 = \frac{a}{a-b+c}.$$

Thus an algorithm for solving 3×3 symmetric games is:

(i)   Look for a saddle point.
(ii)   If there is no saddle point then arrange (by interchanging the last rows and

columns if necessary) that $a > 0$, $b < 0$, $c > 0$.
(iii)   The optimal strategy for both players is given by the above formulae.

1.7 OTHER KINDS OF TWO PERSON ZERO SUM GAMES.

We begin by considering a rather silly game called 'Choose a number'. Two players simultaneously each choose a positive integer. The one choosing the greater number is paid one pound by the other. If both choose the same number the game ends in a tie. Each player has infinitely many pure strategies. This game, obviously, cannot have a solution. However, there are classes of infinite games which do have solutions.

> **Example 1.8** Imagine the finite payoff matrix in a 2-person zero sum game to be replaced by an infinite array of payoffs, one for each point on the unit square, say. Let A choose a point x on the x-axis with $0 \le x \le 1$, and independently B choose a point y on the y-axis with $0 \le y \le 1$. Suppose the payoff to A is given by
>
> $$P(x, y) = -x^2 + y^2 - 4xy + 4x + 2y - 2.$$

This is an infinite game since each player has continuum many pure strategies. Also the payoff kernel $P(x, y)$ is a continuous function of the point (x, y), so we have here an example of a **continuous game on the unit square**. Such games are often further classified by some special property of the function P. In this instance P is a polynomial in two variables, so Example 1.8 is also a **polynomial game**.

This particular example arises from an area-division game as illustrated in Figure 1.9. Two farmers wish to divide an area of farmland OPQRS, by two straight-line boundaries, into four parts. Farmer A chooses x, $0 \le x \le 1$, and independently B chooses y, $0 \le y \le 1$. The land is then divided up as indicated. Naturally both wish to maximise their land area, so we define the payoff kernel as

$P(x, y) = $ (A's total area) - (B's total area).

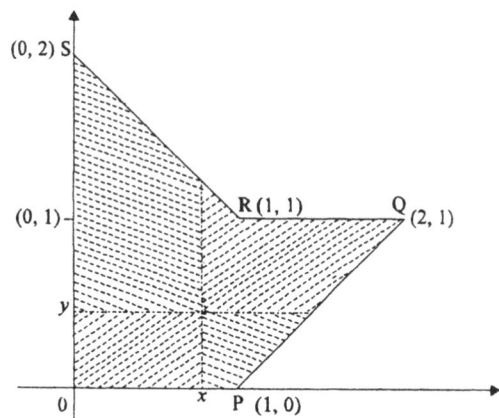

Figure 1.9 An area division game.

Some   elementary   coordinate geometry shows that $P(x, y)$ so defined is exactly the function given in Example 1.8. A wishes to maximise $P(x, y)$, B to minimise $P(x, y)$ (that is, to maximise $-P(x, y)$).

By analogy with matrix games a saddle point for a game on the unit square is a point (x_0, y_0) such that

$$P(x_0, y_0) \geq P(x, y_0) \text{ for all } x, 0 \leq x \leq 1$$
$$\text{and } P(x_0, y_0) \leq P(x_0, y) \text{ for all } y, 0 \leq y \leq 1$$

We observe in passing that the notion of a saddle point in the theory of games differs in two respects from the analogous notion in geometry. In geometry the 'saddle shape' of a surface about some point is independent of 'which way up' the surface is oriented. In game theory, however, for a point to be a saddle point it is necessary that $P(x, y)$ attain there a maximum with respect to its *first* coordinate and a minimum with respect to its *second*. Secondly a saddle point of some surface in geometry is an analytic property associated with the fact that the corresponding partial derivatives vanish at that point. In game theory analytic smoothness is not required. Moreover it is often the case, as with other kinds of extremal problems, that a saddle point turns out to be at the boundary of the domain of definition of $P(x, y)$. For example, the game on the unit square having $P(x, y) = x + y$ has a saddle point at $(1, 0)$, but the partial derivatives do not vanish there.

In Example 1.8, by equating partial derivatives to zero and checking orientation one can discover a saddle point at $(x_0, y_0) = (8/10, 6/10)$. That this *is* a saddle point is clearly revealed if we write

$$P\left(x, \frac{6}{10}\right) = \frac{2}{10} - \left(x - \frac{8}{10}\right)^2,$$

$$P\left(\frac{8}{10}, y\right) = \frac{2}{10} + \left(y - \frac{6}{10}\right)^2.$$

Thus the value of the game is $v = 2/10$.

The solution in pure strategies of this game is exceptional, because games on a square, like matrix games, do not usually possess saddle points.

In general each player must 'weight' the points of the interval. Such distributions of weights (which allocate the probability that the choice will lie in any small interval) are probability density functions; they play a role analogous to that of mixed strategies in matrix games. The payoff $P(x, y)$ then becomes an expected payoff

$$E(p, q) = \iint P(x, y) \, p(x) \, q(y) \; dx \, dy$$

where $p(x)$ is the probability density function of the first player, and $q(y)$ of the second. Yet even when both players employ probability distributions, the minimax theorem does not hold without some mild restrictions on how the payoff $P(x, y)$ varies over the square. For the continuous case, where $P(x, y)$ has no abrupt jumps, the equality of maximin and minimax was proved by Ville [9] in 1938.

In dealing with problems of this type it is usually more convenient to work with distribution functions rather than probability density functions. A **distribution function** on [0,1] is a non-decreasing real valued function F such that $0 \leq F(x) \leq 1$ and $F(1) = 1$. Intuitively, a distribution function is just the integral of the corresponding probability density function, thus

$$F(x) = \int_0^x p(t)\,dt, \quad G(y) = \int_0^y q(t)\,dt,$$

are the distribution functions associated with p and q respectively and contain essentially the same information. This device allows one to consider 'discrete weightings' of points, in which case F and G have jump discontinuities and the corresponding 'probability density functions' are rather hard to describe. The next example illustrates this point.

Example 1.9 Consider the continuous game on the unit square obtained by taking

$$P(x, y) = 16(x - y)^2.$$

It is easy to see this game has no saddle point, although Ville's theorem assures us of the existence of a mixed strategy solution. However, to find the solution of such a game is usually very difficult and it can actually be found only in some special cases.

In this instance it is not hard to verify (see Problem 14) that the optimal strategy for A is described by the distribution function

$$F_0(x) = \begin{cases} \dfrac{1}{2}, & 0 \le x \le 1, \\[2mm] 1, & x = 1 \end{cases}$$

(that is, choose $x = 0$ with probability 1/2 and $x = 1$ with probability 1/2) and that of B is

$$G_0(y) = \begin{cases} 0, & 0 \le y \le \dfrac{1}{2}, \\[2mm] 1, & \dfrac{1}{2} \le y \le 1, \end{cases}$$

(that is, choose $y = 1/2$ with probability 1). Thus the game has value 4. Note that B actually employs a pure strategy although A does not.

The games so far considered in this section have one move for each player and have been infinite because each player has infinitely many pure strategies.

Example 1.10 (The binary game). In this game A and B alternately choose binary digits 0 or 1. A complete play of the game consists of a countable number of such moves or choices. Thus each play consists of an infinite sequence $s(1), s(2), s(3),\dots$ of digits 0 or 1 which determines a real number

$$s = \sum_{n=1}^{\infty} \frac{s(n)}{2^n}.$$

Player A wins one pound from B if s belongs to a certain subset U of the unit interval, while B wins one pound from A if s is not in U.

An alternative way to visualise this game is as an 'intervals game'. A begins by choosing I_1 to be either $[0, 1/2]$ or $[1/2, 1]$, that is, the left or right closed half interval of $[0, 1]$. Next B chooses I_2 to be the left or right closed half interval of I_1 etc. In this way a complete play of the game generates a nested sequence of closed intervals

$$[0, 1] \supset I_1 \supset I_2 \supset \dots ,$$

whose lengths approach zero. Hence, by the Cantor intersection theorem, there exists a unique point s contained in every interval. A wishes to choose the intervals $I_1, I_3, I_5, \dots$ so that s is in U whilst B is attempting to choose $I_2, I_4, I_6, \dots$ so that s is not in U.

Whichever way we visualise it we have here an example of an infinite win-lose game of perfect information with no last move yet whose outcome is quite well defined!

Do there exist optimal pure strategies for either player? The answer depends delicately on the nature of the set U. Intervals games of this general type were studied by Schmidt [10]. For certain classes of such games he showed for example that where U is the set of badly approximable numbers (that is, real numbers α with $|\alpha - p/q| \geq c/q^2$ for some $c > 0$ and all rationals p/q) then *despite the fact U has Lebesgue measure zero, A can always win*. The same is true if U is taken to be the set of numbers non-normal to one or many bases.

A very important branch of the theory of games is the study of *differential games*. Here is an example.

> **Example 1.11** (A pursuit game). A purse snatcher makes the mistake of stealing the purse of a young lady who is expert in Judo. He runs off across a plane campus at uniform speed w_2 (we assume that he runs in such a way that his motion generates a smooth differentiable curve). Our heroine, the quick-witted Judo expert, leaps upon a bicycle conveniently lying nearby and proceeds to chase him at speed w_1. She steers the bicycle by selecting the curvature at each instant, but the radius of curvature of the bicycle is bounded below by ρ. Capture occurs when their distance apart is at most l. We also assume, naturally, that $w_1 > w_2$.

This example is due in essence to Isaacs [11] who founded the topic of differential games. We can model the situation by taking into account the position coordinates of heroine and thief at time t, say $(x_1(t), y_1(t))$ and $(x_2(t), y_2(t))$ respectively, the direction of the handlebar at time t for the heroine, the running direction of the thief and the curvature selected by the heroine at time t. At time $t = t_0$ let their travel directions be φ and ψ respectively as in Figure 1.10(a). After a small time interval h, the heroine is at $(x_1(t_0 + h), y_1(t_0 + h))$ as shown in Figure 1.10(b).

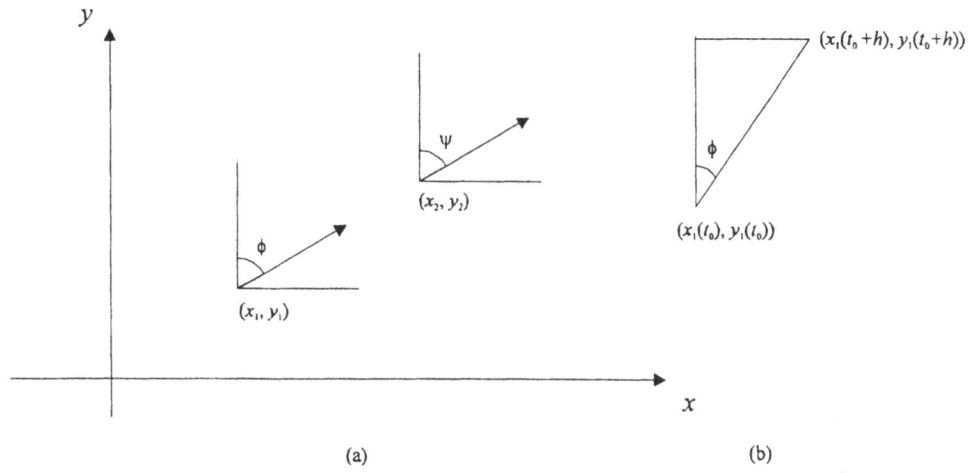

Figure 1.10 Coordinates for the pursuit game.

Her speed being w_1, the distance travelled during this small time interval is $w_1 h$. Thus

$$\frac{y_1(t_0+h) - y_1(t_0)}{w_1 h} = \cos\phi,$$

which, letting $h \rightarrow 0$, gives

$$\frac{dy_1}{dt} = w_1 \cos\phi.$$

Similarly

$$\frac{dx_1}{dt} = w_1 \sin\phi$$

In the same way for the thief we obtain

$$\frac{dx_2}{dt} = w_2 \sin\psi \quad \text{and} \quad \frac{dy_2}{dt} = w_2 \cos\psi.$$

Finally we must take into account the curvature restriction for the heroine and her curvature at time t.

The curvature for the curve described by the motion of the heroine is simply $d\phi/ds$; the radius of curvature is the reciprocal of this quantity. Here s denotes the arc length measured along the curve from some fixed point. Since the radius of curvature is bounded below by ρ we have

$$\left| \frac{d\phi}{ds} \right| \leq \frac{1}{\rho} \quad \text{or} \quad \frac{d\phi}{ds} = \frac{\theta(t)}{\rho},$$

where $|\theta| \leq 1$. Now

$$\frac{d\phi}{dt} = \frac{d\phi}{ds}\frac{ds}{dt} = \frac{d\phi}{ds}w_1 .$$

Thus we have the differential equation

$$\frac{d\phi}{dt} = \frac{w_1}{\rho}\theta(t) , \quad |\theta| \le 1 ,$$

describing the curvature restriction.

The payoff can be described in terms of the total time to capture. Capture occurs when

$$(x_1 - x_2)^2 + (y_1 - y_2)^2 \le l^2 .$$

Isaacs explicitly finds the condition under which the thief can continue indefinitely without being caught. Naturally this depends on the ratio of their speeds and the curvature restriction. If

$$\frac{l}{\rho} > \sqrt{1 - \frac{w_2^2}{w_1^2}} + \sin^{-1}\frac{w_2}{w_1} - 1$$

the thief will ultimately be caught, in which case the thief wishes to maximise the time to capture and the heroine wants to minimise this. If the inequality is reversed the thief can indefinitely be at a distance greater than l and hence never be caught.

The differential equations are often called *kinematic equations* of the system. The angle ψ is under the control of the thief, and the angle ϕ and the parameter θ are under the control of the heroine. Here $-1 \le \theta \le 1$, and the collection of all possible, suitably smooth, functions (ϕ, θ) constitutes the *control space* for the heroine.

If the thief did not run, then the problem would be to minimise the time to capture, which is essentially an optimal control problem. Pontryagin [12] and Bellman [13] have given necessary conditions for the existence of (pure strategy) optimal solutions for such 1-player games under very general conditions on the kinematic equations and the payoff.

For the game theoretic case of at least two players the existence of pure strategy optimal solutions is a very much more difficult question. One can understand why differential games might be technically daunting when one observes that classical non-linear control theory is analogous to playing a game against Nature and that differential games corresponds to a similar analytical study except that Nature is replaced by an one or more intelligent opponents.

Example 1.12 (Fighter-Bomber duel). This is another example of an infinite game. A fighter F has a single air-to-air missile and tries to destroy a bomber B. The strategy of F is a matter of choosing when to fire the missile. Let $a(t)$ be the accuracy of the missile, that is the probability that B is destroyed if F fires at time t; the variable t is normalised to vary over [0, 1]. A mixed strategy for F is a distribution function over [0, 1]. We assume that once F has fired the missile, successfully or otherwise, F flies off home (if it can) and the duel is ended. The bomber is equipped with a machine gun, and in defending itself seeks to destroy the fighter. A pure strategy for B is specified by a function $p(t)$ which measures the intensity with which the machine gun is being fired at time t. We assume that B's total amount of ammunition, if fired at maximum intensity will last for a period $\delta < 1$. Therefore the pure strategies for B consist of all functions p satisfying the restrictions

$$0 \leq p(t) \leq 1 \quad \text{and} \quad \int_0^1 p(t)\,dt = \delta.$$

The accuracy of the bomber's defensive fire is given by a continuous monotone function $r(t)$ with $r(0) = 0$, $r(1) = 1$. This function measures the accuracy in the sense that if B fires with intensity $p(t)$ over a small interval $[t, t+h]$ then the probability that F is destroyed is

$$p(t)\, r(t)\, h + o(h), \text{ as } h \to 0,$$

where $o(h)$ denotes some function which when divided by h tends to zero as $h \to 0$.

Suppose now that $\phi(t, p)$ is the probability that F will survive up to time t when B has chosen to fire with intensity $p(t)$. Then

$$\phi(t+h, p) = \phi(t, p)(1-p(t)r(t)h) + o(h),$$

so that

$$\frac{\phi(t+h, p) - \phi(t, p)}{h} = -\phi(t, p)p(t)r(t) + o(1) .$$

Letting h $\to$ 0 we obtain

$$\phi'(t, p) = -\phi(t, p)p(t)r(t),$$

with the boundary condition $\phi(0, p) = 1$. This differential equation has the solution

$$\phi(t, p) = \exp\left(-\int_0^t p(u)r(u)\,du \right).$$

The payoff function $P(\xi, p)$ when F uses a pure strategy ξ (time of firing the missile) and B uses the strategy p is evaluated as follows. Denote the payoff to F by α if F destroys B, by $-\beta$ if B destroys F, and by zero if neither is destroyed. Then the payoff kernel given as the expected payoff accruing to F is

$$P(\xi, p) = \alpha a(\xi)\, \phi(\xi, p) - \beta(1 - \phi(\xi, p)).$$

The payoff if F adopts a mixed strategy distribution function $x(\xi)$ and B chooses p is obtained by averaging the last equation with respect to $dx(\xi)$, that is

$$P(x, p) = \int_0^1 P(\xi, p)\, dx(\xi) .$$

The differences in form between the last three examples and games on the unit square are apparent. In each case the sets of strategies available to each player, their **strategy spaces**, are of a quite different character. The mathematical machinery required to find optimal strategies (where they exist) for such games is correspondingly diverse. To derive the form of the optimal strategy for F in the Fighter-Bomber duel one needs the Neyman-Pearson lemma, an important result in measure theory (see [14]).

We close this introductory chapter with the following observations. All the 2-person games we have considered, and these have all been non-cooperative games, can be reduced to the normal form

$$\langle X, Y, P_1, P_2 \rangle.$$

Here X denotes the set of all strategies for the first player, known as the first player's strategy space, similarly Y denotes the second player's strategy space, and P_1, P_2 are functions $X \times Y \to \mathbb{R}$ representing the payoffs to the first and second player respectively. If

$$P_1(x, y) + P_2(x, y) = 0$$

for all pairs $x \in X$, $y \in Y$ the game is zero sum and can then be summarised by the triplet

$$\langle X, Y, P_1 \rangle.$$

PROBLEMS FOR CHAPTER 1.

1. Develop in extensive form the game of matching pennies, the rules being that each player *chooses* to show head or tails, A wins one unit if both players make the same choice, and B wins one unit if the choices are different. List the pure strategies of both players and give the normal form.

2. Develop in extensive form the game of matching pennies, the rules being as above except that each player *tosses* the coin. If possible list the players' pure strategies, and give the normal form.

3. The British have decided to attack the American arsenal at Concord, and the Americans know this. They do not know which way the British have chosen to come - whether by land or by sea. The American force is too small to defend both routes; they must choose to defend one or the other and take the consequences.

In fact the British are low on ammunition, and if the two forces meet the British will retreat. This scores one for the Americans. If the forces do not meet, the British reach Concord arsenal and then have plenty of ammunition. Now both sides must plan what to do about the British return from Concord. The Americans can either lay an ambush on the known path of return or move up and attack the British at the arsenal. The British can either leave immediately by day or wait for night.

If the British meet an ambush by day they will be destroyed (score one for the Americans). If the British meet the ambush by night they can filter through with small loss (score one for the British). If the Americans attack the arsenal and the British have already left, the British score one for getting away free. But if the Americans attack and find the British waiting for night, both sides suffer heavy losses and score zero.

Model the above situation as a 2-person zero sum game in extensive form. Since it is arbitrary who is considered to make the first decision assume it to be the Americans (=A). State the pure strategies for both sides and find the payoff matrix.

4. Al and Bill play a game. Al has a real fly and a fake fly. Bill has a fly-swatter. The general idea is: Bill should swat the real fly.

Al chooses either the real fly or the fake fly, puts it on the table and conceals it with his hand. Meanwhile Bill decides whether or not he will swat the fly. At a signal Al removes his hand and Bill swats or not.

If Bill swats he wins one silver thaler (they are both Peace Corps workers in rural Ethopia) if the fly was real and loses one silver thaler if the fly was fake. If Bill does not swat and the fly was real, there is no payment and the game ends because the fly flies off. If Bill does not swat and the fly was fake, they play the game once more - this time for double the stakes, and this time if Bill does not swat the fake fly Bill wins two silver thalers and the game ends.

Develop this game in extensive form and then find the pure strategies and the payoff matrix.

5. Represent the following game in extensive form and reduce it to matrix form. Ann has an Ace and a Queen. Bea has a King and a Joker. The rank of the cards is A>K>Q, but the Joker is peculiar, as will be seen.

Each antes (an initial bet before any move is made) a penny to the pot. Each then selects one of her cards and simultaneously they reveal their selections. If Bea selects the King, the highest card chosen wins the pot and the game ends. if Bea selects the Joker and Ann the Queen, they split the pot and the game ends. If Bea selects the Joker and Ann the Ace, then Ann may either resign (in which case Bea gets the pot) or demand a replay. If Ann demands a replay they each ante another penny to the pot and play again. This time if Bea selects the Joker and Ann the Ace, Bea gets the pot.

6. For the following matrix games find saddle points, if they exist, otherwise try to reduce the games by looking for dominance. Solve the game if you can.

$$(a) \begin{pmatrix} 2 & 4 & 0 & -2 \\ 4 & 8 & 2 & 6 \\ -2 & 0 & 4 & 2 \\ -4 & -2 & -2 & 0 \end{pmatrix} \qquad (b) \begin{pmatrix} 16 & 14 & 6 & 11 \\ -14 & 4 & -10 & -8 \\ 0 & -2 & 12 & -6 \\ 22 & -12 & 6 & 10 \end{pmatrix}$$

$$(c) \begin{pmatrix} 0 & 3 & 6 & 5 \\ 15 & 10 & 8 & 9 \\ 10 & 15 & 11 & 7 \\ 5 & 9 & 4 & 2 \end{pmatrix} \qquad (d) \begin{pmatrix} 2 & -3 & 1 & -4 \\ 6 & -4 & 1 & -5 \\ 4 & 3 & 3 & 2 \\ 2 & -3 & 2 & -4 \end{pmatrix}$$

7. A and B play the following card game with a pack consisting of two cards Hi and Lo. Each player pays an ante of one dollar. Player A then draws a card, looks at it and bets either two dollars or four dollars; B can either 'resign' or 'see'. If B resigns, then A wins the pot. If B decides to 'see', he or she must match A's bet. In this case A will win the pot if holding Hi, otherwise B wins the pot.

Develop the extensive form, list the pure strategies for both players, find the normal form and reduce by dominance as far as you can.

8. Players A and B have been playing the following version of 1-2 Nim: there are two piles of matches (I , I I), one containing one match and the other two. Each player in turn selects a pile and removes at least one match from it. The player taking the last match *wins* one.

After getting bored because the first player A always wins, they decide to spice up the game. They agree that on the first move A will conceal the piles from B and truthfully tell B only how many matches A took. Player B then announces only how many matches B will take, and to compensate for this blind choice B is allowed at this move to take *no* matches if desired.

The piles are then uncovered. If B made an impossible call (that is, asked for two matches when no pile contained two) then B loses one. Otherwise the game is played openly to the end by the original rules.

Develop the game tree to the point where the original rules take over; the outcomes are then readily determined. List the pure strategies relating to this tree and determine the payoff matrix. Reduce by dominance and hence solve the game. Is it fair?

9. Solve the following matrix games, finding an optimal strategy for each player and the value of the game.

(a) $\begin{pmatrix} 4 & -3 \\ 2 & 1 \\ -1 & 3 \\ 0 & -1 \\ -3 & 0 \end{pmatrix}$
   (b) $\begin{pmatrix} -1 & 3 & -5 & 7 & -9 \\ 2 & -4 & 6 & -8 & 10 \end{pmatrix}$

(c) $\begin{pmatrix} -3 & 3 & 0 & 2 \\ -4 & -1 & 2 & -2 \\ 1 & 1 & -2 & 0 \\ 0 & -1 & 3 & -1 \end{pmatrix}$

10. Player A holds two Kings and one Ace. A discards one card, lays the other two face down on the table and announces either "Two Kings" or "Ace King". 'Ace King' is a better hand than 'Two Kings'. Player B may either 'accept' or 'question' A's call. Then the hand is shown and the payoffs are as follows.

If A calls the hand correctly and B accepts, A wins one. If A calls the hand better than it is and B accepts, A wins two. If A calls the hand worse than it is and B accepts, A loses two. if B questions the call, the above payoffs are doubled and reversed.

Develop this game in extensive form, list the pure strategies for each player, give the normal form and solve the game. Is the game fair?

11. Construct a 3×3 matrix game with value $v = \frac{1}{2}$ and such that the set of optimal strategies for the row player is exactly the set

$$\{ (\mu, 1 - \mu, 0); \frac{3}{8} \leq \mu \leq \frac{5}{8}\}.$$

12. A father, mother and daughter decide to hold a certain type of board game family tournament. The game is a 2-person win-lose game. Since the father is the weakest player, he is given the choice of deciding the two players of the first game. The winner of any game is to play the person who did not play in that game, and so on. The first player to win two games wins the tournament. If the daughter is the strongest player, it is intuitive that the father will maximise his probability of winning the tournament if he chooses to play the first game with the mother. Prove that this strategy is indeed optimal. It is assumed that any player's probability of winning an individual game from another player does not change throughout the tournament.
                                        (Third U.S. Mathematical Olympiad.)

13. A and B play a game on the unit square. A chooses a number $x \in [0, 1]$, and B, not knowing A's choice, chooses $y \in [0, 1]$. The payoff to A is calculated by

$$P(x, y) = \frac{1}{2}y^2 - 2x^2 - 2xy + \frac{7}{2}x + \frac{5}{4}y.$$

Solve the game.

After a while A gets bored with always winning and decides to confuse B by playing a mixed strategy, using a probability density function $3\xi^2$, so that the probability A chooses $x \in [\xi, \xi + d\xi]$ is $3\xi^2 d\xi$. Why is this not an outrageously stupid thing for A to do? How much can A expect to lose by doing it?

14. Let $P(x, y)$ be the payoff kernel of a continuous game on the unit square. Since Ville's theorem assures us of the existence of a mixed strategy solution it is clear by definition that F_0, G_0 are optimal distribution functions, for the first and second player respectively, precisely if

$$\int_0^1\int_0^1 P(x, y)\,dF(x)\,dG_0(y) \le \int_0^1\int_0^1 P(x, y)\,dF_0(x)\,dG_0(y) \le \int_0^1\int_0^1 P(x, y)\,dF_0(x)\,dG(y),$$

for all distribution functions F, G on $[0, 1]$.

Use this criterion to verify the solution given in Example 1.9.

15. A and B play a differential game of perfect information. A chooses $y = y(t, x)$, where y is a suitably continuous and differentiable function satisfying $0 \le y \le 1$ for all values of t and x. B chooses $z = z(t, x)$, where again z is suitably well behaved and $0 \le z \le 1$. The kinematic equation is

$$\frac{dx}{dt} = (y - z)^2, \quad x(0) = x_0.$$

Play terminates at $t = T$. A wishes to maximise

$$\int_0^T x\,dt$$

and B wishes to minimise this integral.

Prove that

$$\max_y \min_z \int_0^T x\,dt = x_0 T,$$

and that

$$\min_z \max_y \int_0^T x\,dt = x_0 T + \frac{1}{8}T^2.$$

Thus this differential game has no pure strategy optimal solution.

(Berkovitz [15])

CHAPTER REFERENCES

[1] Borel, E., 'The theory of play and integral equations, with skew symmetrical kernels, On games that involve chance and the skill of the players', and 'On systems of linear forms of skew symmetric determinants and the general theory of play', trans. L. J. Savage, *Econometrica*, **21**, (1953), 97-117.

[2] von Neumann, J. and Morgenstern, O., *Theory of Games and Economic Behaviour*, Princeton University Press (1944 revised 1947), Princeton, N.J.

[3] Copeland, A. H., (review of [2]), *Bull. Amer. Math. Soc.*, **51**, (1945), 498-504.

[4] Thorp, E., *Beat the Dealer*, Random House (1966), New York.

[5] Cutler, W. H., 'An optimal strategy for pot-limit poker', *Amer. Math. Monthly*, **82**, (1975), 368-376.

[6] Hardy G. H., and Wright, E. M., *An Introduction to the Theory of Numbers*, Clarendon Press (Fourth Edition 1962), Oxford.

[7] Williams, J. D., *The compleat strategyst*, McGraw Hill (revised edition 1966), New York.

[8] Kaplansky, I., 'A contribution to von Neumann's theory of games', *Annals of Maths.*, **46**, (1945), 474-479.

[9] Ville, J., 'Sur la théorie générale des jeux au intervient l'habilité des joueurs', *Traité du calcul des probabilités et de ses applications*, **2**(5), (1938), 105-113, ed. by E. Borel and others, Gauthier-Villars & Cie, Paris.

[10] Schmidt, W. M., 'On badly approximable numbers and certain games', *Trans. Amer. Math. Soc.*, **123**, (1966), 178-199.

[11] Isaacs, R., *Differential Games*, J. Wiley and Sons (1965), New York.

[12] Pontryagin, L.S., Boltyanskii, V. G., Gamkrelidze, R. V. and Mishchenko, E. F., *The Mathematical theory of Optimal Processes*, Pergamon Press (1964), New York.

[13] Bellman, R., *Dynamic Programming*, Princeton University Press (1957), Princeton, N.J.

[14] Karlin, S., *The Theory of Infinite Games*, Vol. II, Addison-Wesley (1959), Reading, Mass.

[15] Berkovitz, L. D., 'A differential game with no pure strategy solutions', *Advances in Game Theory*, 175-194 (Ann. Math. Studies No. 52), ed. by M. Dresher, L. S. Shapley and A. W. Tucker, Princeton University Press, Princeton, N.J.

[16] McKinsey, J. C. C., *Introduction to the Theory of Games*, McGraw-Hill (1952), New York.

[17] Widder, D. V., *The Laplace Transform*, Princeton University Press (1941), Princeton, N.J.

[18] Widder, D. V., *Advanced Calculus*, Prentice Hall Inc. (1947), New York.

2

Non-cooperative Games

A man in a lion's den turns to wolves for friendship.

This chapter begins by examining non-cooperative games in general. Later we specialise to 2-person zero sum games and provide the detailed theory used extensively in Chapter 1. In what follows it is often sufficient merely to understand the statements of the theorems, and on a first reading many of the proofs can be omitted, indeed in some instances should be.

2.1 EXTENSIVE FORMS AND EQUILIBRIUM N-TUPLES.

By a **non-cooperative** game is meant a game in which absolutely no preplay communication is permitted between the players and in which players are awarded their due payoff according to the rules of the game. In particular, agreements to share payoffs, even if this were practicable (and in many instances it is not), are specifically forbidden. Thus in a non-cooperative game it is 'all players for themselves'. We do *not* assert that transitory strategic cooperation cannot occur in a non-cooperative game if permitted by the rules. Typically, however, such arrangements to cooperate are not 'binding unto death'. For a requirement of this type would possess the limitation of cooperative games (that agreements are binding) without the possibility of preplay negotiation or profit sharing, at least one of which normally occurs in cooperative games.

An n-person **non-cooperative game** Γ **in extensive form** can be regarded as a graph theoretic tree of vertices (states) and edges (decisions or choices) with certain properties. These properties can be summarised as follows:

(i) Γ has a distinguished vertex called the initial state.

(ii)   There is a payoff function which assigns to each outcome an n-tuple $(P_1, ..., P_n)$, where P_i denotes the payoff to the ith player.

(iii) Each non-terminal vertex of Γ is given one of $n+1$ possible labels according to which player makes the choice at that vertex. If the choice is made by chance the vertex is labelled with an N.

(iv)   Each vertex labelled with an N is equipped with a probability distribution over the edges leading from it.

(v)   The vertices of each player, other than Nature, are partitioned into disjoint subsets known as information sets. A player is presumed to know which information set he or she is in, but not which vertex of the information set. This has the consequence that

    (a)   Any two vertices in the same information set have identical sets of choices (edges) leading from them.

(b)   No vertex can follow another vertex in the same information set.

Player i $(1 \leq i \leq n)$ is said to have **perfect information** in Γ if each information set for this player consists of one element. The game Γ in extensive form is said to have **perfect information** if every player in Γ has perfect information. By a **pure strategy** for player i is meant a function which assigns to each of player i's information sets one of the edges leading from a representative vertex of this set. We denote by S_i the set of all pure strategies for player i. A game in extensive form is **finite** if it has a finite number of vertices.

If Γ has no chance elements the payoff P_i to the ith player is completely determined by an n-tuple $(\sigma_1, ..., \sigma_n)$, where $\sigma_i \in S_i$, that is $P_i = P_i(\sigma_1, ..., \sigma_n)$. If, however, chance elements are involved then $P_i(\sigma_1, ..., \sigma_n)$ is taken to be the statistical expectation of the payoff function of player i, with respect to the probability distributions specified from property (iv), when the pure strategies $(\sigma_1, ..., \sigma_n)$, $\sigma_i \in S_i$, are chosen. A game is **zero sum** if

$$\sum_{i=1}^{n} P_i(\sigma_1, ..., \sigma_n) = 0$$

for all n-tuples $(\sigma_1, ..., \sigma_n)$, $\sigma_i \in S_i$.

We now introduce a solution concept for an n-person non-cooperative game Γ. The extent to which this concept, of equilibrium point, is intuitively satisfying will be discussed in §2.3. A pure strategy n-tuple $(\sigma_1, ..., \sigma_n)$, $\sigma_i \in S_i$, is said to be an **equilibrium point** of Γ if for each i, $1 \leq i \leq n$, and any $\sigma_i' \in S_i$,

$$P_i(\sigma_1, ..., \sigma_i', ..., \sigma_n) \leq P_i(\sigma_1, ..., \sigma_i, ..., \sigma_n)$$

Thus an n-tuple $(\sigma_1, ..., \sigma_n)$ is an equilibrium point if no player has a positive incentive for a *unilateral* change of strategy. In particular this means that once all choices of pure strategies are revealed, no player has any cause for regret.

The bimatrix game

$$\begin{pmatrix} (3,2) & (0,0) \\ (0,0) & (2,3) \end{pmatrix}$$

has equilibria at $(3, 2)$ and $(2, 3)$. In the case of a 2-person zero sum game the notion of a pure strategy equilibrium point is identical to that of a saddle point. We know from Chapter 1 that many 2-person zero sum games do not have saddle points; it is therefore plain that not every n-person game has an equilibrium point in pure strategies. However, it is true that every finite n-person game in extensive form which has perfect information *does* possess a pure strategy equilibrium point. This is a natural generalisation of the Zermelo-von Neumann theorem on 2-person zero sum games of perfect information. Our first object is to prove this theorem.

We **truncate** a finite n-person game Γ having perfect information by deleting the initial vertex and the edges leading from it. Because each information set consists of a single vertex, what remains is a finite number of subgames Γ_1, ..., Γ_r, called the **truncations** of Γ, each having perfect information. We can also consider the **truncation** of a given pure strategy $\sigma_i \in S_i$ by restricting it as a function to the vertices of some truncation of Γ.

Theorem 2.1. A finite n-person non-cooperative game Γ in extensive form which has perfect information possesses an equilibrium point in pure strategies.

Proof. Let x_0 be the initial vertex of Γ and let the other ends of the edges from x_0 be the vertices a_1, ..., a_r. Then a_j, $1 \le j \le r$, are the initial vertices of the games Γ_1, ..., Γ_r respectively, obtained by truncating Γ. See Figure 2.1.

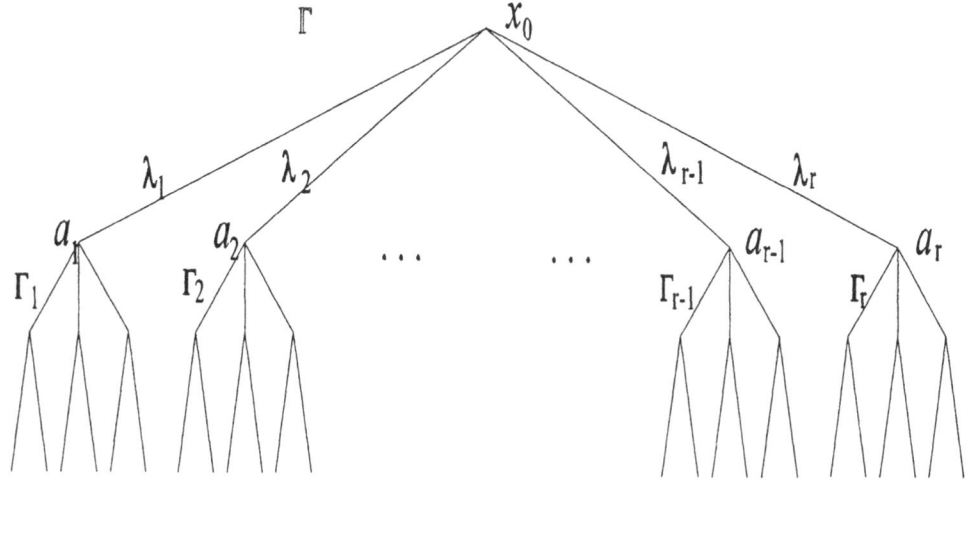

Figure 2.1 Truncating Γ.

Let the longest play in Γ be of length N. We shall prove the theorem by induction on N. Clearly the games Γ_1, ..., Γ_r have length at most $N-1$.

Let $(\sigma_{1j}, ..., \sigma_{nj})$ be pure strategies for each player for the game Γ_j $(1 \le j \le r)$. Let $(\sigma_1, ..., \sigma_n)$ be pure strategies for each player for the game Γ. We write

$$P_i(\sigma_1, ..., \sigma_n), \quad P_i^j(\sigma_{1j}, ..., \sigma_{nj})$$

for the payoffs to player i in Γ and Γ_j respectively.

For games of length zero the theorem is trivial (inaction is equilibrium), so we assume the existence of equilibrium points for games of perfect information with length at most $N-1$, in particular for Γ_1, ..., Γ_r. Let $(\sigma_{1j}^0, ..., \sigma_{nj}^0)$ be such a point for Γ_j, that is, for every i, $1 \le i \le n$,

$$P_i^j(\sigma_{1j}^0, \ldots, \sigma_{ij}', \ldots, \sigma_{nj}^0) \leq P_i^j(\sigma_{1j}^0, \ldots, \sigma_{ij}^0, \ldots, \sigma_{nj}^0) \qquad (2.1)$$

We shall construct an equilibrium point $(\sigma_1^0, \ldots, \sigma_n^0)$ for the game Γ.

Case 1. x_0 is *labelled* N. Let $\lambda_1, \ldots, \lambda_r, 0 \leq \lambda_j \leq 1, \Sigma\lambda_j = 1$, denote the probabilities for the vertices $a_1, \ldots, a_r$ to be selected. Let x be any vertex of Γ. If $x = x_0$ we do not need to define $\sigma_i^0(x)$, nor do we need to define it if x is any other vertex labelled with an N.

Otherwise $x \in \Gamma_j$ for some j and is labelled with an i, $1 \leq i \leq n$. We then define

$$\sigma_i^0(x) = \sigma_{ij}^0(x).$$

For any pure strategies $(\tau_1, \ldots, \tau_n)$ of Γ we denote the restriction of τ_i to Γ_j by $\tau_i | \Gamma_j$. We plainly have

$$P_i(\tau_1, \ldots, \tau_n) = \sum_{j=1}^r \lambda_j P_i^j(\tau_1 | \Gamma_j, \ldots, \tau_n | \Gamma_j),$$

and $\sigma_i^0 | \Gamma_j = \sigma_{ij}^0$.

Thus for $1 \leq i \leq n$,

$$P_i(\sigma_1^0, \ldots, \sigma_n^0) = \sum_{j=1}^r \lambda_j P_i^j(\sigma_{1j}^0, \ldots, \sigma_{ij}^0, \ldots, \sigma_{nj}^0)$$

$$\geq \sum_{j=1}^r \lambda_j P_i^j(\sigma_{1j}^0, \ldots, \sigma_{ij}', \ldots, \sigma_{nj}^0)$$

from (2.1). But

$$\sum_{j=1}^r \lambda_j P_i^j(\sigma_{1j}^0, \ldots, \sigma_{ij}', \ldots, \sigma_{nj}^0) = P_i(\sigma_1^0, \ldots, \sigma_i', \ldots, \sigma_n^0)$$

so that, for each i, $1 \leq i \leq n$,

$$P_i(\sigma_1^0, \ldots, \sigma_i', \ldots, \sigma_n^0) \leq P_i(\sigma_1^0, \ldots, \sigma_i^0, \ldots, \sigma_n^0),$$

that is, $(\sigma_1^0, \ldots, \sigma_n^0)$ is an equilibrium point for Γ.

Case 2. x_0 is *labelled with a player index*. Without loss of generality we can suppose x_0 is labelled with a 1. If $x = x_0$ we define $\sigma_1^0(x)$ to be that choice of $j = \alpha$ for which

$$\max_{1 \leq j \leq r} P_i^j(\sigma_{1j}^0, \ldots, \sigma_{nj}^0)$$

is attained, that is, $\sigma_1^0(x_0) = \alpha$. For any other vertex $x \neq x_0$, $\sigma_i^0(x)$ is defined, where necessary, as in Case 1. Then

$$P_1(\sigma_1^0, \ldots, \sigma_n^0) = P_1^\alpha(\sigma_{1\alpha}^0, \ldots, \sigma_{n\alpha}^0) \geq P_1^j(\sigma_{1j}^0, \ldots, \sigma_{nj}^0)$$

for $1 \leq j \leq r$.

Since $(\sigma_{1j}^0, \ldots, \sigma_{nj}^0)$ is an equilibrium point for Γ_j,

$$P_1^j(\sigma_{1j}', \sigma_{2j}^0, \ldots, \sigma_{nj}^0) \leq P_1^j(\sigma_{1j}^0, \ldots, \sigma_{nj}^0).$$

Now any pure strategy σ_1' for player 1 in Γ will truncate to some pure strategy σ_{1j}' in Γ_j for any j, $1 \leq j \leq r$. Thus

$$P_1(\sigma_1', \sigma_2^0, \ldots, \sigma_n^0) = P_1^j(\sigma_{1j}', \sigma_{2j}^0, \ldots, \sigma_{nj}^0).$$

where $\sigma_1'(x_0) = j$. Hence

$$P_1(\sigma_1', \sigma_2^0, \ldots, \sigma_n^0) \leq P_1^\alpha(\sigma_{1\alpha}^0, \ldots, \sigma_{i\alpha}^0, \ldots, \sigma_{n\alpha}^0) = P_1(\sigma_1^0, \ldots, \sigma_n^0). \tag{2.2}$$

If $i \neq 1$ since $(\sigma_{1\alpha}^0, \ldots, \sigma_{n\alpha}^0)$ is an equilibrium point for Γ_α

$$P_i(\sigma_1^0, \ldots, \sigma_i', \ldots, \sigma_n^0) = P_i^\alpha(\alpha_1^0 | \Gamma_\alpha, \ldots, \sigma_i' | \Gamma_\alpha, \ldots, \sigma_n^0 | \Gamma_\alpha)$$

$$= P_i^\alpha(\sigma_{1\alpha}^0, \ldots, \sigma_{i\alpha}', \ldots, \sigma_{n\alpha}^0) \leq P_i^\alpha(\sigma_{1\alpha}^0, \ldots, \sigma_{i\alpha}^0, \ldots, \sigma_{n\alpha}^0)$$

$$= P_i(\sigma_1^0, \ldots, \sigma_i^0, \ldots, \sigma_n^0)$$

since $\sigma_1^0(x_0) = \alpha$. Hence if $i \neq 1$,

$$P_i(\sigma_1^0, \ldots, \sigma_i', \ldots, \sigma_n^0) \leq P_i(\sigma_1^0, \ldots, \sigma_i^0, \ldots, \sigma_n^0). \tag{2.3}$$

But (2.2) and (2.3) together assert that $(\sigma_1^0, \ldots, \sigma_n^0)$ is an equilibrium point for Γ, and this completes the proof.

For a finite 2-person win-lose game (where the only possible outcome is for one player to win and the other to lose) which has perfect information, a very much simpler proof is possible. Let Γ be such a game with at most N moves for each player, and let A_i, B_i be the sets of possible i^{th} moves for players A and B respectively.

Consider the statement '$\exists\, a \in A_1 \,\forall\, b \in B_1$ A wins' which we read as 'there exists a in A_1 such that for every b in B_1 A wins' (it is convenient to omit the phrase 'such that' when using the logical symbols). Since the game Γ has perfect information A knows the state on A's first move, thus the statement '$\exists\, a \in A_1 \ldots$' implies not only that there exists a first move for A with a certain property (namely it will win against all of B's first moves) but that A *can determine which move it is.*

This being said we are now able to write down the statement that A has a winning pure strategy. It is:

$$\exists\, a \in A_1 \;\forall\, b \in B_1 \;\exists\, a \in A_2 \; ... \; \forall\, b \in B_N \; A \text{ wins.}$$

To logically negate a formal statement like this it is only necessary to interchange the symbols '∃' and '∀' everywhere. Thus the statement that A does *not* have a winning pure strategy is:

$$\forall\, a \in A_1 \;\exists\, b \in B_1 \;\forall\, a \in A_2 \; ... \; \exists\, b \in B_N \; A \text{ does not win.}$$

However, because Γ is a win-lose game, this is just the statement that B has a winning pure strategy. Hence, either A or B has a winning pure strategy; that is, the game is strictly determined.

Our next objective is to show that if Γ is no longer finite the above results are false; that is, there are such games which are not strictly determined. In some respects Problem 15 of Chapter 1 illustrates this point. However, in that case there were analytic requirements which may be regarded as limiting the choice of pure strategies. The example we shall use is in fact a 2-person win-lose game having perfect information, the binary game of Example 1.10. We shall show that the winning set $U \subseteq [0, 1]$ of the first player can be chosen in such a way that neither player has a winning pure strategy. The proof rests firmly on the Axiom of Choice:

> If S is a set, and the elements of S are not ϕ (the empty set) and are mutually disjoint, then there is a set T which has precisely one element in common with each element of S.

Thus this axiom conflicts with the above interchange of ∃ and ∀, to form a negation, when extended to infinite N. Some writers have advocated an alternative to the Axiom of Choice by advancing the interchange of ∃ and ∀, to form a negation, over any collection of sets as itself a possible axiom. Although this would have the pleasant consequence that every game of perfect information would be strictly determined, we shall adopt the conventional course and assume the Axiom of Choice.

Theorem 2.2 (Gale and Stewart [1]). There is an infinite game with no chance elements and having perfect information which is not strictly determined.

Before embarking upon the proof the reader should know that we shall make cavalier use of transfinite ordinals. Be not deterred. it is no more necessary to understand the set-theoretic construction of transfinite ordinals in order to make free use of them than is the case with real numbers.

Proof. Let S_A, S_B denote the sets of pure strategies for the first and second player respectively. A pair $\sigma \in S_A$, $\tau \in S_B$ determines a complete play of the binary game and hence a unique real number $(\sigma, \tau) \in [0, 1]$. We have, from the specification of the game, $(\sigma, \tau) \in U$ if A wins.

Denote the cardinality of any set S by $|S|$.

Since the game has perfect information, $\sigma \in S_A$ is just a function from the countable set of vertices labelled A to the set $\{0, 1\}$. The set of all such functions has cardinality $2^{\aleph_0}$. Similarly for $\tau \in S_B$.

Thus

$$|S_A| = |S_B| = 2^{\aleph_0}.$$

We can therefore index the respective strategies for each player by σ_β, τ_β, $\beta < \alpha$, where α is the smallest ordinal preceded by at least $2^{\aleph_0}$ ordinals.

Next we observe that for any pure strategy $\tau_0 \in S_B$ player A has $2^{\aleph_0}$ pure strategies σ which will give different values of (σ, τ_0); that is, if $S_{\tau_0} = \{(\sigma, \tau_0); \sigma \in S_A\} \subseteq [0, 1]$ then

$$|S_{\tau_0}| = 2^{\aleph_0}. \tag{2.4}$$

This amounts to saying that if every second binary digit of some number in $[0, 1]$ is already fixed (by τ_0) we can still produce $2^{\aleph_0}$ distinct numbers by varying the remaining digits (that is, by varying σ). Similarly if $S_{\sigma_0} = \{(\sigma_0, \tau); \tau \in S_B\} \subseteq [0, 1]$ then

$$|S_{\sigma_0}| = 2^{\aleph_0}. \tag{2.5}$$

In order to specify a set U with the required peculiar property we now proceed to construct two sets V and W of real numbers in $[0, 1]$. We first choose real numbers $s_0 \in S_{\tau_0}$, $t_0 \in S_{\sigma_0}$ so that $t_0 \neq s_0$, which is plainly possible by (2.4) and (2.5). Here σ_0, τ_0 are respectively the first elements in S_A, S_B according to the ordinal indexing defined above. We continue the construction inductively on the ordinal index β. If s_γ, t_γ have been chosen for all $\gamma < \beta < \alpha$, the set $\{t_\gamma; \gamma < \beta\}$ has fewer than $2^{\aleph_0}$ members so $S_{\tau\beta} \setminus \{t_\gamma; \gamma < \beta\}$ is not empty. Choose one of its elements and call it s_β. Similarly $S_{\sigma\beta} \setminus \{s_\gamma; \gamma \leq \beta\}$ is not empty. Call one of its members t_β. By the Axiom of Choice we can define sets

$$V = \{s_\gamma; \gamma < \alpha\}, \quad W = \{t_\gamma; \gamma < \alpha\}.$$

Moreover, V and W are disjoint, that is, for any γ and β, $s_\gamma \neq t_\beta$. For if $\gamma \leq \beta$, then $t_\gamma \in S_{\sigma\beta} \setminus \{s_\delta; \delta \leq \beta\}$, so $t_\beta \notin \{s_\delta; \delta \leq \beta\}$, and therefore $t_\beta \neq s_\gamma$. On the other hand if $\gamma > \beta$ then $s_\gamma \in S_{\tau\gamma} \setminus \{t_\delta; \delta < \gamma\}$, so $s_\gamma \notin \{t_\delta; \delta < \gamma\}$, and therefore $s_\gamma \neq t_\beta$.

We now choose U in any manner provided only that $V \subseteq U$ and $W \subseteq [0, 1] \setminus U$. To show that the game so defined is not strictly determined it suffices to show that neither player has a winning pure strategy. Let $\sigma \in S_A$. Then σ corresponds to some index β, say. By construction there exists $t_\beta \in S_{\sigma\beta}$ and hence there exists $\tau \in S_B$ such that $(\sigma_\beta, \tau) = t_\beta \in W \subseteq [0, 1] \setminus U$. Thus if A uses the arbitrary pure strategy σ there is a pure strategy τ for B such that B wins. However, the situation is symmetrical. In exactly the same way it follows that for any $\tau \in S_B$ there exists $\sigma \in S_A$ such that $(\sigma, \tau) \in U$; that is, A wins, and the theorem is proved.

2.2 NORMAL FORMS AND MIXED STRATEGY EQULIBRIA.

Although not all finite n-person non-cooperative games have pure strategy equilibria we can ask about the situation if mixed strategies are permitted. Is it true that in terms of mixed strategies every such game has an equilibrium point? This question was answered affirmatively in 1951 by J. F. Nash [2]. His result, which generalises the von Neumann minimax theorem, is the main objective of this section and certainly provides one of the strongest arguments in favour of 'equilibrium points' as a solution concept for n-person non-coooperative games. Because the proof involves an appeal to some moderately advanced mathematics and is also of a non-constructive nature we shall give a technically rather simpler proof of the von Neumann minimax theorem in §2.5.

Denote by the index set I the set of players. For an n-person game $I = \{1, 2, ..., n\}$. Let x_i be an arbitrary mixed strategy for the ith player, that is, a probability distribution on the set S_i of that player's pure strategies. The probability assigned by x_i to some $\sigma_i \in S_i$ is denoted by $x_i(\sigma_i)$. The set of all mixed strategies of player i is denoted by X_i. We assume, since the game is non-cooperative, that the mixed strategies of all the players $1, 2, ..., n$, viewed as probability distributions, are jointly independent, that is, the probability $x(\sigma)$ of arriving at the pure strategy n-tuple $\sigma = (\sigma_1, ..., \sigma_n)$, $\sigma_i \in S_i$ is assumed to be $x(\sigma) = x_1(\sigma_1)x_2(\sigma_2)...x_n(\sigma_n)$.

In terms of pure strategies the payoff to player i is given by P_i, where

$$P_i : S = S_1 \times ... \times S_n \to \mathbb{R} ,$$

that is, P_i is a function which maps each $\sigma = (\sigma_1, ..., \sigma_n) \in S$ to a real number. If mixed strategies x_i, $i \in I$, are used the payoff will be the statistical expectation of P_i with respect to the distribution $x(\sigma)$, namely

$$P_i(x_1, ..., x_n) = \sum_{\sigma \in S} P_i(\sigma)x(\sigma),$$

$$P_i(x_1, ..., x_n) = \sum_{\sigma_1 \in S_1} \cdots \sum_{\sigma_n \in S_n} P_i(\sigma_1, ..., \sigma_n) \prod_{j=1}^{n} x_j(\sigma_j). \qquad (2.6)$$

It is convenient to introduce the following notation. For the strategy n-tuple $x = (x_1, ..., x_n)$ we write

$$x|x_i' = (x_1, ..., x_{i-1}, x_i', x_{i+1}, ..., x_n).$$

This means that the player i has in x replaced the strategy x_i by x_i'. We note in passing that (2.6) implies

$$P_i(x \| \sigma_i) = \sum_{\sigma_1 \in S_1} \cdots \sum_{\sigma_{i-1} \in S_{i-1}} \sum_{\sigma_{i+1} \in S_{i+1}} \cdots \sum_{\sigma_n \in S_n} P_i(\sigma) \prod_{\substack{j=1 \\ j \neq i}}^{n} x_j(\sigma_j), \qquad (2.7)$$

where $\sigma = (\sigma_1, ..., \sigma_n)$.

Definition. A **non-cooperative game** Γ **in normal form** is a collection

$$\Gamma = \langle I, \{X_i\}_{i \in I}, \{P_i\}_{i \in I} \rangle,$$

in which the set of players is I, the set of strategies for player i is X_i, and the payoff to player i is given by

$$P_i : \prod_{i \in I} X_i \to \mathbf{R}$$

Here the sets X_i could be taken to be sets of pure or mixed strategies. If, as intended, we take the X_i to consist of mixed strategies then Γ is called the **mixed extension** of the original pure strategy game (where X_i is replaced by S_i throughout), and the payoff functions P_i are given by (2.6). We shall assume throughout that $I = \{1, 2, ..., n\}$ for some integer $n \geq 2$.

If the sets X_i are all subsets of a finite dimensional vector space the game Γ is said to be a **finite** n-person game. This definition is consistent with that given in §2.1 for games in extensive form. For if in a game Γ player i has a finite number m_i of pure strategies, then for this player a typical mixed strategy is

$$\mathbf{x}_i = (x_1, ..., x_{m_i}), \ 0 \leq x_j \leq 1, \ \sum_{j=1}^{m_i} x_j = 1,$$

where $x_j = \mathbf{x}_i(\sigma_i^j)$ is the probability assigned by $\mathbf{x}_i$ to the jth pure strategy $\sigma_i^j \in S_i$. The set of all such $\mathbf{x}_i$ is an (m_i-1)-dimensional simplex in m_i-dimensional Euclidean space. We denote this geometrical representation of X_i by $\Sigma^{(i)}$. Figure 2.2 illustrates $\Sigma^{(i)}$ in the case $m_i = 3$.

A mixed strategy n-tuple $x = (x_1, ..., x_n)$ is called **admissible for player** i if

$$P_i(x \| x_i') \leq P_i(x)$$

for every mixed strategy $x_i' \in X_i$. This means that if the other players use the strategies specified by x, then player i has no positive incentive to alter the strategy x_i. If x is admissible for every player then it is an equilibrium point; that is:

Definition. A mixed strategy n-tuple $x = (x_1, ..., x_n)$, $x_i \in X_i$, is an **equilibrium point** of an n-person non-cooperative game Γ if for each i, $1 \leq i \leq n$, and any $x_i' \in X_i$,

$$P_i(x \| x_i') \leq P_i(x). \qquad (2.8)$$

Theorem 2.3. A mixed strategy n-tuple $x = (x_1, ..., x_n)$ is an equilibrium point of a finite game Γ if and only if for each player index i,

$$P_i(x\|\sigma_i) \le P_i(x) \qquad (2.9)$$

for every pure strategy $\sigma_i \in S_i$.

Proof. If x is an equilibrium point of Γ, for each i, $1 \le i \le n$, the inequality (2.9) is immediate from (2.8) since a pure strategy is a particular case of a mixed strategy.

To prove that the condition is sufficient to ensure that x is an equilibrium point, choose an arbitrary mixed strategy $x_i{'} \in X_i$, multiply (2.9) by $x_i{'}(\sigma_i)$, and sum over all $\sigma_i \in S_i$. We obtain

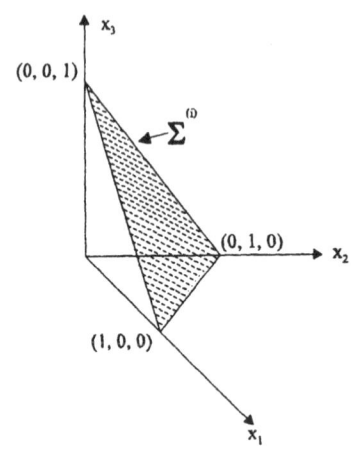

Figure 2.2 Set of mixed strategies $\Sigma^{(i)}$ ($m_i = 3$)

$$\sum_{\sigma_i \in S_i} P_i(x\|\sigma_i)x_i{'}(\sigma_i) \le \sum_{\sigma_i \in S_i} x_i{'}(\sigma_i) P_i(x).$$

From (2.7) and (2.6) this gives

$$P_i(x\|x_i{'}) \le P_i(x) \sum_{\sigma_i \in S_i} x_i{'}(\sigma_i),$$

which is (2.8) since

$$\sum_{\sigma_i \in S_i} x_i{'}(\sigma_i) = 1$$

This theorem gives an effective procedure for checking a possible equilibrium point.

In order to prove our main theorem we require

Lemma 2.1. For any mixed strategy n-tuple $x = (x_1, \ldots, x_n)$ each player i, $1 \le i \le n$, possesses a pure strategy σ_i^k such that

$$x_i(\sigma_i^k) > 0 \text{ and } P_i(x\|\sigma_i^k) \le P_i(x).$$

Proof. Suppose the conclusion is false, then for all pure strategies $\sigma_i \in S_i$ with $x_i(\sigma_i) > 0$ we have $P_i(x\|\sigma_i) > P_i(x)$. Hence for all such strategies σ_i,

$$P_i(x\|\sigma_i) x_i(\sigma_i) > P_i(x) x_i(\sigma_i). \qquad (2.10)$$

For all other $\sigma_i \in S_i$ we have $x_i(\sigma_i) = 0$ so that

$$P_i(x \| \sigma_i) \; x_i(\sigma_i) \; = \; P_i(x) \; x_i(\sigma_i) \; = \; 0 \,. \tag{2.11}$$

Summing (2.10) and (2.11) over all pure strategies $\sigma_i \in S_i$ we obtain, since $x_i(\sigma_i) \neq 0$ for at least one $\sigma_i \in S_i$,

$$\sum_{\sigma_i \in S_i} P_i(x \| \sigma_i) \; x_i(\sigma_i) \; > \sum_{\sigma_i \in S_i} P_i(x) \; x_i(\sigma_i)$$

which on using (2.6) and (2.7) gives

$$P_i(x) > P_i(x) \sum_{\sigma_i \in S_i} x_i(\sigma_i) \; = \; P_i(x) \,,$$

a contradiction.

We recall that a set Σ in Euclidean space is called **convex** if the line segment joining any two points of Σ lies entirely in Σ; that is if $\mathbf{x}, \mathbf{y} \in \Sigma$, then for any real λ, $0 \le \lambda \le 1$, the point $\lambda \mathbf{x} + (1 - \lambda)\mathbf{y}$ is also in Σ. Figure 2.3 illustrates the idea in 2-dimensional space.

It is simple to show that a finite Cartesian product of convex sets $\Sigma^{(i)}$ in m_i-dimensional space is a convex set in $(m_1 + m_2 \ldots + m_n)$-dimensional space. For example the Cartesian product of a square and a line segment is a cube.

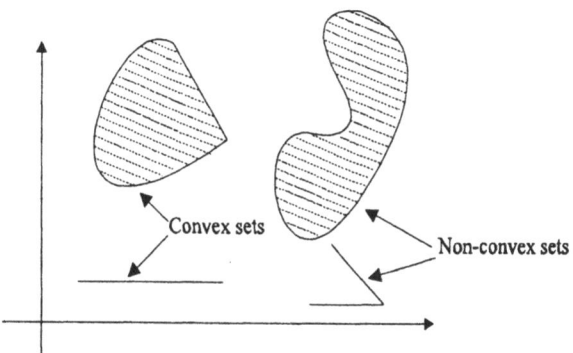

Theorem 2.4 (Nash). Any finite n-person non-coooperative game Γ has at least one mixed strategy equilibrium point.

Proof. If player i in the game Γ possesses m_i pure strategies then, as remarked earlier, the set X_i of his or her mixed strategies can be represented geometrically by an $(m_i\text{-}1)$-dimensional simplex in m_i-dimensional space which we denote by $\Sigma^{(i)}$. Each $\Sigma^{(i)}$ is obviously a convex closed bounded subset of m_i-dimensional space. Hence the Cartesian product $\Sigma^{(1)} \times \ldots \times \Sigma^{(n)}$ is a convex closed bounded $(m_1 + \ldots + m_n - n)$-dimensional subset of $(m_1 + \ldots + m_n)$-dimensional Euclidean space. Moreover any mixed strategy n-tuple $x = (x_1, \ldots, x_n)$ can be viewed as a point in $\Sigma^{(1)} \times \ldots \times \Sigma^{(n)}$.

Figure 2.3 Convex and non-convex sets.

For any such $x \in \Sigma^{(1)} \times \ldots \times \Sigma^{(n)}$ and any pure strategy $\sigma_i^j \in S_i$ for player i we define

$$\phi_{ij}(x) \; = \; \max \; \{0, \; P_i(x \| \sigma_i^j) \, - \, P_i(x) \} \,. \tag{2.12}$$

Plainly $\phi_{ij}(x) \geq 0$, and these functions measure the increase in the payoff to player i when the mixed strategy n-tuple x is being used if this player replaces the strategy x_i by the pure strategy σ_i^j. Any possible decrease is not indicated by ϕ_{ij} since in this case the function vanishes.

For each i, j, $1 \leq i \leq n$, $1 \leq j \leq m_i$ we consider the numbers

$$\frac{x_i(\sigma_i^j) + \phi_{ij}(x)}{1 + \sum_{j=1}^{m_i} \phi_{ij}(x)} \tag{2.13}$$

Each such number is plainly non-negative, and summing over all $\sigma_i^j \in S_i$ we have

$$\sum_{j=1}^{m_i} \frac{x_i(\sigma_i^j) + \phi_{ij}(x)}{1 + \sum_{j=1}^{m_i} \phi_{ij}(x)} = 1 ,$$

since

$$\sum_{j=1}^{m_i} x_i(\sigma_i^j) = 1 .$$

Hence for fixed x and i the number in (2.13) could be taken as a probability attached to the pure strategy σ_i^j. For $1 \leq j \leq m_i$ the set of all m_i of these numbers is then a particular mixed strategy for player i.

Since the numbers in (2.13) are constructed for each i, their totality as x is fixed, and i varies, $1 \leq i \leq n$, determines a mixed strategy n-tuple $(y_1, ..., y_n)$. Plainly $y = (y_1, ..., y_n)$ is a function of the original mixed strategy n-tuple $x = (x_1, ..., x_n)$, that is $y = f(x)$. Thus the function f maps the convex closed bounded set $\Sigma^{(1)} \times ... \times \Sigma^{(n)}$ into itself.

We next establish that f is a *continuous* vector valued function of the variable x viewed as a point in $(m_1 + ... + m_n)$-dimensional Euclidean space, for the payoff function $P_i(x)$ is determined by the $m_1 m_2 ... m_n$ numbers $P_i(\sigma_1, ..., \sigma_n)$, $\sigma_i \in S_i$, according to (2.6). Thus $P_i(x)$ is actually a polynomial in the components $x_i(\sigma_i)$ of x. Hence each $P_i(x)$ is a continuous function of the vector variable x. Thus each $\phi_{ij}(x)$ defined by (2.12) is also a continuous function of x. Now

$$y_i = \frac{x_i(\sigma_i^j) + \phi_{ij}(x)}{1 + \sum_{j=1}^{m_i} \phi_{ij}(x)} ,$$

$x_i(\sigma_i^j)$ is just a component of x, $\phi_{ij}(x)$ is a continuous function of x, and

$$1 + \sum_{j=1}^{m_i} \phi_{ij}(x) \neq 0.$$

It therefore follows that y_i is a continuous function of x. Hence $y = (y_1, ..., y_n) = f(x)$ is a continuous function of x.

From the above we conclude that the conditions of *Brouwer's fixed point theorem* are satisfied (cf. Appendix 1). This theorem asserts that a continuous mapping f of a convex closed bounded subset of a finite-dimensional Euclidean space into itself possesses at least one fixed point, that is, a point x^0 such that $f(x^0) = x^0$. Thus there exists a mixed strategy n-tuple $x^0 = (x_1^0, ..., x_n^0)$ such that for all $i, j, 1 \leq i \leq n, 1 \leq j \leq m_i$,

$$x_i^0(\sigma_i^j) = \frac{x_i^0(\sigma_i^j) + \phi_{ij}(x^0)}{1 + \sum_{j=1}^{m_i} \phi_{ij}(x^0)}. \tag{2.14}$$

From Lemma 2.1 and (2.12) there exists for any player i a pure strategy σ_i^k such that $x_i^0(\sigma_i^k) > 0$ and $\phi_{ik}(x^0) = 0$. For this particular strategy (2.14) becomes

$$x_i^0(\sigma_i^k) = \frac{x_i^0(\sigma_i^k)}{1 + \sum_{j=1}^{m_i} \phi_{ij}(x^0)},$$

whence

$$x_i^0(\sigma_i^k) \sum_{j=1}^{m_i} \phi_{ij}(x^0) = 0.$$

However, since $x_i^0(\sigma_i^k) > 0$, this implies

$$\sum_{j=1}^{m_i} \phi_{ij}(x^0) = 0.$$

But since, for each j, $\phi_{ij}(x^0) \geq 0$ we conclude

$$\phi_{ij}(x^0) = 0, \qquad (1 \leq i \leq n, \ 1 \leq j \leq m_i).$$

From (2.12) this tells us that for each player index i

$$P_i(x^0 \| \sigma_i^j) \leq P(x^0)$$

for all j, $1 \leq j \leq m_i$, that is, for all pure strategies $\sigma_i^j \in S_i$. Hence from Theorem 2.3 $x^0 = (x_1^0, ..., x_n^0)$ is an equilibrium n-tuple for Γ and the proof is complete.

It is worth repeating that Nash's theorem is 'non-effective' in the sense that it does not provide an algorithm whereby one can compute an equilibrium n-tuple for the game Γ. This is because Brouwer's fixed point theorem does not reveal how to find a fixed point but merely guarantees its existence. Unfortunately no general algorithm for computing a mixed strategy equilibrium point if $n \geq 3$ appears to be known at present.

However, in games with a small number of players and where each player has only a few pure strategies it is usually possible to produce an analysis. We illustrate by sketching the process for 2×2 bimatrix games. The details can be found in the excellent text of Vorob'ev [3].

Let

		Player 2		
		y	$1\text{-}y$	a_{ij}= payoff to 1
Player 1	x	(a_{11}, b_{11})	(a_{12}, b_{12})	b_{ij}= payoff to 2
	$1\text{-}x$	(a_{21}, b_{21})	(a_{22}, b_{22})	

be an arbitrary 2×2 bimatrix game. Since only two pure strategies are available to each player the sets of mixed strategies for players 1 and 2 respectively can be described as

$$X = \{ \ (x, 1 - x); \ 0 \leq x \leq 1 \ \},$$
$$Y = \{ \ (y, 1 - y); \ 0 \leq y \leq 1 \ \}.$$

With this notation the payoff functions $P_i((x, 1\text{-}x), (y, 1\text{-}y))$ $(i = 1, 2)$ are just functions of the ordered pair (x, y), that is, $P_i((x, 1\text{-}x), (y, 1\text{-}y)) = P_i(x, y)$. Thus a point (x, y) in the unit square uniquely represents a mixed strategy 2-tuple. A little algebra reveals that the admissible mixed strategy 2-tuples *for player 1* are just those points in the unit square of the form

(a) $(0, y)$ if $Ay - a < 0,$
or (b) (x, y) if $Ay - a = 0,$
or (c) $(1, y)$ if $Ay - a > 0,$

where $A = a_{11} - a_{12} - a_{21} + a_{22}$, $a = a_{22} - a_{12}$. Similarly the admissible mixed strategy 2-tuples *for player 2* are just those points in the unit square which satisfy

(a$'$) $(x, 0)$ if $Bx - b < 0,$
or (b$'$) (x, y) if $Bx - b = 0,$
or (c$'$) $(x, 1)$ if $Bx - b > 0,$

where $B = b_{11} - b_{12} - b_{21} + b_{22}$, $b = b_{22} - b_{21}$. For an equilibrium point we require that the mixed strategy 2-tuple be admissible for *both players*. Taking the intersection of the above two sets we find that if a mixed strategy equilibrium exists at all it is given by

$$(x, y) = (b/B, a/A) \in [0, 1] \times [0, 1].$$

Thus, for example, the bimatrix game

$$\begin{pmatrix} (4,\ 1) & (0,\ 0) \\ (0,\ 0) & (1,\ 4) \end{pmatrix}$$

has pure strategy equilibria given by $((1, 0), (1, 0))$ and $((0, 1), (0, 1))$ corresponding to the payoff entries $(4, 1)$ and $(1, 4)$ respectively, and a mixed strategy equilibrium given by $((^4/_5, ^1/_5),$ $(^1/_5, ^4/_5))$. This last 'solution' has a certain attraction since it gives the same payoff of $^4/_5$ to both players. However, this is merely a reflection of the symmetry of the game, and in general such equilibria will not occur.

2.3 DISCUSSION OF EQUILIBRIA.

The notion of an 'equilibrium point' as a solution concept for non-zero sum, non-cooperative games was introduced without comment, and it is now time to examine the idea more carefully. To a mathematician the theorems proved on equilibria have a satisfying feel. They neatly generalise the results on 2-person zero sum games described in the first chapter. However, we now call a halt to this wave of mathematical euphoria and look at some of the basic problems inherent in non-zero sum, non-cooperative games.

Because such games often have more than one equilibrium point we are frequently unable to obtain an intuitively satisfying solution. In these circumstances we cannot say, for example, that to each player the game has a certain 'value', for the same player might obtain different payoffs at different equilibria. Thus each player may have his or her preferred equilibrium, but if each insists upon his or her own preferred choice the consequences could well be disastrous for all. The way is therefore open, especially when the game is played many times, for a *tacit* agreement between players where each employs strategies which rely on the other's good sense (see [4]).

To pursue this line of thought further, consider the game

		Player 2	
		τ_1	τ_2
Player 1	σ_1	(-20,-20)	(15,-15)
	σ_2	(-15, 15)	(10, 10)

Here both of the pure strategy combinations (σ_1, τ_2) and (σ_2, τ_1) are equilibrium points (there is also a mixed strategy equilibrium which gives payoffs $(-^5/_2, -^5/_2)$). Although player 1 prefers the former and player 2 the latter, once either of these strategy combinations is employed neither player will want to move away from it unilaterally. In these circumstances if each player chooses a strategy corresponding to his or her preferred equilibrium the outcome will be (σ_1, τ_1) which is the worst possible outcome for both players. Yet given that each player has full knowledge of the opponent's possible strategies and corresponding payoffs, it hardly seems likely that players would want to pursue a policy of maximising their respective equilibrium payoffs knowing that this would lead to the worst outcome for all.

On the other hand it does seem credible that each player may individually adopt a cautious strategy which, granted tacit cooperation of the other, would result in the payoff pair (10, 10) generated by the pure strategy combination (σ_2, τ_2). But (σ_2, τ_2) is *not* an equilibrium point: each player will be tempted to switch to his or her first pure strategy. The relentless pursuit of individual rationality for a game such as this can, evidently, be unsatisfactory.

The preceding discussion prompts us to ask whether behaviour according to a *tacit* collective rationality, even in non-cooperative games, would be fruitful, given that the players have complete information about each of the others' payoffs. Consider, for example, the well known *Prisoners' Dilemma* attributed to A. W. Tucker. In this game two men are in prison serving sentences for previous convictions. They are held in separate cells. The D.A. suspects that these two men were responsible for an armed robbery which he has so far been unable to clear up. However, the evidence is inadequate to prove his suspicions, so he approaches each prisoner and offers him the opportunity to confess and give evidence against the other. He promises that if one confesses and the other does not, the former will get one year's remission on his present sentence (that is, early parole) and the latter will get a nine-year sentence. The prisoners know that if both confess they will each get a four-year sentence, but that if neither confesses the robbery charges will be dropped although both will get one year for carrying a concealed weapon. This interprets into the bimatrix game

	C_2	D_2	
C_1	(-4,-4)	(1,-9)	C = Confess
D_1	(-9, 1)	(-1,-1)	D = Deny

Clearly the row player will prefer C_1 over D_1, and the column player will prefer C_2 over D_2. Thus from the viewpoint of *individual rationality* both players should confess. For either to depart from this strategy would be foolhardy since it may result in a nine-year sentence, while his partner would be released a year earlier. Indeed (C_1, C_2) is the *only* equilibrium point.

The trouble is that both would be better off to deny the charges. From the viewpoint of *collective rationality* the payoff pair (-1, -1) is preferred to (-4, -4). Yet each is tempted to break any tacit agreement to deny the charges and *knows* the other player *is also tempted* to defect and confess. The dilemma is whether each player is to trust the other; that is, the dilemma confronting each is whether to operate according to an individual rationality or a collective rationality.[10]

When a game is played *only once* it seems plausible that individual rationality is more likely to operate. If, however, the game is played *many* times then it is arguable that collective rationality is more likely to come into play. For although the prisoners cannot communicate

---

[10]This is precisely the dilemma of a labour union considering an inflationary wage claim.

directly, they can signal their willingness to cooperate with each other by playing cooperative strategies in the initial plays of the game in order to establish their good intentions. Nevertheless these observations in no way ease the logical difficulty of providing a solution which is satisfactory under *all* assumptions.

For the reasons indicated, the role of equilibria in the theory of non-zero sum, non-cooperative games is controversial.

2.4 PRELIMINARY RESULTS FOR 2-PERSON ZERO SUM GAMES.

In the remainder of this chapter the theory of 2-person zero sum games described in Chapter 1 is established (although we shall not prove Ville's theorem). We consider such games in their normal form $\langle X, Y, P \rangle$, where X denotes the set of strategies for player 1, Y denotes the set of strategies for player 2, and P is a real-valued function $P: X \times Y \to \mathbb{R}$, which represents the payoff to player 1.

A real valued function f defined on a convex subset S of Euclidean space is called a **convex function** if

$$f(\lambda x + (1 - \lambda)y) \le \lambda f(x) + (1 - \lambda) f(y)$$

for all $x, y \in S$ and $0 \le \lambda \le 1$. A function f is called **concave** if $-f$ is convex. Figure 2.4 is intended to illustrate these ideas when S is a disc. The main feature is that a convex function lies *below* each point of the chord joining any two points on the graph.

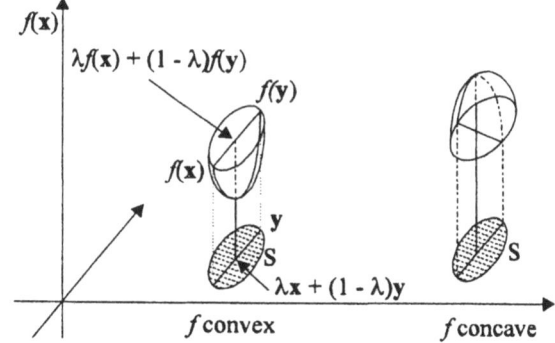

The examples of §1.7 demonstrate the enormous diversity of strategy sets X, Y which can arise in these games. Likewise the function P can take many different forms. We can, however, single out two cases as being of special interest.

Figure 2.4 Convex and concave functions.

(i) X, Y are sets of mixed strategies which are convex closed bounded subsets of Euclidean space, and $P(x, y)$ is linear in each variable separately.

(ii) X, Y are sets of pure strategies which are convex closed bounded subsets of Euclidean space, and $P(x, y)$ is a continuous function.

The first corresponds to finite matrix games, for which we have the von Neumann minimax theorem

$$\max_{x \in X} \min_{y \in Y} P(x, y) = \min_{y \in Y} \max_{x \in X} P(x, y).$$

The second generalises the notion of a continuous game on the unit square and can, in general, be solved only by defining probability distribution functions on the sets X, Y and passing to the mixed strategy extension. For continuous games on the unit square the earliest proof that such solutions exist was given by Ville, who was also the first to exhibit an example which demonstrated the impossibility of proving a general mixed strategy minimax theorem if P is merely supposed to be a bounded function.

Given X, Y, sets of pure or mixed strategies, what are the minimal requirements for a minimax theorem to hold? Firstly some kind of continuity assumption on $P(x, y)$, which in turn requires that X, Y be at least topological spaces. In addition, convexity assumptions invariably play a critical role, which requires that X and Y be *linear* topological spaces. With suitable further assumptions on $P(x, y)$ it is sometimes possible to prove

$$\sup_{x \in X} \ \inf_{y \in Y} \ P(x, y) = \inf_{y \in Y} \ \sup_{x \in X} \ P(x, y) = v.$$

This is essentially equivalent to guaranteeing the existence of ε-optimal strategies for any $\varepsilon > 0$. A pair of strategies x_0 and y_0 are said to be ε-**optimal** if

$$P(x_0, y) \geq v - \varepsilon \quad \text{for all } y \in Y,$$

and

$$P(x, y_0) \leq v + \varepsilon \quad \text{for all } x \in X.$$

If in addition the *compactness* of the strategy spaces, with respect to the appropriate topology imposed upon them, is assumed, then inf and sup can be replaced by min and max respectively. (Recall that in Euclidean space 'compact' is equivalent to 'closed bounded'.) A typical result of this kind is:

Theorem. Let X and Y be convex closed bounded subsets of Euclidean spaces. Let $P: X \times Y \to \mathbb{R}$ be continuous, convex in **y** for each **x** and concave in **x** for each **y**. Then

$$\max_{x \in X} \ \min_{y \in Y} \ P(x, y) = \min_{y \in Y} \ \max_{x \in X} \ P(x, y).$$

This can be proved (see, for example, [5]) by use of the Kakutani fixed point theorem (see Appendix 1) and covers the case of finite matrix games because there the payoff kernel $P(x, y)$ is *linear* in each variable separately and so trivially convex in **y** and concave in **x**. The hypothesis on P is a very reasonable one, for we can see intuitively that concavity in **x** and convexity in **y** should force the existence of a saddle point (in the case of matrix games this is of course in terms of mixed strategies).

Before we begin the discussion of matrix games we prove some preliminary results in the general context of topological spaces. If the notion of a topological space is unfamiliar, think of X and Y as subsets of Euclidean spaces. In this case the proofs can be simplified.

Lemma 2.2. Let $P(x, y)$ be a real-valued function on $X \times Y$, then

$$\sup_{x \in X} \inf_{y \in Y} P(x, y) \leq \inf_{y \in Y} \sup_{x \in X} P(x, y) .$$

Moreover if X and Y are both compact subsets of topological spaces and P is continuous, then sup and inf are attained and so can be replaced by max and min respectively.

Proof. For any $y \in Y$

$$P(x, y) \leq \sup_{x \in X} P(x, y) \text{ for every } x \in X .$$

Hence

$$\inf_{y \in Y} P(x, y) \leq \inf_{y \in Y} \sup_{x \in X} P(x, y) \text{ for every } x \in X$$

and the result follows.

For the second part if X and Y are compact subsets of their respective spaces then $X \times Y$ is a compact subset of the product space. The image of a compact set under a continuous function is again a compact set. Thus $P(X \times Y)$ is a compact, hence closed bounded, subset of $\mathbb{R}$, and the conclusion follows.

Our next aim is to establish in Theorem 2.5 a very useful criterion for the equality of maximin and minimax. To this end we require:

Lemma 2.3. If X and Y are both compact subsets of topological spaces and $P: X \times Y \to \mathbb{R}$ is continuous, then

$$m(y) = \max_{x \in X} P(x, y) : Y \to \mathbb{R}$$

is continuous.

Proof. The definition of pointwise continuity in a topological space is: m is continuous at y_0 if given a neighbourhood A of $m(y_0)$ then $m^{-1}(A) = \{ y \in Y; m(y) \in A \}$ is a neighbourhood of y_0 (that is, contains an open set containing y_0). We shall show $m(y)$ is continuous at each $y_0 \in Y$.

Since $P(x, y_0)$ is continuous on the compact set X there exists $x_0 \in X$, not necessarily unique, such that $m(y_0) = P(x_0, y_0)$. As P is continuous in y, $\forall \; \varepsilon > 0$ there exists an open neighbourhood $N(y_0) \subseteq Y$ such that

$$| P(x_0, y_1) - P(x_0, y_0) | < \varepsilon \;\; \forall \; y_1 \in N(y_0) .$$

Hence $P(x_0, y_1) > P(x_0, y_0) - \varepsilon$ so that

$$m(y_1) = \max_{x \in X} P(x, y_1) \geq P(x_0, y_1) > m(y_0) - \varepsilon \quad \forall \ y \in N(y_0) \ .$$

To establish a similar inequality in the other direction we proceed as follows. Given $y_0 \in Y$ and $\varepsilon > 0$ we can, since P is continuous on $X \times Y$, define for each $x \in X$ an open neighbourhood $N(x) \times N(y_0, x) \subseteq X \times Y$ of (x, y_0) so that the value of P at every point in the neighbourhood differs from $P(x, y_0)$ by less than ε. Plainly $\{ N(x); x \in X \}$ is an open cover of X and hence, since X is compact, contains a finite subcover $\{ N(x^{(i)}); 1 \leq i \leq n \}$ of X. Then

$$\bigcap_{i=1}^{n} N(y_0, x^{(i)}) = N(y_0) , \text{ say,}$$

is an open neighbourhood of y_0. Let $y_1 \in N(y_0)$. Since P is continuous on the compact set X, there exists $x_1 \in X$, not necessarily unique, such that $m(y_1) = P(x_1, y_1)$. Suppose $x_1 \in N(x^{(j)})$ then $(x_1, y_1) \in N(x^{(j)}) \times N(y_0)$ so that

$$P(x_1, y_1) < P(x^{(j)}, y_0) + \varepsilon \quad \forall \ y_1 \in N(y_0) ,$$

that is,

$$m(y_1) < P(x^{(j)}, y_0) + \varepsilon \quad \forall \ y_1 \in N(y_0),$$

so that

$$m(y_1) < \max_{x \in X} P(x, y_0) + \varepsilon \quad \forall \ y_1 \in N(y_0),$$

that is,

$$m(y_1) < m(y_0) + \varepsilon \quad \forall \ y_1 \in N(y_0).$$

We have now shown that $\forall \ \varepsilon > 0$ if $A = (\ m(y_0) - \varepsilon, \ m(y_0) + \varepsilon \ )$, there exists an open neighbourhood $N(y_0)$ such that $m(N(y_0)) \subseteq A$, hence

$$N(y_0) \subseteq m^{-1}(m(N(y_0))) \subseteq m^{-1}(A) .$$

Thus $m(y)$ is continuous at y_0, and the proof is complete.

We remark that the lemma is true with max replaced by min (simply replace P in the proof by $-P$). Moreover, the proof is greatly simplified if we assume that X and Y are compact subsets of *metric* spaces, for we can then use the theorem that a continuous function from a compact metric space to a metric space (namely $\mathbb{R}$) is uniformly continuous.

Theorem 2.5. Let X and Y be compact subsets of topological spaces, and P be a continuous function $P: X \times Y \to \mathbb{R}$. Then

$$\max_{x \in X} \min_{y \in Y} P(x, y) = \min_{y \in Y} \max_{x \in X} P(x, y) \tag{2.15}$$

if, and only if, there exist $x_0 \in X$, $y_0 \in Y$ and $v \in \mathbb{R}$ such that

$$P(x_0, y) \geq v \quad \forall \ y \in Y \text{ and } P(x, y_0) \leq v \quad \forall \ x \in X . \qquad (2.16)$$

Proof. We first assume (2.15) and show that (2.16) holds. By Lemma 2.3,

$$\max_{x \in X} P(x, y)$$

is continuous. Since Y is compact it follows that there exists $y_0 \in Y$ so that

$$\max_{x \in X} P(x, y_0) = \min_{y \in Y} \max_{x \in X} P(x, y) = v , \text{ say.}$$

Similarly, since

$$\min_{y \in Y} P(x, y)$$

is continuous and X is compact, there exists $x_0 \in X$ so that

$$\min_{y \in Y} P(x_0, y) = \max_{x \in X} \min_{y \in Y} P(x, y) = v ,$$

where the last equality follows from (2.15). Then $P(x_0, y) \geq v$ and $P(x, y_0) \leq v$ as required.

We now assume (2.16) and show that (2.15) holds. Since $P(x_0, y) \geq v \ \forall \ y \in Y$ we have

$$\min_{y \in Y} P(x_0, y) \geq v \text{ and } \max_{x \in X} \min_{y \in Y} P(x, y) \geq v .$$

Similarly

$$\max_{x \in X} P(x, y_0) \leq v \text{ and } \min_{y \in Y} \max_{x \in X} P(x, y) \leq v .$$

Hence by Lemma 2.2

$$\max_{x \in X} \min_{y \in Y} P(x, y) = \min_{y \in Y} \max_{x \in X} P(x, y)$$

which is (2.15).

2.5 THE MINIMAX THEOREM FOR MATRIX GAMES.

For an $m \times n$ matrix game we identify the pure strategies for player 1 with the standard basis vectors $\sigma_1, ..., \sigma_m$, where $\sigma_i = (0, ...,0, 1, 0, ...,0)$ and the 1 occurs as the ith component. The set of all strategies, pure and mixed, X_1 for player 1 can then be represented geometrically by the $(m-1)$-dimensional simplex

$$\Sigma^{(1)} = \{ \mathbf{x} = \sum_{i=1}^{m} x_i \sigma_i; \; x_i \geq 0 \; (1 \leq i \leq m), \; \sum_{i=1}^{m} x_i = 1 \}$$

that is the convex set spanned by the m linearly independent points $\sigma_1, ..., \sigma_m$ (cf. Figure 2.2). Similarly we can take the pure strategies for player 2 to be the standard basis $\tau_1, ..., \tau_n$, and associate X_2 with the corresponding simplex $\Sigma^{(2)}$. The payoff kernel then takes the form

$$P(\mathbf{x}, \mathbf{y}) = \sum_{i=1}^{m} \sum_{j=1}^{n} x_i g_{ij} y_j ,$$

where $g_{ij} = P(\sigma_i, \tau_j)$, $1 \leq i \leq m$, $1 \leq j \leq n$, is the payoff matrix. This derives from (2.6) or equivalently from the observation that the probability of the payoff g_{ij} when the mixed strategies $\mathbf{x} \in X_1$, $\mathbf{y} \in X_2$ are employed is $x_i y_j$.

Consider the payoff matrix

$$G = \begin{pmatrix} g_{11} & \cdots & g_{1n} \\ & & \\ \cdot & & \\ \cdot & & \\ \cdot & & \\ g_{m1} & \cdots & g_{mn} \end{pmatrix} = (\mathbf{g}_1, \mathbf{g}_2, ..., \mathbf{g}_n),$$

say, where $\mathbf{g}_j$ $(1 \leq j \leq n)$ denotes the column vector $(g_{1j}, ..., g_{mj})$, which we write as a row vector for convenience.

We have seen that the simplices $\Sigma^{(1)}$ and $\Sigma^{(2)}$ represent the strategies of players 1 and 2 respectively. The proof of the minimax theorem which follows rests on the idea of replacing the simplices $\Sigma^{(1)}$, $\Sigma^{(2)}$ by two other objects and then applying a simple result from convexity theory.

Let Γ be the convex set in $\mathbf{R}^m$ spanned by the vectors $\mathbf{g}_j$. Formally,

$$\Gamma = \{ \sum_{j=1}^{n} y_j \mathbf{g}_j; \; y_j \geq 0 \; (1 \leq j \leq n), \; \sum_{j=1}^{n} y_j = 1 \} .$$

In Figure 2.5 Γ is illustrated in the case $m = 2$, $n = 3$.

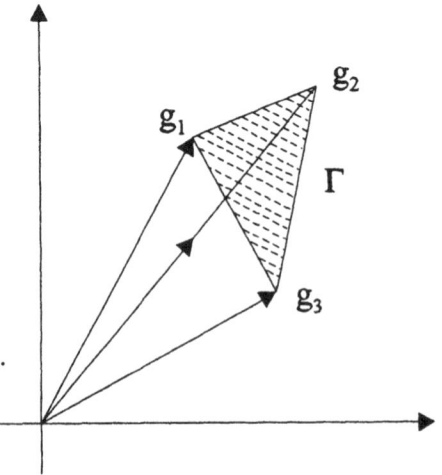

Figure 2.5 The set Γ $(m = 2, n = 3)$.

Then strategies $\mathbf{y} = (y_1, ..., y_n) \in \Sigma^{(2)}$ can be mapped to points in Γ by writing

$$p_i = \sum_{j=1}^{n} g_{ij} y_j \ (1 \le i \le m) \ (\mathbf{p} = G\mathbf{y}),$$

where $\mathbf{p} = (p_1, ..., p_m) \in \Gamma$. This sets up the map $\Sigma^{(2)} \to \Gamma$.

The mapping for $\Sigma^{(1)}$ is a trifle more involved. We recall that if $\mathbf{a} \in \mathbf{R}^m$, $\mathbf{a} \ne \mathbf{0}$ and $\alpha \in \mathbf{R}$, the set

$$H = \{ \ \mathbf{x} \in \mathbf{R}^m; \ \mathbf{x}.\mathbf{a} = \alpha \ \}$$

is called a **hyperplane** in $\mathbf{R}^m$. Any hyperplane in $\mathbf{R}^m$ determines two closed half-spaces

$$H_1 = \{ \ \mathbf{x} \in \mathbf{R}^m; \ \mathbf{x}.\mathbf{a} \ge \alpha \ \},$$
$$H_2 = \{ \ \mathbf{x} \in \mathbf{R}^m; \ \mathbf{x}.\mathbf{a} \le \alpha \ \}.$$

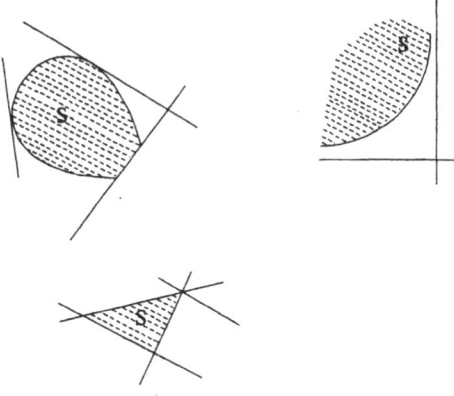

A hyperplane H is a **supporting** hyperplane (or briefly a support plane: see Figure 2.6) to a convex set S if

either $S \subseteq H_1$ and $\inf_{\mathbf{x} \in S} \mathbf{x}.\mathbf{a} = \alpha$,

or $S \subseteq H_2$ and $\sup_{\mathbf{x} \in S} \mathbf{x}.\mathbf{a} = \alpha$.

Figure 2.6 Support planes to S ($m = 2$).

Our next step is to construct a convex set O_{λ_0}. The points of $\Sigma^{(1)}$ will then be associated with support planes to O_{λ_0}. We let

$$O_\lambda = \{ \mathbf{x} = (x_1, ..., x_m) \in \mathbf{R}^m$$
$$x_i \le \lambda \ (1 \le i \le m) \ \},$$

where λ is any real number. We call λ Γ-**admissible** if $O_\lambda \cap \Gamma \ne \phi$ (see Figure 2.7).

If Ω denotes the set of all Γ-admissible λ, then $\Omega \ne \phi$, since $N \in \Omega$ provided $N > g_{ij}$ for all i, j, and Ω is bounded below by $-N$ for sufficiently large N. Hence we can define

$$\lambda_0 = \inf_{\lambda \in \Omega} \lambda.$$

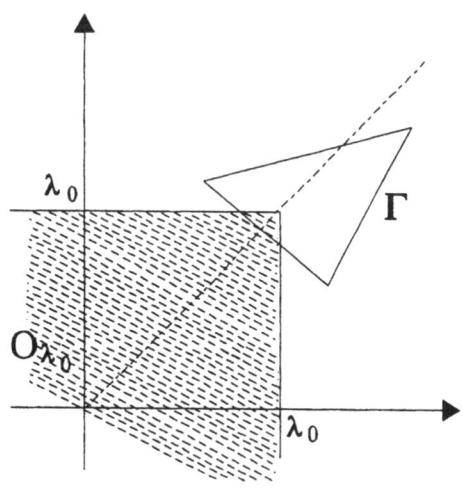

Figure 2.7 Γ and O_λ ($m = 2$). Because $O_\lambda \cap \Gamma \ne \varphi$ λ is Γ-admissible.

The situation is illustrated in Figure 2.8.

I claim that λ_0 is itself Γ-admissible; that is, $\lambda_0 \in \Omega$. This is essentially because both O_{λ_0} and

Γ are closed sets and hence contain their boundary points. Thus $O_{\lambda_0} \cap \Gamma$ is just the set of common boundary points. To give a formal proof let λ_ν ($\nu = 1, 2, 3, ...$) be a sequence of points of Ω such that $\lambda_\nu \to \lambda_0$ as $\nu \to \infty$. Each λ_ν is Γ-admissible, and so there exists $\mathbf{p}_\nu \in \Gamma$ such that $\mathbf{p}_\nu \in O_{\lambda_\nu} \cap \Gamma$.

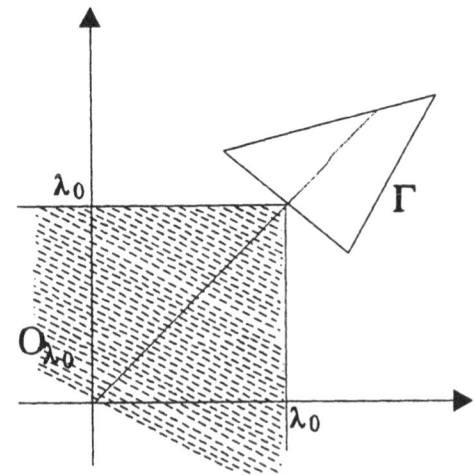

Since Γ is compact, the Bolzano-Weierstrass theorem in $\mathbf{R}^m$ (every bounded infinite sequence has a convergent subsequence) implies the existence of a subsequence $\mathbf{p}_{\nu_j}$ of $\mathbf{p}_\nu$ and a $\mathbf{p}_0 \in \Gamma$ such that $\mathbf{p}_{\nu_j} \to \mathbf{p}_0$ as $j \to \infty$. It remains to show $\mathbf{p}_0 \in O_{\lambda_0}$.

Since $\mathbf{p}_{\nu_j} = (p_{1j}, ..., p_{mj}) \in O_{\lambda \nu_j}$ we have

$$\max_{1 \le i \le m} p_{ij} \le \lambda_{\nu_j},$$

and since $\lambda_{\nu_j} \to \lambda_0$, $p_{ij} \to p_{i0}$, as $j \to \infty$, it follows that

$$\max_{1 \le i \le m} p_{i0} \le \lambda_0,$$

Figure 2.8 Γ and $O_{\lambda 0}$ ($m = 2$).

where $\mathbf{p}_0 = (p_{10}, ..., p_{m0})$, that is, $\mathbf{p}_0 \in O_{\lambda_0}$. Thus O_{λ_0} touches Γ with no overlap, that is $\mathbf{p}_0 \in O_{\lambda_0} \cap \Gamma$, but the two sets have no interior points in common.

We next characterise those support planes H to O_{λ_0} whose normals μ are directed into the half space *not* containing O_{λ_0}; see Figure 2.9.

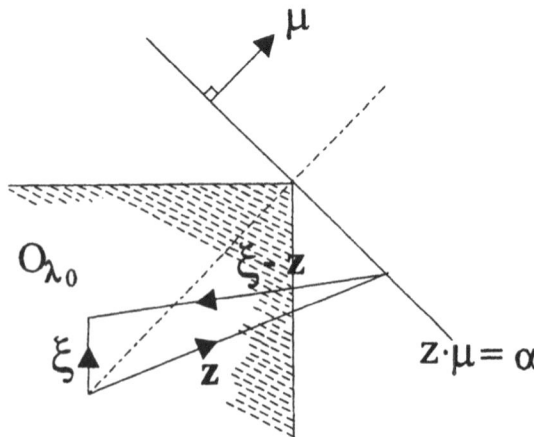

Figure 2.9 Support plane to $O_{\lambda 0}$.

Let H be defined by $\mathbf{z}.\mu = \alpha$. This is a support plane for O_{λ_0}, with μ directed away from O_{λ_0}.

if and only if *firstly*

$$(\xi - z).\mu \le 0 \text{ for every } \xi \in O_{\lambda_0} \text{ (and some } z \in H),$$

that is

$$\sum_{i=1}^{m} \xi_i \mu_i \le z.\mu = \alpha \text{ for every } \xi \in O_{\lambda_0}, \tag{2.17}$$

where $\xi = (\xi_1, ..., \xi_m)$, $\mu = (\mu_1, ..., \mu_m)$, and *secondly* there exists some $\xi_0 \in O_{\lambda_0}$ for which equality holds in (2.17). The components μ_i are characterised by

$$\mu_i \ge 0 \ (1 \le i \le m), \ \sum_{i=1}^{m} \mu_i > 0, \ \alpha = \lambda_0 \sum_{i=1}^{m} \mu_i. \tag{2.18}$$

For firstly $\mu_i \ge 0$, since the point $(\lambda_0, ..., \lambda_0, -N, \lambda_0, ..., \lambda_0) \in O_{\lambda_0}$ (the $-N$ is in the ith position) and so by (2.17)

$$\lambda_0 \mu_1 + ... + (-N)\mu_i + ... + \lambda_0 \mu_m \le \alpha,$$

hence $\mu_i \ge 0$ by taking N sufficiently large. Secondly *any* support plane to O_{λ_0} must contain the point $\xi = (\lambda_0, ..., \lambda_0)$ so that

$$\sum_{i=1}^{m} \lambda_0 \mu_i = \alpha \text{ that is, } \alpha = \lambda_0 \sum_{i=1}^{m} \mu_i.$$

Conversely any hyperplane H for which μ satisfies (2.18) is a support plane to O_{λ_0}. Let us write

$$x_i = \frac{\mu_i}{\sum_{i=1}^{m} \mu_i}.$$

Then support planes to O_{λ_0} with normals directed away from O_{λ_0} can be characterised by

$$\sum_{i=1}^{m} x_i \xi_i = \lambda_0, \ x_i \ge 0 \ (1 \le i \le m), \ \sum_{i=1}^{m} x_i = 1.$$

This sets up a 1-1 map from strategies $x \in \Sigma^{(1)}$ to support planes to O_{λ_0}.

We shall need the following simple

Lemma 2.4. If $x \in \Sigma^{(1)}$ and $b \in \mathbb{R}^m$ then

$$\max_{x \in \Sigma^{(1)}} \sum_{i=1}^{m} x_i b_i = \max_{1 \le i \le m} b_i.$$

Proof.

$$\sum_{i=1}^{m} x_i b_i \leq (\max_{1 \leq i \leq m} b_i) \sum_{i=1}^{m} x_i = \max_{1 \leq i \leq m} b_i,$$

since $\Sigma x_i = 1$. Moreover the upper bound is attained, for let $b_j = \max b_i$, then $(0, ..., 0, 1, 0, ...0) \in \Sigma^{(1)}$, where the 1 is in the jth position, is such a point.

We can now prove the minimax theorem for matrix games.

Theorem 2.6 (von Neumann).

$$\max_{x \in \Sigma^{(1)}} \min_{y \in \Sigma^{(2)}} P(x, y) = \min_{y \in \Sigma^{(2)}} \max_{x \in \Sigma^{(1)}} P(x, y),$$

where

$$P(x, y) = \sum_{i=1}^{m} \sum_{j=1}^{n} x_i g_{ij} y_j.$$

Proof. Construct O_{λ_0}, Γ as above. Since O_{λ_0}, Γ are non-overlapping convex sets then, by a standard result from convexity theory, there exists a hyperplane H^0 which separates the interior points of these sets. Since O_{λ_0} and Γ actually touch, this plane H^0 must also be a support plane to O_{λ_0}. Let the normal to H^0 be $x^0 = (x_1^0, ..., x_m^0)$, where

$$x_i^0 \geq 0 \; (1 \leq i \leq m), \text{ and } \sum_{i=1}^{m} x_i = 1.$$

Then

$$\sum_{i=1}^{m} \xi_i x_i^0 \leq \lambda_0 \text{ for every } \xi \in O_{\lambda_0},$$

and

$$\sum_{i=1}^{m} \xi_i x_i^0 \geq \lambda_0 \text{ for every } \xi \in \Gamma. \tag{2.19}$$

If $y \in \Sigma^{(2)}$ then the vector ξ given by

$$\xi_i = \sum_{j=1}^{n} g_{ij} y_j$$

is in Γ. Hence by (2.19)

$$\sum_{i=1}^{m} x_i^0 \sum_{j=1}^{n} g_{ij} y_j \geq \lambda_0 \, ,$$

that is,

$$P(\mathbf{x}^0, \mathbf{y}) \geq \lambda_0 \text{ for every } \mathbf{y} \in \Sigma^{(2)} \, .$$

Conversely if $\mathbf{y}^0 = (y_1^0, \dots, y_n^0)$ denotes a strategy corresponding to a point in $O_{\lambda_0} \cap \Gamma$, then by definition of O_{λ_0}

$$\sum_{j=1}^{n} g_{ij} y_j^0 \leq \lambda_0 \ (1 \leq i \leq m) \, .$$

Hence by Lemma 2.4 with

$$b_i = \sum_{j=1}^{n} g_{ij} y_j^0$$

we have

$$\sum_{i=1}^{m} x_i \sum_{j=1}^{n} g_{ij} y_j^0 \leq \lambda_0 \, ,$$

that is

$$P(\mathbf{x}, \mathbf{y}^0) \leq \lambda_0 \text{ for every } \mathbf{x} \in \Sigma^{(1)} \, .$$

The theorem now follows from Theorem 2.5 with $\mathbf{x}^0 = \mathbf{x}_0$, $\mathbf{y}^0 = \mathbf{y}_0$ and $\lambda_0 = v$.

The above proof of the theorem is unsymmetrical in the sense that $\mathbf{x}$ strategies correspond to hyperplanes and $\mathbf{y}$ strategies to points of $\mathbb{R}^m$. However, a completely parallel analysis can be developed which associates $\mathbf{y}$ strategies with hyperplanes and $\mathbf{x}$ strategies with points in $\mathbb{R}^n$. Thus the analog of Γ would be the convex set spanned by the row vectors of G, and so on. This is a reflection of the fact that there is an underlying duality theorem which occurs in linear programming theory (see §3.1 and §3.3).

2.6 PROPERTIES OF MATRIX GAMES.

From Theorem 2.6 we know there exist strategies $\mathbf{x}' \in \Sigma^{(1)}$, $\mathbf{y}' \in \Sigma^{(2)}$ such that

$$P(\mathbf{x}, \mathbf{y}') \leq P(\mathbf{x}', \mathbf{y}') \leq P(\mathbf{x}', \mathbf{y}) \tag{2.20}$$

for all strategies $\mathbf{x} \in \Sigma^{(1)}$, $\mathbf{y} \in \Sigma^{(2)}$. We call such strategies $\mathbf{x}'$, $\mathbf{y}'$ **optimal**. For any other optimal strategies $\mathbf{x}''$, $\mathbf{y}''$ we have from (2.20), on putting $\mathbf{x} = \mathbf{x}''$, $\mathbf{y} = \mathbf{y}''$,

$$P(\mathbf{x}'', \mathbf{y}') \le P(\mathbf{x}', \mathbf{y}') \le P(\mathbf{x}', \mathbf{y}'') . \qquad (2.21)$$

Similarly, on applying the definition of optimality to $\mathbf{x}''$, $\mathbf{y}''$ and putting $\mathbf{x} = \mathbf{x}'$, $\mathbf{y} = \mathbf{y}'$ we have

$$P(\mathbf{x}', \mathbf{y}'') \le P(\mathbf{x}'', \mathbf{y}'') \le P(\mathbf{x}'', \mathbf{y}') . \qquad (2.22)$$

Combining (2.21) and (2.22) we obtain in particular

$$P(\mathbf{x}', \mathbf{y}') = P(\mathbf{x}'', \mathbf{y}'') .$$

This common **value** v, say, therefore depends only on the game Γ, and we write $v = v(\Gamma)$. The triplet $\mathbf{x}'$, $\mathbf{y}'$, v is called a **solution** of Γ. We have proved

Theorem 2.7. If $\mathbf{x}^0 \in \Sigma^{(1)}$, $\mathbf{y}^0 \in \Sigma^{(2)}$, and $w \in \mathbb{R}$ have the properties

$$P(\mathbf{x}, \mathbf{y}^0) \le w \le P(\mathbf{x}^0, \mathbf{y}) \qquad (2.23)$$

for all strategies $\mathbf{x} \in \Sigma^{(1)}$, $\mathbf{y} \in \Sigma^{(2)}$, then $w = v(\Gamma)$ and $\mathbf{x}^0$, $\mathbf{y}^0$, w is a solution of Γ.

The converse is of course immediate from the definition of a solution.

Although we do not intend to give a detailed analysis of the sets of optimal strategies for each player, it is comparatively simple to prove.

Theorem 2.8. The set of optimal strategies for each player in a matrix game is a convex closed bounded set.

Proof. We give the proof for player 1. The convexity comes immediately from the fact that $P(\mathbf{x}, \mathbf{y})$ is linear in each variable separately. For if $v = v(\Gamma)$ and $\mathbf{x}'$, $\mathbf{x}''$ are optimal strategies for player 1, then

$$P(\mathbf{x}', \mathbf{y}) \ge v, \ P(\mathbf{x}'', \mathbf{y}) \ge v \text{ for every } \mathbf{y} \in \Sigma^{(2)} .$$

Thus, for $0 \le \lambda \le 1$,

$$P(\lambda \mathbf{x}' + (1 - \lambda)\mathbf{x}'', \mathbf{y}) = \lambda P(\mathbf{x}', \mathbf{y}) + (1 - \lambda)P(\mathbf{x}'', \mathbf{y})$$

$$\ge \lambda v + (1 - \lambda)v = v ,$$

for every $\mathbf{y} \in \Sigma^{(2)}$. Hence $\lambda \mathbf{x}' + (1 - \lambda)\mathbf{x}''$, which is obviously a strategy since $\Sigma^{(1)}$ is convex, is actually an optimal strategy for player 1.

The set of optimal strategies is certainly bounded, since $\Sigma^{(1)}$ is bounded. If $\mathbf{x}^0$ is a limit point of the set of optimal strategies, then the continuity of $P(\mathbf{x}, \mathbf{y})$ gives

$$P(\mathbf{x}^0, \mathbf{y}) \geq v \text{ for every } \mathbf{y} \in \Sigma^{(2)},$$

hence $\mathbf{x}^0$ is itself an optimal strategy and so the set is closed.

For an $m \times n$ matrix game $G = (g_{ij})$ let α_i $(1 \leq i \leq m)$ denote the ith row of the matrix G and β_j $(1 \leq j \leq n)$ the jth column. Observe that the expected payoff if player 1 employs any mixed strategy $\mathbf{x} \in \Sigma^{(1)}$ against player 2's pure strategy τ_j is

$$P(\mathbf{x}, \tau_j) = \mathbf{x} . \beta_j \quad (1 \leq j \leq n). \tag{2.24}$$

Similarly for $\mathbf{y} \in \Sigma^{(2)}$

$$P(\sigma_i, \mathbf{y}) = \mathbf{y} . \alpha_i \quad (1 \leq i \leq m).$$

Theorem 2.7 is the basis of the 'solution check scheme' explained in Chapter 1. This technique amounted to evaluating mixed strategies $\mathbf{x}^0$, $\mathbf{y}^0$ against all of the opponent's pure strategies, that is,

Corollary 2.1. If $\mathbf{x}^0 \in \Sigma^{(1)}$, $\mathbf{y}^0 \in \Sigma^{(2)}$, and $w \in \mathbb{R}$ have the properties

$$\mathbf{y}^0 . \alpha_i \leq w \leq \mathbf{x}^0 . \beta_j \quad (1 \leq i \leq m, \ 1 \leq j \leq n),$$

then $w = v(\Gamma)$ and $\mathbf{x}^0$, $\mathbf{y}^0$, w is a solution of Γ.

Proof. Since $\tau_1, \ldots, \tau_n$ is the standard basis and $P(\mathbf{x}, \mathbf{y})$ is linear in $\mathbf{y}$, we have, for any $\mathbf{y} = (y_1, \ldots, y_n) \in \Sigma^{(2)}$,

$$P(\mathbf{x}^0, \mathbf{y}) = y_1 P(\mathbf{x}^0, \tau_1) + \ldots + y_n P(\mathbf{x}^0, \tau_n)$$

$$= y_1 \mathbf{x}^0 . \beta_1 + \ldots + y_n \mathbf{x}^0 . \beta_n \qquad \text{(from (2.24))}$$

$$\geq y_1 w + \ldots + y_n w = \left(\sum_{j=1}^{n} y_j \right) w,$$

that is

$$P(\mathbf{x}^0, \mathbf{y}) \geq w \text{ for every } \mathbf{y} \in \Sigma^{(2)},$$

since $\Sigma y_j = 1$. Similarly

$$P(\mathbf{x}, \mathbf{y}^0) \leq w \text{ for every } \mathbf{x} \in \Sigma^{(1)}.$$

Hence condition (2.23) of Theorem 2.7 is satisfied, and so $w = v(\Gamma)$, and $\mathbf{x}^0$, $\mathbf{y}^0$, w is a solution.

Another result we had occasion to use in the first chapter was the following intuitively apparent observations. If a pure strategy played against an opponent's optimal strategy yields less than the value of the game, then this pure strategy will not appear as a component of any optimal mixed strategy (cf. part (ii) of the folowing theorem). To be precise we have

Theorem 2.9. Let $v = v(\Gamma)$.

(i) If x^0 is an optimal strategy for player 1 and for some j', $1 \le j' \le n$,

$$x^0 . \beta_{j'} > v , \tag{2.25}$$

then for any optimal strategy $y^0 = (y_1^0, ..., y_n^0) \in \Sigma^{(2)}$ we have $y_{j'}^0 = 0$.

(ii) If y^0 is an optimal strategy for player 2 and for some i', $1 \le i' \le m$,

$$y^0 . \alpha_{i'} < v ,$$

then for any optimal strategy $x^0 = (x_1^0, ..., x_m^0) \in \Sigma^{(1)}$ we have $x_{i'}^0 = 0$.

Proof. We prove (i); the proof of (ii) is similar. Since x^0 is optimal we have

$$x^0 . \beta_j \ge v \ (1 \le j \le n) . \tag{2.26}$$

For $j \ne j'$ multiply each inequality (2.26) by y_j^0 and add. We obtain

$$\sum_{j \ne j}^{n} y_j^0 P(x^0, \tau_j) \ge \left(\sum_{j \ne j} y_j^0 \right) v . \tag{2.27}$$

If $y_{j'}^0 \ne 0$ (2.25) gives

$$y_{j'}^0 x^0 . \beta_{j'} > y_{j'}^0 v ,$$

that is,

$$y_{j'}^0 P(x^0, \tau_{j'}) > y_{j'}^0 v . \tag{2.28}$$

Adding (2.27) and (2.28) we have

$$\sum_{j=1}^{n} y_j^0 P(x^0, \tau_j) > \left(\sum_{j=1}^{n} y_j^0 \right) v , \tag{2.29}$$

that is,

$$P(x^0, y^0) > v ,$$

since $\Sigma y_j^0 = 1$. This is a contradiction since $P(\mathbf{x}^0, \mathbf{y}^0) = v$. Hence $y_j.^0 = 0$ and the result is proved.

The converse of this theorem is also true (see [3] §1.20.10), although somewhat harder to prove, namely: if $v = v(\Gamma)$ and for every optimal strategy $\mathbf{y}^0 = (y_1^0, ..., y_n^0)$ of player 2 $y_j.^0 = 0$, then there exists an optimal strategy $\mathbf{x}^0 = (x_1^0, ..., x_m^0) \in \Sigma^{(1)}$ such that $P(\mathbf{x}^0, \tau_j.) = \mathbf{x}^0.\beta_j. > v$ (a similar converse applies to part (ii)).

We next turn our attention to results on domination of strategies; the definition of domination was given in §1.5. However, we first introduce some convenient notation.

For any subset S of Euclidean space we denote by [S] the smallest convex set which contains S; [S] is known as the **convex hull** of S.

Thus for matrix games

$$\Sigma^{(1)} = [S_1], \ \Sigma^{(2)} = [S_2], \tag{2.30}$$

where $S_1 = \{\sigma_1, ..., \sigma_m\}$, $S_2 = \{\tau_1, ..., \tau_n\}$ denote the pure strategy sets of the respective players.

The principal result on domination assumed in Chapter 1 was that if a pure strategy is eliminated by domination then any solution of the reduced game leads to a solution of the original game. This translates to the absurdly complicated

Theorem 2.10. If, for player 1, a strategy $\mathbf{x}' \in [S_1 \backslash \{\sigma_k\}]$ dominates the pure strategy σ_k, then any solution $\mathbf{x}^0$, $\mathbf{y}^0$, v, with

$$\mathbf{x}^0 = \sum_{i \neq k} x_i^0 \sigma_i, \ \mathbf{y}^0 = \sum_{j=1}^{n} y_j^0 \tau_j,$$

of the reduced game $\langle [S_1 \backslash \{\sigma_k\}], [S_2], P \rangle$ gives rise to a solution $\mathbf{X}^0$, $\mathbf{Y}^0$, v of the original game $\langle [S_1], [S_2], P \rangle$ with

$$\mathbf{X}^0 = \sum_{i=1}^{m} x_i^0 \sigma_i, \ (x_k^0 = 0), \ \mathbf{Y}^0 = \mathbf{y}^0 .$$

A similar result holds for player 2.

For convenience no distinction has been drawn between P and its restriction to a smaller set than $\Sigma^{(1)} \times \Sigma^{(2)}$.

Proof. Since $\mathbf{x}^0$, $\mathbf{y}^0$, v is a solution of $\langle [S_1 \backslash \{\sigma_k\}], [S_2], P \rangle$ we have in particular

$$P(\mathbf{x}^0, \mathbf{Y}) \geq v \text{ for every } \mathbf{Y} \in \Sigma^{(2)} .$$

Since P is linear in the first variable and $x_k^0 = 0$ we have

$$P(\mathbf{X}^0, \mathbf{Y}) = \sum_{i=1}^{m} x_i^0 P(\sigma_i, \mathbf{Y}) = \sum_{i \neq k} x_i^0 P(\sigma_i, \mathbf{Y}) = P(\mathbf{x}^0, \mathbf{Y}).$$

Thus

$$P(\mathbf{X}^0, \mathbf{Y}) \geq v \text{ for every } \mathbf{Y} \in \Sigma^{(2)}. \tag{2.31}$$

Again since $\mathbf{x}^0$, $\mathbf{y}^0$, v is a solution of $\langle \, [S_1 \backslash \{\sigma_k\}], \, [S_2], \, P \, \rangle$ and $\mathbf{Y}^0 = \mathbf{y}^0$ we have

$$P(\mathbf{x}, \mathbf{Y}^0) = P(\mathbf{x}, \mathbf{y}^0) \leq v \tag{2.32}$$

for every $\mathbf{x} \in [S_1 \backslash \{\sigma_k\}]$.

Put $\mathbf{x}' = (x_1', .., x_m')$ where $x_i' \geq 0 \, (1 \leq i \leq m)$, $\Sigma x_i' = 1$ and $x_k' = 0$ since $\mathbf{x}' \in [S_1 \backslash \{\sigma_k\}]$. Now for any $\mathbf{X} \in [S_1] = \Sigma^{(1)}$

$$P(\mathbf{X}, \mathbf{Y}^0) = \sum_{i=1}^{m} x_i P(\sigma_i, \mathbf{Y}^0)$$

$$= x_k P(\sigma_k, \mathbf{Y}^0) + \sum_{i \neq k} x_i P(\sigma_i, \mathbf{Y}^0).$$

Since $\mathbf{x}'$ dominates σ_k we have $P(\sigma_k, \mathbf{Y}) \leq P(\mathbf{x}', \mathbf{Y})$ for every $\mathbf{Y} \in [S_2]$. Taking $\mathbf{Y} = \mathbf{Y}^0$ in the last equation gives

$$P(\mathbf{X}, \mathbf{Y}^0) \leq x_k P(\mathbf{x}', \mathbf{Y}^0) + \sum_{i \neq k} x_i P(\sigma_i, \mathbf{Y}^0).$$

Since $\mathbf{Y}^0 = \mathbf{y}^0$ we can write this as

$$P(\mathbf{X}, \mathbf{Y}^0) \leq x_k P(\mathbf{x}', \mathbf{y}^0) + \sum_{i \neq k} x_i P(\sigma_i, \mathbf{y}^0)$$

$$\leq x_k \sum_{i=1}^{m} x_i' P(\sigma_i, \mathbf{y}^0) + \sum_{i \neq k} x_i P(\sigma_i, \mathbf{y}^0),$$

on using the linearity of P in the first variable. However, since $\mathbf{x}' \in [S_1 \backslash \{\sigma_k\}]$ we have $x_k' = 0$, and the last inequality becomes

$$P(\mathbf{X}, \mathbf{Y}^0) \leq x_k \sum_{i \neq k} x_i' P(\sigma_i, \mathbf{y}^0) + \sum_{i \neq k} x_i P(\sigma_i, \mathbf{y}^0)$$

that is

$$P(\mathbf{X}, \mathbf{Y}^0) \leq \sum_{i \neq k} (x_k x_i' + x_i) P(\sigma_i, \mathbf{y}^0)$$

or

$$P(\mathbf{X}, \mathbf{Y}^0) \leq P(\sum_{i \neq k} (x_k x_i' + x_i) \sigma_i, \mathbf{y}^0). \tag{2.33}$$

However, each $x_k x_i' + x_i \geq 0$, and

$$\sum_{i \neq k} (x_k x_i' + x_i) = x_k \sum_{i \neq k} x_i' + \sum_{i \neq k} x_i$$

$$= x_k + \sum_{i \neq k} x_i = \sum_{i=1}^{m} x_i = 1,$$

hence

$$\sum_{i \neq k} (x_k x_i' + x_i)\sigma_i \in [S_1 \setminus \{\sigma_k\}].$$

Whence from (2.32) and (2.33)

$$P(\mathbf{X}, \mathbf{Y}^0) \leq v \text{ for every } \mathbf{X} \in \Sigma^{(1)}. \tag{2.34}$$

From (2.31) and (2.34) we see that the conditions of Theorem 2.7 are satisfied with $w = v$. Hence the value of $\Gamma = \langle \Sigma^{(1)}, \Sigma^{(2)}, P \rangle$ is indeed v, and $\mathbf{X}^0$, $\mathbf{Y}^0$ are optimal strategies.

It is worth remarking that even if $\mathbf{x}'$ involves σ_k with positive probability the result still holds provided (obviously!) $\mathbf{x}' \neq \sigma_k$, that is $x_k' \neq 1$. To prove this we observe that if $\mathbf{x}'$ dominates σ_k and $x_k' \neq 1$ it is possible to replace $\mathbf{x}'$ by another strategy $\mathbf{x}''$, say, which has $x_k'' = 0$ but which also dominates σ_k. The new strategy $\mathbf{x}''$ will therefore satisfy the hypothesis of the theorem. We put

$$x_i'' = \begin{cases} \dfrac{x_i'}{(1 - x_k')}, & \text{if } i \neq k \\ 0, & \text{if } i = k. \end{cases}$$

Clearly $x_i'' \geq 0$ $(1 \leq i \leq m)$ and

$$\sum_{i=1}^{m} x_i'' = \sum_{i \neq k} \frac{x_i'}{1 - x_k'} = \frac{1}{1 - x_k'} \sum_{i \neq k} x_i' = \frac{1 - x_k'}{1 - x_k'} = 1$$

Thus $\mathbf{x}'' \in \Sigma^{(1)}$. It remains to show that $\mathbf{x}''$ dominates σ_k.

Since $\mathbf{x}'$ dominates σ_k we have

$$P(\mathbf{x}', \mathbf{y}) \geq P(\sigma_k, \mathbf{y}) \text{ for every } \mathbf{y} \in \Sigma^{(2)}.$$

In particular for pure strategies $\mathbf{y} = \tau_j$ we have

$$P(\mathbf{x}', \tau_j) \geq P(\sigma_k, \tau_j) \quad (1 \leq j \leq n),$$

that is,

$$\mathbf{x}' \cdot \beta_j \geq g_{kj} \qquad (1 \leq j \leq n).$$

Dividing through by $1 - x_k' > 0$ and using the definition of $\mathbf{x}''$, this can be written as

$$\mathbf{x}'' \cdot \beta_j + \frac{x_k'}{1 - x_k'} g_{kj} \geq \frac{g_{kj}}{1 - x_k'},$$

which reduces to

$$\mathbf{x}'' \cdot \beta_j \geq g_{kj},$$

or

$$P(\mathbf{x}'', \tau_j) \geq P(\sigma_k, \tau_j) \quad (1 \leq j \leq n).$$

For any strategy $\mathbf{y} = (y_1, ..., y_n) \in \Sigma^{(2)}$ we multiply both sides of this inequality by y_j and add the resulting inequalities for $j = 1, 2, ..., n$. On using the linearity of $P(\mathbf{x}, \mathbf{y})$ in $\mathbf{y}$ we obtain

$$P(\mathbf{x}'', \mathbf{y}) \geq P(\sigma_k, \mathbf{y}) \text{ for every } \mathbf{y} \in \Sigma^{(2)},$$

that is $\mathbf{x}''$ dominates σ_k as required.

Combining this observation with the previous theorem we obtain

Corollary 2.2. If, for player 1, the strategy $\mathbf{x}' \in \Sigma^{(1)}$ dominates the pure strategy $\sigma_k \in \Sigma^{(1)}$ and $\mathbf{x}' \neq \sigma_k$, then there exists an optimal strategy $\mathbf{x}^0 = (x_1^0, ..., x_m^0) \in \Sigma^{(1)}$ such that $x_k^0 = 0$. A similar result holds for player 2.

Our final theorem on dominance is that if a pure strategy is *strictly* dominated then it will not appear in any optimal mixed strategy.

Theorem 2.11. If, for player 1, the pure strategy σ_k is strictly dominated by a (pure or mixed) strategy $\mathbf{x}' \in \Sigma^{(1)}$, then in any optimal strategy $\mathbf{x}^0 = (x_1^0, ..., x_m^0) \in \Sigma^{(1)}$ we have $x_k^0 = 0$. Similarly for player 2.

Proof. Since $\mathbf{x}'$ strictly dominates σ_k we have
$$P(\sigma_k, \mathbf{y}) < P(\mathbf{x}', \mathbf{y}) \text{ for every } \mathbf{y} \in \Sigma^{(2)}$$

In particular if $\mathbf{y} = \mathbf{y}^0$, an optimal strategy for player 2,

$$P(\sigma_k, y^0) < P(x', y^0) \le v,$$

since y^0 is optimal. Thus

$$y^0 . \alpha_k < v.$$

The conclusion is now immediate from Theorem 2.9(ii) with $i' = k$.

The following result on symmetric games was also used in Chapter 1. Recall that an $m \times n$ matrix game (g_{ij}) is symmetric if $m = n$ and $g_{ji} = -g_{ij}$ for all pairs i, j.

Theorem 2.12. In a symmetric matrix game Γ any strategy which is optimal for one player is also optimal for the other. Moerover $v(\Gamma) = 0$.

Proof. Since $m = n$ we can indentify the mixed strategy sets $\Sigma^{(1)}$, $\Sigma^{(2)}$ and write $\Sigma^{(1)} = \Sigma^{(2)} = \Sigma$, say.

Let x^0, y^0 be optimal strategies for players 1 and 2 respectively. Then

$$P(x, y^0) \le P(x^0, y^0) \le P(x^0, y) \tag{2.35}$$

for all $x, y \in \Sigma$. Now

$$P(x, y) = \sum_{i=1}^{m} \sum_{j=1}^{m} x_i g_{ij} y_j = \sum_{j=1}^{m} \sum_{i=1}^{m} x_j g_{ji} y_i$$

$$= \sum_{j=1}^{m} \sum_{i=1}^{m} x_j (-g_{ij}) y_i = - \sum_{i=1}^{m} \sum_{j=1}^{m} y_i g_{ij} x_j$$

$$= -P(y, x).$$

Substituting into (2.35), multiplying through by -1, and rearranging gives

$$P(y, x^0) \le P(y^0, x^0) \le P(y^0, x) \text{ for all } y, x \in \Sigma.$$

However, this is exactly the statement that y^0 is optimal for player 1 and x^0 is optimal for player 2. Moreover,

$$v(\Gamma) = P(x^0, y^0) = - P(y^0, x^0) = - v(\Gamma),$$

so that $v(\Gamma) = 0$.

2.7 SIMPLIFIED 2-PERSON POKER.

In the study of games poker, in its varied forms, has become a popular source of models for mathematical analysis. Various simple pokers have been investigated by von Neumann ([6], 186-219), Bellman and Blackwell [7], and Nash and Shapley [8] to mention but a few.

We terminate our discussion of 2-person zero sum matrix games with a detailed analysis of a simplified version of poker. This study is due to H. W. Kuhn [9].

As actually played, poker is far too complex a game to permit a complete analysis at present; however, this complexity is computational, and the restrictions that we will impose serve only to bring the numbers involved within a reasonable range. The only restriction that is not of this nature consists in setting the number of players at two. The simplifications, though radical, enable us to compute *all* optimal strategies for both players. In spite of these modifications, however, it seems that Simplified Poker retains many of the essential characteristics of the usual game.

An *ante* of one unit is required of each of the two players. They obtain a fixed *hand* at the beginning of a play by drawing one card apiece from a pack of three cards (rather than the $^{52}C_5 = 2,598,960$ hands possible in poker) numbered 1, 2, 3. Then the players choose alternatively either to *bet* one unit or *pass* without betting. Two successive bets or passes terminate a play, at which time the player holding the higher card wins the amount wagered previously by the other player. A player passing after a bet also ends a play and loses his or her ante.

Thus thirty possible plays are permitted by the rules. First of all, there are six possible deals; for each deal the action of the players may follow one of five courses which are described in the following table:

| | First round | | second round | Payoff |
	Player 1	Player 2	Player 1	
(1)	pass	pass	-	1 to holder of higher card
(2)	pass	bet	pass	1 to player 2
(3)	pass	bet	bet	2 to holder of higher card
(4)	bet	pass	-	1 to player 1
(5)	bet	bet	-	2 to holder of higher card

We code the pure strategies available to the players by ordered triples (x_1, x_2, x_3) and (y_1, y_2, y_3) for players 1 and 2 respectively $(x_i = 0, 1, 2; y_j = 0, 1, 2, 3)$. The instructions contained in x_i are for card i and are deciphered by expanding x_i in the binary system, the first figure giving directions for the first round of betting, the second giving directions for the second, with 0 meaning pass and 1 meaning bet. For example, $(x_1, x_2, x_3) = (2, 0, 1) = (10, 00, 01)$ means player 1 should bet on a 1 in the first round, always pass with a 2, and wait until the second round to bet on a 3.

Similarly, to decode y_j, one expands in the binary system, the first figure giving directions when confronted by a pass, the second when confronted by a bet, with 0 meaning pass and 1 meaning bet. Thus $(y_1, y_2, y_3) = (2, 0, 1) = (10, 00, 01)$ means that player 2 should pass except when holding a 1 and confronted by a pass, or holding a 3 and confronted by a bet.

In terms of this description of the pure strategies, the payoff to player 1 is given by the following

x_i $\backslash$ y_j	$0 = 00$	$1 = 01$	$2 = 10$	$3 = 11$
$0 = 00$	± 1	± 1	-1	-1
$1 = 01$	± 1	± 1	± 2	± 2
$2 = 10$	1	± 2	1	± 2

where the ambiguous sign is + if $i > j$ and - if $i < j$.

From the coding of pure strategies it is clear that player 1 has 27 pure strategies, and player 2 has 64 pure strategies. Fortunately poker sense indicates a method of reducing this unwieldy number of strategies.

Obviously, no player will decide either to bet on a 1 or pass with a 3 when confronted by a bet. For player 1 (2) this heuristic argument recognises the domination of certain rows (columns) of the game matrix by other rows (columns). As we know, we may drop the dominated rows (columns) without changing the value of the game, and any optimal strategy for the matrix thus reduced will be optimal for the original game. However, since the domination is not strict, these pure strategies could appear in some optimal mixed strategy. For the careful solver who may wish to find all of the optimal strategies, complementary arguments may be made to show that the pure strategies dropped are actually superfluous in this game.

Now that these strategies have been eliminated new dominations appear. First, we notice that if player 1 holds a 2, he or she may as well pass in the first round, deciding to bet in the second if confronted by a bet, as bet originally. On either strategy he or she will lose the same amount if player 2 holds a 3; on the other hand, player 2 may bet on a 1 if confronted by a pass but certainly will not if confronted by a bet. Secondly player 2 may as well pass as bet when holding a 2 and confronted by a pass, since player 1 will now answer a bet only when holding a 3.

We are now in a position to describe the game matrix composed of those strategies not eliminated by the previous heuristic arguments. Each entry is computed by adding the payoffs for the plays determined by the six possible deals for each pair of pure strategies, and thus this matrix is six times the actual game matrix.

(y_1, y_2, y_3)	(0, 0, 3)	(0, 1, 3)	(2, 0, 3)	(2, 1, 3)
(x_1, x_2, x_3)				
(0, 0, 1)	0	0	-1	-1
(0, 0, 2)	0	1	-2	-1
(0, 1, 1)	-1	-1	1	1
(0, 1, 2)	-1	0	0	1
(2, 0, 1)	1	-2	0	-3
(2, 0, 2)	1	-1	-1	-3
(2, 1, 1)	0	-3	2	-1
(2, 1, 2)	0	-2	1	-1

One easily verifies that the following mixed strategies are optimal for this game matrix (and hence for simplified poker):

Player 1: (A) $\frac{2}{3}$ (0, 0, 1) + $\frac{1}{3}$ (0, 1, 1)

(B) $\frac{1}{3}$ (0, 0, 1) + $\frac{1}{2}$ (0, 1, 2) + $\frac{1}{6}$ (2, 0, 1)

(C) $\frac{5}{9}$ (0, 0, 1) + $\frac{1}{3}$ (0, 1, 2) + $\frac{1}{9}$ (2, 1, 1)

(D) $\frac{1}{2}$ (0, 0, 1) + $\frac{1}{3}$ (0, 1, 2) + $\frac{1}{6}$ (2, 1, 2)

(E) $\frac{2}{5}$ (0, 0, 2) + $\frac{7}{15}$ (0, 1, 1) + $\frac{2}{15}$ (2, 0, 1)

(F) $\frac{1}{3}$ (0, 0, 2) + $\frac{1}{2}$ (0, 1, 1) + $\frac{1}{6}$ (2, 0, 2)

(G) $\frac{1}{2}$ (0, 0, 2) + $\frac{1}{3}$ (0, 1, 1) + $\frac{1}{6}$ (2, 1, 1)

(H) $\frac{4}{9}$ (0, 0, 2) + $\frac{1}{3}$ (0, 1, 1) + $\frac{2}{9}$ (2, 1, 2)

(I) $\frac{1}{6}$ (0, 0, 2) + $\frac{7}{12}$ (0, 1, 2) + $\frac{1}{4}$ (2, 0, 1)

(J) $\frac{5}{12}$ (0, 0, 2) + $\frac{1}{3}$ (0, 1, 2) + $\frac{1}{4}$ (2, 1, 1)

(K) $\frac{1}{3}$ (0, 0, 2) + $\frac{1}{3}$ (0, 1, 2) + $\frac{1}{3}$ (2, 1, 1)

(L) $\frac{2}{3}$ (0, 1, 2) + $\frac{1}{3}$ (2, 0, 2)

Player 2: (A) $\frac{1}{3}$ (0, 0, 3) + $\frac{1}{3}$ (0, 1, 3) + $\frac{1}{3}$ (2, 0, 3)

(B) $\frac{2}{3}$ (0, 0, 3) + $\frac{1}{3}$ (2, 1, 3)

These strategies yield the value of simplified poker as $-\frac{1}{18}$.

A striking simplification of the solution is achieved if we return to the extensive form of the game. We do this by introducing *behaviour parameters* to describe the choices remaining available to the players after we have eliminated the superfluous strategies. We define:

Player 1:

α = probability of bet with 1 in first round.
β = probability of bet with 2 in second round.
γ = probability of bet with 3 in first round.

Player 2:

ξ = probability of bet with 1 against a pass.
η = probability of bet with 2 against a bet.

In terms of these parameters, player 1's basic optimal strategies fall into seven sets:

Basic strategies	(α,	β,	γ)
A	(0,	$1/3$,	0)
C	($1/9$,	$4/9$,	$1/3$)
E	($2/15$,	$7/15$,	$2/5$)
B, D, F, G	($1/6$,	$1/2$,	$1/2$)
H	($2/9$,	$5/9$,	$2/3$)
I, J	($1/4$,	$5/12$,	$3/4$)
K, L	($1/3$,	$2/3$,	1)

Thus, in the space of these behaviour parameters, the five dimensions of optimal mixed strategies for player 1 collapse onto the one parameter family of solutions:

$$\alpha = \frac{\gamma}{3},$$

$$\beta = \frac{\gamma}{3} + \frac{1}{3},$$

$$0 \leq \gamma \leq 1.$$

These may be described verbally by saying that player 1 may pass on a 3 in the first round with arbitrary probability, but then he or she must bet on a 1 in the first round one third as often, while the probability of betting on a 2 in the second round is one third more than the probability of betting on a 1 in the first round.

On the other hand, we find that player 2 has the single solution:

$$(\xi, \eta) = (\frac{1}{3}, \frac{1}{3}),$$

which instructs him or her to bet one third of the time when holding a 1 and confronted by a pass, and to bet one third of the time when holding a 2 and confronted by a bet.

The presence of *bluffing* and *underbidding* in these solutions is noteworthy (*bluffing* means betting with a 1; *underbidding* means passing on a 3). All but the extreme strategies for player

1, in terms of the behaviour parameters, involve both bluffing and underbidding, while player 2's single optimal strategy is to bluff with constant probability $^1/_3$ (underbidding is not available to player 2). These results compare favourably with the presence of bluffing in the von Neumann example [6], while bluffing is not available to player 2 in the continuous variant considered by Bellman and Blackwell [7].

The sensitive nature of bluffing and underbidding in this example is exposed by varying the ratio of the bet to the ante. Consider the games described by the same rules in which the ante is a positive real number a and the bet is a positive real number b. We will state the solutions in terms of the behaviour parameters without proof. This one parameter family of games falls naturally into four intervals.

Case 1: $0 < b < a$

$$\text{Player 1: } (\alpha, \beta, \gamma) = (\frac{b}{2a + b}, \frac{2a}{2a + b}, 1)$$

$$\text{Player 2: } (\xi, \eta) = (\frac{b}{2a + b}, \frac{2a - b}{2a + b})$$

Remarks: Both players have a unique optimal mode of behaviour. Player 1 never underbids, and bluffs with probability $b/(2a + b)$; player 2 bluffs with probability $b/(2a + b)$. The value of this game is $-b^2/6(2a + b)$.

Case 2: $0 < b = a$

This is our original game. The value is $-b/18$.

Case 3: $0 < a < b < 2a$

$$\text{Player 1: } (\alpha, \beta, \gamma) = (0, \frac{2a - b}{2a + b}, 0),$$

$$\text{Player 2: } \xi = \frac{b}{2a + b}, \quad \frac{b}{2a + b} \leq \eta \leq \frac{a + b}{2a + b}.$$

Remarks: Player 1 never bluffs and always underbids; player 2 bluffs with probability $b/(2a + b)$. The value of this game is $-b(2a - b)/6(2a + b)$.

Case 4: $0 < 2a = b$

This game has a saddle point in which player 1 never bluffs, always underbids, and never bets on a 2, while player 2 bets only and always with a 3. The strategy for player 1 is unique, while player 2 can vary his or her strategy considerably. The value of this game is 0.

It is remarkable that player 1 has a negative expectation for a play, that is, a disadvantage that is plausibly imputable to being forced to take the initiative. (Compare von Neumann's variant, in which the possession of the initiative seems to be an advantage.)

2.8 CONTINUOUS GAMES ON THE UNIT SQUARE.

Examples 1.8 and 1.9 were instances of this type of game. Whilst we do not intend to embark upon a proof of Ville's theorem that such games can be solved in terms of mixed strategies (that is, distribution functions over [0, 1]) nevertheless we shall state a generalisation of this result due to Glicksberg and show that in some sense this is best possible. Detailed treatment of Ville's theorem can be found in McKinsey ([10], Chapter 10) or Vorob'ev ([3], Chapter 2).

We begin by developing some terminology. For any function $f: X \to \mathbf{R}$, with X is a topological space, we say f is **upper (lower) semi-continuous** if for every $c \in \mathbf{R}$ the set

$$\{ x \in X ; f(x) < c \} \quad (\{ x \in X ; f(x) > c \})$$

is open. If f is both upper and lower semi-continuous then it is plainly continuous.

Theorem (Glicksberg [11]). If $P : [0, 1] \times [0, 1] \to \mathbf{R}$ is upper (lower) semi-continuous on S = [0, 1] × [0, 1] then

$$\sup_{F} \inf_{G} \iint_{S} P(x, y) \, dF(x) \, dG(y) = \inf_{G} \sup_{F} \iint_{S} P(x, y) \, dF(x) \, dG(y) ,$$

where F, G range over the set of distribution functions on [0, 1].

The following example, of a game on the unit square which does *not* have a value (even) in terms of mixed strategies, shows that the condition of semi-continuity cannot be dropped.

Example 2.1 (Sion and Wolfe [12]). Let $0 \le x \le 1, 0 \le y \le 1$ and define

$$P(x, y) = \begin{cases} -1, & \text{if } x < y < x + \dfrac{1}{2} , \\ 0, & \text{if } x = y \text{ or } y = x + \dfrac{1}{2} , \\ +1, & \text{otherwise} . \end{cases}$$

Figure 2.10 shows how P varies over the unit square. Now

$$\{ (x, y) \in [0, 1] \times [0, 1]; P(x, y) > -1 \}$$

is the union of the two shaded closed triangles in Figure 2.10, and this is not an open subset of [0, 1] × [0, 1]. So P is not lower semi-continuous. Similarly, P is not upper semi-continuous.

We shall show that for this game

$$\sup_{F} \inf_{G} \iint_{S} P(x, y) \, dF(x) \, dG(y) = \frac{1}{3} , \tag{2.36}$$

whilst

$$\inf_{G} \sup_{F} \iint_{S} P(x,y)\,dF(x)\,dG(y) = \frac{3}{7} \tag{2.37}$$

Consequently the game has no value.

The verification of (2.36) is in two parts. Let F be any distribution function on $[0, 1]$. We begin by showing that

$$\int_{0}^{1} P(x, y_F)\,dF \le \frac{1}{3} \tag{2.38}$$

for some $y_F \in [0, 1]$. For any (Borel) subset $B \subseteq [0, 1]$ we write

$$E_F(B) = \int_{B} dF .$$

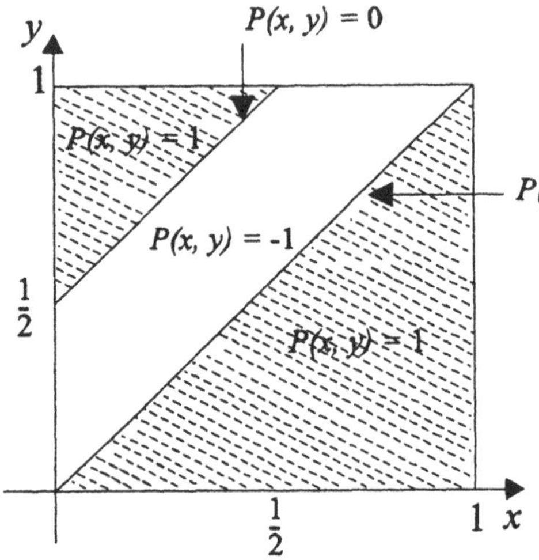

Figure 2.10 Variation of $P(x, y)$.

If $E_F([0, \frac{1}{2})) \le \frac{1}{3}$, let $y_F = 1$. Then

$$\int_{0}^{1} P(x, y_F)\,dF = \int_{0}^{1} P(x, 1)\,dF = \int_{0}^{\frac{1}{2}} 1\,dF + \int_{\frac{1}{2}}^{1} (-1)\,dF$$

$$= E_F([0, \tfrac{1}{2})) - E_F((\tfrac{1}{2}, 1)) .$$

But $E_F((\frac{1}{2}, 1)) \ge 0$ since it represents a probability. Hence

$$\int_{0}^{1} P(x, y_F)\,dF \le E_F([0, \tfrac{1}{2})) \le \tfrac{1}{3} .$$

If $E_F([0, \frac{1}{2})) > \frac{1}{3}$ choose $\delta > 0$ so that $E_F([0, \frac{1}{2} - \delta)) \ge \frac{1}{3}$ and let $y_F = \frac{1}{2} - \delta$. Then

$$\int_{0}^{1} P(x, y_F)\,dF = \int_{0}^{1} P(x, \tfrac{1}{2} - \delta)\,dF = \int_{0}^{\frac{1}{2} - \delta} P(x, \tfrac{1}{2} - \delta)\,dF + \int_{\frac{1}{2} - \delta}^{1} P(x, \tfrac{1}{2} - \delta)\,dF$$

and so

$$\int_0^1 P(x, y_F) dF = \int_0^{\frac{1}{2} - \delta} (-1) dF + \int_{\frac{1}{2} - \delta}^1 dF$$

$$= - E_F([0, \frac{1}{2} - \delta)) + E_F((\frac{1}{2} - \delta, 1]) .$$

Now $E_F((\frac{1}{2} - \delta, 1]) \le \frac{2}{3}$ since $E_F([0, \frac{1}{2} - \delta)) \ge \frac{1}{3}$. Hence

$$\int_0^1 P(x, y_F) dF \le -\frac{1}{3} + \frac{2}{3} = \frac{1}{3} ,$$

which verifies (2.38). Now choose a distribution function G_0, defined so that $E_{G_0}(\{y_F\}) = 1$. Then

$$\iint_S P(x, y) dF dG_0 = \int_0^1 P(x, y_F) dF \le \frac{1}{3} .$$

Hence

$$\inf_G \iint_S P(x, y) dF dG \le \frac{1}{3} ,$$

and so

$$\sup_F \inf_G \iint_S P(x, y) dF dG \le \frac{1}{3} . \qquad (2.39)$$

On the other hand, if F is chosen to be F_0, defined so that

$$E_{F_0}(\{0\}) = E_{F_0}(\{\frac{1}{2}\}) = E_{F_0}(\{1\}) = \frac{1}{3} ,$$

then for every $y \in [0, 1]$

$$\int_0^1 P(x, y) dF_0 = \frac{1}{3}(P(0, y) + P(\frac{1}{2}, y) + P(1, y)) \ge \frac{1}{3} .$$

Hence for all distribution functions G on $[0, 1]$

$$\iint_S P(x, y) dF_0 dG \ge \frac{1}{3} \quad (\int_0^1 dG = 1) ,$$

so that

$$\inf_{G} \iint_S P(x,y)\,dF_0\,dG \ge \frac{1}{3}\,,$$

whence

$$\sup_{F} \inf_{G} \iint_S P(x,y)\,dF\,dG \ge \frac{1}{3}\,. \tag{2.40}$$

Equation (2.36) now follows from (2.39) and (2.40).

The verification of (2.37) proceeds similarly, and we content ourselves with a sketch. The details are filled in as above.

Let G be any distribution function on $[0, 1]$. We begin by showing that

$$\int_0^1 P(x_G, y)\,dG \ge \frac{3}{7} \tag{2.41}$$

for some $x_G \in [0, 1]$. If $E_G([0, 1)) \ge \frac{3}{7}$, let $x_G = 1$. If $E_G([0, 1)) < \frac{3}{7}$ then $E_G([1]) > \frac{4}{7}$ and two subcases arise. If $E_G([0, \frac{1}{2})) \le \frac{1}{7}$, let $x_G = 0$. If $E_G([0, \frac{1}{2})) > \frac{1}{7}$ choose $\delta > 0$ so that $E_G([0, \frac{1}{2}-\delta)) \ge \frac{1}{7}$ and let $x_G = \frac{1}{2}-\delta$. The verification of (2.41) is now routine. Now choose a distribution function F_0, defined so that $E_{F_0}(\{x_G\}) = 1$. Then

$$\iint_S P(x,y)\,dF_0\,dG = \int_0^1 P(x_G, y)\,dG \ge \frac{3}{7}\,.$$

Hence

$$\sup_{F} \iint P(x,y)\,dF\,dG \ge \frac{3}{7}\,,$$

and so

$$\inf_{G} \sup_{F} \iint P(x,y)\,dF\,dG \ge \frac{3}{7}\,. \tag{2.42}$$

On the other hand, if G is chosen to be G_0, defined so that

$$E_{G_0}(\{\tfrac{1}{4}\}) = \frac{1}{7}\,, \quad E_{G_0}(\tfrac{1}{2}) = \frac{2}{7}\,, \quad E_{G_0}(\{1\}) = \frac{4}{7}\,,$$

then for every $x \in [0, 1]$

$$\int_0^1 P(x,y)\,dG_0 = \frac{1}{7}\left(P(x,\tfrac{1}{4}) + 2P(x,\tfrac{1}{2}) + 4P(x,1) \right) \le \frac{3}{7}\,.$$

Hence for all distribution functions F on $[0, 1]$

$$\iint_S P(x,y)\,dF\,dG \leq \frac{3}{7} \quad \left(\int_0^1 dF = 1 \right),$$

so that

$$\sup_F \iint_S P(x,y)\,dF\,dG_0 \leq \frac{3}{7},$$

whence

$$\inf_G \sup_F \iint_S P(x,y)\,dF\,dG \leq \frac{3}{7} \qquad (2.43)$$

Equation (2.37) now follows from (2.42) and (2.43), and the proof that the game has no value is complete.

This game can be viewed as a 'Colonel Blotto' game. The title 'Blotto games' has come to describe a variety of military deployment problems. The original version goes back at least to Caliban in *The Weekend Book* (June 1924). A typical example is the following. Blotto has four units of armed force with which to oppose an enemy of three units. Between the opposing forces is a mountain range with four passes, and in each pass is a fort. War is declared in the evening, and the issue will be decided at first light. Decisions are based on who outnumbers the other at each pass. One point is scored for each fort taken, and one point for each unit taken. For technical reasons Blotto has available only three pure strategies:

I   One unit to each pass.
II   Three units to one pass and one unit to another.
III   Two units to two passes.

His opponent likewise has three pure strategies:

I   One unit to each pass.
II   All three units to one pass.
III   Two units to one pass and one to another.

Since neither Blotto nor the enemy know one another's intentions we can assume that, once each has selected a pure strategy, the particular line-up of forces in the passes is decided by chance. Routine calculation of the payoffs then gives the matrix

$$\begin{pmatrix} 1 & 1 & 0 \\ \dfrac{1}{2} & \dfrac{1}{2} & \dfrac{3}{4} \\ 2 & -\dfrac{1}{2} & 1 \end{pmatrix}.$$

where payoffs are to Blotto. The techniques of Chapter 1 give a solution as

Blotto's optimal strategy : $(\frac{1}{5}, \frac{4}{5}, 0)$,
Enemy's optimal strategy : $(0, \frac{3}{5}, \frac{2}{5})$,
Value = $\frac{3}{5}$.

Such problems can occur in a variety of different guises, for example in the most efficient distribution of spare parts. However, in practical applications the matrix tends to be so large (for example 2000 × 1000) that a computer is required to solve the game. The techniques appropriate to such situations are discussed in Chapter 3.

It is sometimes useful to consider 'continuous' analogs of 'Blotto games'. This amounts to the assumption that the forces involved are infinitely divisible (that is, can be split in any ratios) and gives rise to continuum many pure strategies.

Example 2.1 can be considered as a 'Blotto game' as follows. Player 1 (Blotto) must assign a force x, $0 \leq x \leq 1$, to the attack of one of two mountain passes, and $1 - x$ to the other. Player 2 must assign a force y, $0 \leq y \leq 1$, to the defense of the first pass, and $1 - y$ to the other, at which is also located an extra stationary force of ½. A player scores 1, from the other, at each pass, if his or her force at that pass exceeds the opponent's, and scores nothing if they are equal there. Define

$$\operatorname{sgn} x = \begin{cases} 1, & \text{if } x > 0, \\ 0, & \text{if } x = 0, \\ -1, & \text{if } x < 0. \end{cases}$$

The payoff is then

$$B(x,y) = \operatorname{sgn}(x - y) + \operatorname{sgn}((1 - x) - (\frac{3}{2} - y)) .$$

It is easily checked that if $P(x, y)$ is the payoff function of Example 2.1, then

$$P(x,y) = 1 + B(x,y) .$$

Thus the present game is insoluble, even in mixed strategies, having minorant value $-\frac{2}{3}$ and majorant value $-\frac{4}{7}$. Curiously enough the 'Blotto game' played with forces x, $a - x$ for player 1 and y, $b - y$ (with no reserves) for player 2, always has a value [13].

In general, the determination of optimal strategies for continuous games on the unit square is very complicated. Solving such games is rather like solving differential equations; a combination of good guesswork, adroit use of previously learned techniques and ability to exploit the special features of the problem.

Particular classes of continuous functions $P(x, y)$ have been singled out for detailed study. Thus for example, **polynomial games** are those for which $P(x, y)$ is a polynomial in the variables x y. For these games some theoretical knowledge is available [14]; in particular there are optimal strategies for both players which involve only a finite number of pure strategies. Unfortunately the explicit solution of a given game involves severe computational problems. The general approach is to reduce the solution of the game to the solution of certain systems of algebraic equations, linear in some cases, non-linear in others.

By the Weierstrass approximation theorem any continuous function on the closed unit square can be uniformly approximated by polynomials. Early hopes that the solution of polynomial games would pave the way to the analysis of all continuous games were unfounded. Polynomial games of large degree can exhibit solution sets of great complexity which are very difficult to describe qualitatively, let alone calculate. Moreover the slow convergence of polynomials to continuous functions means that polynomial games are impractical as approximations to continuous games.

Another class of continuous games is those with **bell-shaped kernels** (for a precise definition see [5]). These include and generalise properties of the well known probability density function

$$p(\xi) = \frac{1}{\sqrt{\pi}} e^{-\xi^2} .$$

Such analytic functions occur as payoff kernels, for example in pursuit games in which the pursuer does not know the precise location of the quarry. For these games it can be shown that the optimal strategies of both players are unique and involve only a finite number of pure strategies.

A continuous game is called **convex** if for each fixed x, $0 \leq x \leq 1$, $P(x, y)$ is a convex function of y. For convex games there is an elegant and fairly detailed theory. It can be shown (for example [3], §2.11) that player 2 possesses a pure optimal strategy whilst player 1 has an optimal strategy which involves at most two pure strategies (cf. Example 1.9). Since we know this much about the form of the optimal strategies, such games are easily solved.

To illustrate the difficulties which can arise in any theory of continuous games consider the *Cantor distribution function* $C(\xi)$. The graph of $C(\xi)$ can be constructed as follows. Define $C(0) = 0$, $C(1) = 1$. Then divide the interval $[0, 1]$ into three equal subintervals, and in the middle one ($1/3 \leq \xi \leq 2/3$) put $C(\xi) = \frac{1}{2}(C(0) + C(1))$. The process is iterated as follows. Divide each maximal interval $[a, b]$ in which $C(\xi)$ has not been defined into three equal subintervals and in the middle one put $C(\xi) = \frac{1}{2}(C(a) + C(b))$. For example $C(\xi) = \frac{1}{4}$ for $1/9 \leq \xi \leq 2/9$, $C(\xi) = 3/4$ for $7/9 \leq \xi \leq 8/9$, etc. This function has some unusual properties: it is continuous everywhere, and differentiable except on an uncountable set of Lebesgue measure zero. Moreover, the derivative vanishes wherever it is defined, but $C(\xi)$ is a perfectly acceptable distribution function.

Now specify an *arbitrary* pair of strategies (distribution functions on $[0, 1]$) $F_0(x)$, $G_0(y)$ for players 1 and 2 respectively. It can be shown [15] that there exists a function $P(x, y)$, *analytic* in both variables, for which F_0, G_0 are the unique optimal strategies in the corresponding game. Worse still, if by way of illustration we take F_0, G_0 both equal to the infamous Cantor distribution, then there exists a *rational* function $P(x, y)$ which is continuous on the unit square and with F_0, G_0 as unique optimal strategies! In fact

$$P(x,y) = (y - \tfrac{1}{2}) \left(\frac{1 + (x - \tfrac{1}{2})(y - \tfrac{1}{2})^2}{1 + (x - \tfrac{1}{2})^2(y - \tfrac{1}{2})^4} - \frac{1}{1 + (\tfrac{x}{3} - \tfrac{1}{2})^2(y - \tfrac{1}{2})^4} \right)$$

(see [5], Chapter 7).

PROBLEMS FOR CHAPTER 2.

1. Discuss the relative merits of the 'solutions' given by the pure strategy equilibria of the non-cooperative games

(a)

	τ_1	τ_2
σ_1	(4,-300)	(10,6)
σ_2	(8, 8)	(5,4)

(b)

	τ_1	τ_2
σ_1	(4,-300)	(10,6)
σ_2	(12, 8)	(5,4)

2. Let (x', y'), (x'', y'') $(x', x'' \in \Sigma^{(1)}, y', y'' \in \Sigma^{(2)})$ be equilibrium points in a 2-person zero sum matrix game. Prove that (x', y''), (x'', y') are also equilibrium points. Give an example to show that this result is false for 2-person non-zero sum games. (In general if for any two equilibrium points $x = (x_1, ..., x_n)$, $y = (y_1, ..., y_n)$, and any $i \in I$ $x\|y_i$ is also an equilibrium point, we say the equilibria form a **rectangular** set.)

3. Consider the non-cooperative game

$$\Gamma = \langle\, I, \{S_i\}_{i \in I}, \{P_i\}_{i \in I} \,\rangle,$$

where $I = \{1, 2, ..., n\}$, the pure strategy set $S_i = \{1, 2, ..., m\}$ for every $i \in I$, and

$$P_i(\sigma_1, ..., \sigma_n) = \begin{cases} a_{ik} > 0, & \text{if } \sigma_1 = ... = \sigma_n = k, \\ 0 & , \text{ otherwise.} \end{cases}$$

Show that the only *pure* strategy n-tuples $(\sigma_1, ..., \sigma_n)$ which are *not* equilibrium points are those with exactly $n - 1$ of the σ_i equal. Deduce that the set of equilibrium points of Γ is not rectangular.

4. Prove the following generalisation of Theorem 2.9. Let Γ be a finite n-person non-cooperative game. If the mixed strategy x_i for player i appears in an equilibrium point $x = (x_1, ..., x_n)$, and if for some pure strategy σ_i^j for this player $P_i(x\|\sigma_i^j) < P_i(x)$, then $x_i(\sigma_i^j) = 0$.

5. Find all equilibrium points of the games

(a)

$$\begin{pmatrix} (-10, 5) & (2, -2) \\ (1, -1) & (-1, 1) \end{pmatrix}$$

(b)

$$\begin{pmatrix} (1, 2) & (0, 0) \\ (0, 0) & (2, 1) \end{pmatrix}$$

6. Consider the non-cooperative n-person game in which each player $i \in I$ has exactly two pure strategies; either $\sigma_i = 1$ or $\sigma_i = 2$. The payoff is

$$P_i(\sigma_1,\ldots,\sigma_n) = \sigma_i \prod_{j \neq i} (1 - \delta(\sigma_i, \sigma_j)), \quad (i \in I) ,$$

where δ is the Kronecker δ given by

$$\delta(\sigma_i, \sigma_j) = \begin{cases} 1, & \text{if } \sigma_i = \sigma_j , \\ 0, & \text{otherwise.} \end{cases}$$

If player i uses a mixed strategy in which pure strategy 1 is chosen with probability p_i ($i \in I$), prove that $(p_1, \ldots, p_n)$ defines an equilibrium point if and only if

$$\prod_{j \neq i} (1 - p_j) = 2 \prod_{j \neq i} p_j \text{ for every } i \in I .$$

Deduce that a mixed strategy equilibrium is given by

$$p_i = \cfrac{1}{\left(1 + 2^{\frac{1}{(n-1)}} \right)}, \quad \forall \, i \in I ,$$

and that for $n = 2, 3$ this is the only equilibrium point.

7. A 2-person zero sum matrix game (g_{ij}) is called **diagonal** if $m = n$ and

$$g_{ij} = \begin{cases} g_i > 0, & \text{if } i = j , \\ 0 & , \text{otherwise} . \end{cases}$$

A diagonal game can be interpreted as a variation of *Hide and Seek*. Player 2 hides one of m possible objects in one of m possible locations in such a way that the object of vaule g_i is hidden in location i (that is, player 2 chooses a column). Player 1 then searches for the object (by choosing a row).

Prove that any diagonal game has a positive value, and that if $x^0 = (x_1{}^0, \ldots, x_m{}^0)$ is an optimal strategy for player 1 then $x_j{}^0 > 0$ ($1 \leq j \leq m$). Use this information and Theorem 2.9 to prove that the value of the game is

$$v = \cfrac{1}{\displaystyle\sum_{k=1}^{m} \frac{1}{g_k}} ,$$

and that the unique optimal strategies $x^0 = (x_1{}^0, \ldots, x_m{}^0)$, $y^0 = (y_1{}^0, \ldots, y_m{}^0)$ for players 1 and 2 respectively are given by

$$x_i^0 = \frac{1}{g_i} \frac{1}{\sum\limits_{k=1}^{m} \frac{1}{g_k}}, \quad y_j^0 = \frac{1}{g_j} \frac{1}{\sum\limits_{k=1}^{m} \frac{1}{g_k}}, \quad (1 \le i \le m, \ 1 \le j \le m).$$

8. If $A = (a_{ij})$ is an $m \times n$ matrix game with value v show that the game with matrix $B = (b_{ij})$, where $b_{ij} = a_{ij} + k$ ($1 \le i \le m, 1 \le j \le n$) has value $v + k$ and that the optimal strategies of each player are unchanged.

9. An $m \times m$ matrix is called a **latin square** if each row and each column contains all the integers from 1 to m, for example

$$\begin{pmatrix} 1 & 3 & 2 & 4 \\ 2 & 4 & 3 & 1 \\ 3 & 1 & 4 & 2 \\ 4 & 2 & 1 & 3 \end{pmatrix}.$$

Show that an $m \times m$ matrix game whose matrix is a latin square has the value $(m + 1)/2$. Generalise the result.

10. Solve the infinite matrix game

$$\begin{pmatrix} d & 2d & \frac{1}{2} & 2d & \frac{1}{4} & \cdots \\ d & 1 & 2d & \frac{1}{3} & 2d & \cdots \end{pmatrix}$$

where $d > 0$.

11. Ann and Bill play the following form of simplified poker. There are three cards numbered 1, 2 and 3 respectively. The ranking of the cards is $1 < 2 < 3$. The pack is shuffled, each player antes $1 to the pot, and Ann is dealt a card face down. Ann looks at her card and announces 'Hi' or 'Lo'. To go Hi costs her $2 to the pot, and Lo costs her $1. Next Bill is dealt one of the remaining cards face down. Bill looks at his card and can then 'Fold' or 'See'. If Bill folds, the pot goes to Ann. If Bill sees he must match Ann's bet, and then the pot goes to the holder of the higher card if Ann called Hi, or the lower card if Ann called Lo.

Draw up the extensive form of this game and show that Ann has 8 pure strategies and Bill 64. (Do not attempt to construct the matrix form at this stage.)

Careful inspection of the extensive form enables certain of these strategies to be discarded to reduce the game to 2×4. Explain why, and list the remaining pure strategies for each player. What formal reasons could be used to justify this reduction?

Find the matrix form of the 2×4 game and solve it.

12. Let $P(x, y)$ be the payoff kernel of a continuous game on the unit square. If $P(x, y)$ is strictly concave in x for each fixed y and strictly convex in y for each fixed x, it is intuitively apparent that the game has a saddle point and so is soluble in pure strategies. Prove this by using the following approach.

For any value of x there exists a *unique* value of y which minimises $P(x, y)$ (by strict convexity). Let this value of x be $\phi(x)$. Thus

$$P(x, \phi(x)) = \min_{y \in [0, 1]} P(x, y) .$$

Analogously we may define $\psi(y)$ so that

$$P(\psi(x), y) = \max_{x \in [0, 1]} P(x, y) .$$

(i) Prove ϕ and ψ are continuous.

Hence the composition $\psi \circ \phi (x) = \psi(\phi(x))$ is continuous.

(ii) Now apply the Brouwer fixed point theorem to $\psi \circ \phi : [0, 1] \to [0, 1]$ and obtain the result.

13. Consider the symmetric game on the unit square with

$$P(x, y) = \begin{cases} K(x, y) , & \text{if } x < y , \\ 0 , & \text{if } x = y , \\ -K(y, x) , & \text{if } x > y , \end{cases}$$

where K is suitably smooth. We do *not* assume $P(x, y)$ is continuous and so *there is no guarantee that optimal strategies exist.* The object of this question is to establish two necessary conditions which a special type of distribution function must satisfy to be an optimal mixed strategy.

If the game does have a value it is plain it will be zero, and the optimal stragieges will be the same for both players. Let us *assume* that there exists an optimal mixed strategy F (that is, a distribution function on $[0, 1]$) which is continuous on some interval (a, b) with $0 \le a < b \le 1$, has a continuous derivative F' on $[0, 1]$ with $F' > 0$ on (a, b), and $F' = 0$ outside (a, b).

(i) Show that $E(F, y)$, the expected payoff, is a continuous function of y. Deduce that $E(F, y) = 0$ if $y \in (a, b)$.

(ii) It follows from (i) that

$$\frac{\partial}{\partial y} E(F, y) = 0 \text{ if } y \in (a, b) .$$

By differentiating under the sign of integration deduce that F must satisfy the integral equation

$$-2K(y,y)F'(y) = \int_a^y \frac{\partial K(x,y)}{\partial y} F'(x)dx - \int_y^b \frac{\partial K(y,x)}{\partial y} F'(x)dx$$

for all $y \in (a, b)$.

14. *A silent duel.* Two men are engaged in a duel. At time $t = 0$ they begin to walk towards each other at a constant rate; they will meet at time $t = 1$ if nothing intervenes. Each man has a pistol with one bullet and may fire at any time. The duel is silent, which means that a player does not know that his opponent has fired (unless of course ...). If one hits the other the duel is ended and the payoff is +1 to the winner and -1 to the loser. If neither hits the other or both are hit simultaneously the payoff to both is zero. Assume that if a player fires at time t he has a probability t of hitting his opponent.

Show that the payoff kernel is given by

$$P(x,y) = \begin{cases} x - y + xy, & \text{if } x < y \\ 0 & , \text{if } x = y \\ x - y - xy, & \text{if } x > y . \end{cases}$$

Using the results of the previous question find a continuous distribution F which is an optimal mixed strategy for both players. [Hint: Differentiate the integral equation with respect to y and solve the resulting differential equation for F, then determine a and b.]

15. *A noisy duel.* Consider a duel as in the previous question except that a player knows if his opponent fires and misses. Determine the payoff function and solve the game. [Hint: It has a saddle point.]

CHAPTER REFERENCES

[1] Gale, D. and Stewart, F. M., 'Infinite games with perfect information'. *Contributions to the Theory of Games*, II, (Ann. Math. Studies no. 28) 245-266 (1953), Princeton.

[2] Nash, J. F. 'Non-cooperative games'. *Annals of Math.*, **54**, 286-295 (1951).

[3] Vorob'ev, N. N., *Game Theory, Lectures for Economists and Systems Scientists*, Springer-Verlag (1977), New York.

[4] Schelling, T. C., *The Strategy of Conflict,* Harvard University Press (1960).

[5] Karlin, S., *The Theory of Infinite Games,* Vol. II, Addison-Wesley (1959), Reading, Mass.

[6] von Neumann, J. and Morgenstern, O., *Theory of Games and Economic Behavior,* Princeton University Press (1947), Princeton.

[7] Bellman, R. and Blackwell, D., 'Some two-person games involving bluffing'. *Proc. N.A.S.,* **35**, 600-605 (1949).

[8] Nash, J. F. and Shapley, L. S., 'A simple three-person poker game'. *Contributions to the Theory of Games,* I, (Ann. Math. Studies No. 24) 105-116 (1950), Princeton.

[9] Kuhn, H. W. 'A simplified two-person poker'. *Contributions to the Theory of Games,* I, (Ann. Math. Studies No. 24) 97-103 (1950), Princeton.

[10] McKinsey, J. C. C., *Introduction to the Theory of Games,* McGraw-Hill (1952), New York.

[11] Glicksberg, I. L., *Minimax theorem for upper and lower semicontinuous payoffs.* The RAND Corporation, Research Memorandum RM-478, October 1950.

[12] Sion, M. and Wolfe, P., 'On a game without a value'. *Contributions to the Theory of Games,* III, (Ann. Math. Studies No. 39) 299-306 (1957), Princeton.

[13] Gross, O. A. and Wagner, R. A., *A continuous Colonel Blotto game.* The RAND Corporation, Research Memorandum RM-408, June 1950.

[14] Dresher, M., Karlin, S. and Shapley, L. S., 'Polynomial games'. *Contributions to the Theory of Games,* I, (Ann. Math. Studies No. 24) 161-180 (1950), Princeton.

[15] Glicksberg, I. L. and Gross, O. A., 'Notes on Games over the Square I'. *Contributions to the Theory of Games,* I, (Ann. Math. Studies No. 24) 173-182 (1950), Princeton.

[16] Vorob'ev, N. N., 'Equilibrium points in bimatrix games'. *Teor. Veroyatnost i Primenen.,* **3**, 318-331 (1958).

[17] Kuhn, H. W., 'An algorithm for equilibrium points in bimatrix games'. *Proc. Nat. Acad. Sci. U.S.A.,* **47**, 1657-1662 (1961).

[18] Lemke, C. E. and Howson, J. T., 'Equilibrium points of bimatrix games'. *J. Soc. Indust. Appl. Math.,* **12**, 413-423 (1964).

[19] Apostol, T. M., *Mathematical Analysis,* Addison-Wesley (1963), Reading, Mass.

3

Linear Programming and Matrix Games

Duality of aspect is a mirror for observers.

3.1 INTRODUCTION.

In the early 1950's great strides were made in the technique which came to be called **linear programming**. Linear programming is concerned with the problem of planning a complex of interdependent activities in the best possible way. Examples of its application can be found in determining optimal product mix under given purchase and selling price, machine productivities and capacities; problems of optimum storage, shipment and distribution of goods; minimising set-up times in machine shop operations and problems relating to optimal labor allocation. More precisely, linear programming consists of maximising or minimising a *linear* function, known as the **objective function**, subject to a finite set of *linear* constraints. Many real-life problems fall naturally within this context; in other cases it may be possible to approximate the objective function by a linear function or non-linear constraints by linear ones. The use of linear approximations as a means of dealing with important problems has a respectable history in economics as well as in the natural sciences, including mathematics. It is to be found in economics in the *tableau économique* of Quesnay, the general equilibrium systems of Walras and Cassel, and the input-output analyses of Leontief.

Much of the stimulus in the development of the mathematical theory of linear programming seems to have been derived from the early work of von Neumann and others in game theory; no doubt this worked both ways. However, it was Dantzig's general formulation of the problem and development of the simplex algorithm which made possible the now widespread application of the earlier theoretical work.

Consider, for example, the *Diet problem*, proposed by Stiegler in 1945 and which helped to stimulate the development of linear programming.

> **Example 3.1** (The diet problem). The problem is to find the minimal cost of subsistence. We assume that dieticians have determined that a healthy diet must contain minimal amounts b_j of nutrients N_j $(1 \leq j \leq n)$. For example N_1 could be vitamin C, in which case present knowledge seems to suggest b_1 should be about 30 milligrams/day. The range of foods to be included are those available that are not forbidden by dietary custom. Let us suppose that there are m different foods $F_1, ..., F_m$ which can be used. If an extremely expensive food is included in the list then the solution should provide that very little or none of this food be included in the diet.

Let

$$a_{ij} = \text{the number of units of } N_j \text{ in } F_i,$$
$$c_i = \text{the cost of one unit of } F_i.$$

The problem is to choose x_i units of F_i ($1 \leq i \leq m$) so that the cost of the diet,

$$c_1 x_1 + \ldots + c_m x_m$$

is a minimum, subject to the constraints that the diet must contain at least b_j units of N_j ($1 \leq j \leq n$), that is the x_i must satisfy

$$a_{1j} x_1 + a_{2j} x_2 + \ldots + a_{mj} x_m \geq b_j \qquad (1 \leq j \leq n) .$$

The remaining constraints

$$x_1 \geq 0, \ x_2 \geq 0, \ \ldots x_m \geq 0$$

are just commonsense ones in that it is impracticable to consume a negative quantity of food.

It is obviously convenient to develop some matrix notation. In what follows all vectors are row vectors, and if

$$A = \begin{pmatrix} a_{11} & a_{12} & \cdots & a_{1n} \\ a_{21} & a_{22} & \cdots & a_{2n} \\ \cdots & \cdots & \cdots & \cdots \\ a_{m1} & a_{m2} & \cdots & a_{mn} \end{pmatrix}$$

is an $m \times n$ matrix, then

$$A^T = \begin{pmatrix} a_{11} & a_{21} & \cdots & a_{m1} \\ a_{12} & a_{22} & \cdots & a_{m2} \\ \cdots & \cdots & \cdots & \cdots \\ a_{1n} & a_{2n} & \cdots & a_{mn} \end{pmatrix}$$

is an $n \times m$ matrix called the **transpose** of A. For the vector $y = (y_1, \ldots, y_n)$ we have

$$y^T = \begin{pmatrix} y_1 \\ \cdot \\ \cdot \\ \cdot \\ y_n \end{pmatrix}$$

If $u = (u_1, \ldots, u_m)$, $v = (v_1, \ldots, v_m)$, are two vectors of the same dimension we write $u \geq v$ if

$u_i \geq v_i$ $(1 \leq i \leq m)$; and $\mathbf{0} = (0, 0, ..., 0)$ so that $\mathbf{u} \geq \mathbf{0}$ means $u_i \geq 0$ $(1 \leq i \leq m)$.

With this notation we see that the diet problem is an example of the following class of linear programming problems.

Pure minimum problem (**Primal**):

$$\left. \begin{array}{l} \text{Minimise } \mathbf{x}\,\mathbf{c}^{\,T} \\ \text{subject to } \mathbf{x}\,A \geq \mathbf{b} \text{ and } \mathbf{x} \geq \mathbf{0} \,. \end{array} \right\} \qquad (3.1)$$

Here $\mathbf{c} = (c_1, ..., c_m)$, $\mathbf{x} = (x_1, ..., x_m)$, $A = (a_{ij})$ $(1 \leq i \leq m, 1 \leq j \leq n)$, and $\mathbf{b} = (b_1, ..., b_n)$.

Corresponding to any linear program there is a related linear program called the dual program. For the pure minimum problem the dual is the

Pure maximum problem (**Dual**):

$$\left. \begin{array}{l} \text{Maximise } \mathbf{y}\,\mathbf{b}^T \\ \text{subject to } A\,\mathbf{y}^T \leq \mathbf{c}^T \text{ and } \mathbf{y} \geq \mathbf{0} \,. \end{array} \right\} \qquad (3.2)$$

It is easy to check that the dual of the dual is the original problem, for we can write the maximum problem in the form

$$\text{Minimise } -\mathbf{y}\,\mathbf{b}^T$$
$$\text{subject to } -\mathbf{y}\,A^T \geq -\mathbf{c} \text{ and } \mathbf{y} \geq \mathbf{0} \,.$$

If we now form the dual of this minimum problem we find it is

$$\text{Maximise } -\mathbf{x}\,\mathbf{c}^T$$
$$\text{subject to } -A^T\mathbf{x}^T \leq -\mathbf{b}^T \text{ and } \mathbf{x} \geq \mathbf{0} \,,$$

which can be rewritten as the pure minimum problem we started with.

The dual problem often has an interesting interpretation in terms of the original problem.

> **Example 3.2.** A hospital administrator, who also happens to have shares in a pharmaceutical firm, hears about the diet problem. He decides that the hospital board can be persuaded to substitute pills for food in their patients' diet provided that the pills cost no more than the food. The patients of course are too ill to object. The administrator's problem is to choose the price y_j of one unit of the N_j pills so that the receipts of the pharmaceutical firm, based on the minimal nutritional requirements of the diet, are maximised. This is equivalent to maximising
>
> $$y_1 b_1 + y_2 b_2 + ... + y_n b_n$$
>
> subject to the constraints

$$a_{i1} y_1 + a_{i2} y_2 + \ldots + a_{in} y_n \le c_i \quad (1 \le i \le m)$$

which say that the equivalent pill form of each food is not more expensive than the food itself. The remaining constraints

$$y_1 \ge 0, \ldots y_n \ge 0$$

merely express the condition that the prices be non-negative.

Expressed in matrix form the administrator's problem is

$$\text{Maximise } \mathbf{y}\,\mathbf{b}^T$$
$$\text{subject to } A\,\mathbf{y}^T \le \mathbf{c}^T \text{ and } \mathbf{y} \ge \mathbf{0} \,,$$

which is precisely the dual of the diet problem!

Consider now the problem of player 1 in solving an $m \times n$ matrix game $A = (a_{ij})$. There is no loss of generality in assuming that $a_{ij} > 0$ for all i and j, since adding the same quantity to all the payoff entries does not alter the strategic structure of the game.[11] Hence the value of the game is strictly positive. Player 1 wishes to find a $\mathbf{p} = (p_1, \ldots, p_m)$ such that

$$p_1 a_{1j} + p_2 a_{2j} + \ldots + p_m a_{mj} \ge v \quad (1 \le j \le n),$$

where $v > 0$ is as large as possible and $0 \le p_i \le 1$, $\Sigma p_i = 1$. If we change variables by putting

$$x_i = \frac{p_i}{v} \quad (1 \le i \le m)$$

then the condition $\Sigma p_i = 1$ becomes

$$\sum_{i=1}^{m} x_i = \frac{1}{v}.$$

Thus maximising v is equivalent to *minimising* Σx_i. If we write $\mathbf{J}_m = (1, 1, \ldots, 1)$, where $\mathbf{J}_m$ is m-dimensional, and $\mathbf{J}_n = (1, 1, \ldots, 1)$, where $\mathbf{J}_n$ is n-dimensional, we can restate player 1's problem as

$$\left. \begin{array}{l} \text{Minimise } \mathbf{x}\,\mathbf{J}_m^T \\ \text{subject to } \mathbf{x}A \ge \mathbf{J}_n \text{ and } \mathbf{x} \ge \mathbf{0} \end{array} \right\} \tag{3.3}$$

Thus player 1's problem is a special case of the pure minimum problem.

---

[11]Chapter 2, Problem 8.

Now consider the problem for player 2. This player has to find $\mathbf{q} = (q_1, \ldots, q_n)$, $0 \le q_j \le 1$, $\Sigma q_j = 1$ so that

$$q_1 a_{i1} + q_2 a_{i2} + \ldots + q_n a_{in} \le u \quad (1 \le i \le m)$$

where $u > 0$ is as small as possible. Setting

$$y_j = \frac{q_j}{u} \quad (1 \le j \le n)$$

and proceeding as before we can restate player 2's problem as

$$\left. \begin{array}{l} \text{Maximise } \mathbf{y}\,\mathbf{J}_n^T \\[2mm] \text{subject to } A\,\mathbf{y}^T \le \mathbf{J}_m^T \text{ and } \mathbf{y} \ge \mathbf{0} \end{array} \right\} . \tag{3.4}$$

It comes as no surprise to find that player 2's problem is the dual of player 1's.

Definition. A vector satisfying the constraints of a linear program is called a **feasible vector**. If a feasible vector exists for a linear program we call the program **feasible**.

Proposition 3.1. Both linear programs associated with a matrix game $A = (a_{ij})$, $a_{ij} > 0$ for all i, j are feasible.

Proof. That (3.4) is feasible is trivial since $\mathbf{y} = \mathbf{0}$ satisfies the constraints. To show (3.3) is feasible, for any k, $1 \le k \le m$, put

$$a_{kl} = \min_{1 \le j \le n} a_{kj} > 0 .$$

Then a feasible vector for (3.3) is $\mathbf{x} = a_{kl}^{-1} (0, \ldots, 0, 1, 0, \ldots, 0)$, where the 1 is in the kth entry, for $\mathbf{x} \ge \mathbf{0}$ and

$$\mathbf{x} A = a_{kl}^{-1} (a_{k1}, a_{k2}, \ldots, a_{kn}) \ge \mathbf{J}_n .$$

Theorem 3.1. Let $\mathbf{x}$, $\mathbf{y}$ satisfy the constraints of (3.1) and (3.2) respectively. Then

$$\mathbf{y}\,\mathbf{b}^T \le \mathbf{x}\,\mathbf{c}^T .$$

Proof. We have

$$\mathbf{y}\,\mathbf{b}^T = \sum_{j=1}^n y_j b_j \le \sum_{j=1}^n y_j \left(\sum_{i=1}^m x_i a_{ij} \right) \quad (\text{since } \mathbf{x} A \ge \mathbf{b})$$

$$= \sum_{i=1}^m x_i \left(\sum_{j=1}^n y_j a_{ij} \right) \le \sum_{i=1}^m x_i c_i \quad (\text{since } A\,\mathbf{y}^T \le \mathbf{c}^T)$$

$$= \mathbf{x}\,\mathbf{c}^T$$

Corollary 3.1. Suppose x^*, y^* satisfy the constraints of (3.1) and (3.2) respectively, and $y^* b^T = x^* c^T$. Then x^* is a solution to (3.1) and y^* is a solution to (3.2).

In general a linear program need not have a sloution. One of two things may go wrong. The constraints may be inconsistent so that there are no feasible vectors for the program, in which case the program is said to be **infeasible**. Or it may be that the desired minimum (or maximum) does not exist even if the program is feasible, since the objective function may take arbitrarily small (or large) values in the constraint set. For example the program

$$\text{Maximise } 3x$$
$$\text{subject to } -x \leq 1 \text{ and } x \geq 0$$

does not have a solution, as the objective function $3x$ can be made arbitrarily large. Such a program is said to be **unbounded**. Theorem 3.1 has a second corollary.

Corollary 3.2. If one of the pure linear programs is unbounded then its dual is infeasible.

Proof. Suppose (3.1) is feasible, so there exists x which satisfies the constraints of this program. Then Theorem 3.1 shows that the objective function $y b^T$ of (3.2) is bounded above by $x c^T$. Hence (3.2) is not unbounded. Thus if (3.2) is unbounded then (3.1) is infeasible. Similarly if (3.1) is unbounded then (3.2) is infeasible.

The converse of this corollary is false. That is, if a program is infeasible it does not follow that its dual is unbounded. It can happen that both programs are infeasible.

Example 3.3 The dual programs

$$\text{Minimise } x_1 - 3x_2$$
$$\text{subject to}$$
$$x_1 - x_2 \geq 6$$
$$-x_1 + x_2 \geq 1 \text{ and } x_1 \geq 0, x_2 \geq 0 ,$$

and

$$\text{Maximise } 6y_1 + y_2$$
$$\text{subject to}$$
$$y_1 - y_2 \leq 1$$
$$-y_1 + y_2 \leq -3 \text{ and } y_1 \geq 0, y_2 \geq 0 ,$$

are both obviously infeasible.

Thus if a program is infeasible its dual may be unbounded, or it may also be infeasible. Our first objective will be to prove (Theorem 3.4) that these are the only possibilities. It then follows that there are only four possible relationships between the dual pair (3.1) and (3.2). They can both be infeasible, or one can be infeasible while the other is unbounded or both may be feasible and bounded. That a feasible bounded program possesses a solution is not a trivial matter since the constraint set, although closed, may be unbounded. However, in Theorem 3.5, our second objective, the question of existence, is settled by showing that if a program and its dual are both feasible then both possess solutions and take the same value, that is if $\mathbf{x}^*$ is a solution of (3.1) and $\mathbf{y}^*$ a solution of (3.2), then $\mathbf{x}^* \mathbf{c}^T = \mathbf{y}^* \mathbf{b}^T$. Putting these facts together: if one program has a solution it must be feasible and bounded, hence by Corollary 3.2 and Theorem 3.4 the dual is also feasible and bounded, but then Theorem 3.5 asserts that both programs possess solutions and take the same value. We are thus led to the *duality theorem of linear programming*, Theorem 3.6.

3.2 PRELIMINARY RESULTS.

Lemma 3.1 (Separating hyperplane lemma). If $\mathbf{C} \subseteq \mathbf{R}^n$ is a closed bounded convex set and $\mathbf{b} \notin \mathbf{C}$ then there exists $\mathbf{y} \in \mathbf{R}^n$ such that

$$\mathbf{y}\,\mathbf{b}^T > \mathbf{y}\,\mathbf{z}^T \text{ for every } \mathbf{z} \in \mathbf{C} \ .$$

Proof. Choose a closed bounded ball S with radius R and centre $\mathbf{b}$ such that $\mathbf{C} \cap S$ is closed bounded. Hence the continuous function $f(\mathbf{z}) = (\mathbf{b} - \mathbf{z})(\mathbf{b} - \mathbf{z})^T$ attains its bounds in $\mathbf{C} \cap S$. In particular there exists $\mathbf{z}_0 \in \mathbf{C} \cap S$ such that

$$f(\mathbf{z}_0) = \min_{\mathbf{z} \,\in\, \mathbf{C} \cap S} f(\mathbf{z}) \ .$$

Moreover $0 < f(\mathbf{z}_0) \leq R^2$ since $f(\mathbf{z}) = |\mathbf{b} - \mathbf{z}|^2$, $\mathbf{b} \notin \mathbf{C}$ and $\mathbf{z}_0 \in S$. Hence

$$0 < f(\mathbf{z}_0) \leq f(\mathbf{z}) \text{ for every } \mathbf{z} \in \mathbf{C} , \tag{3.5}$$

for if $\mathbf{z} \in \mathbf{C} \cap S$ then $f(\mathbf{z}) \geq f(\mathbf{z}_0)$, and if $\mathbf{z} \notin S$ then $f(\mathbf{z}) > R_2 \geq f(\mathbf{z}_0)$.

Let $\mathbf{z}$ be any point of $\mathbf{C}$, then since $\mathbf{C}$ is convex, $\lambda \mathbf{z} + (1 - \lambda)\mathbf{z}_0 \in \mathbf{C}$. Hence by (3.5)

$$f(\mathbf{z}_0) \leq f(\lambda \mathbf{z} + (1-\lambda)\mathbf{z}_0) \ ,$$

that is,

$$(\mathbf{b} - \mathbf{z}_0)(\mathbf{b} - \mathbf{z}_0)^T \leq (\mathbf{b} - \lambda\mathbf{z} - (1 - \lambda)\mathbf{z}_0)(\mathbf{b} - \lambda\mathbf{z} - (1 - \lambda)\mathbf{z}_0)^T .$$

If we put $\mathbf{k} = \mathbf{b} - \mathbf{z}_0$ this last inequality can be written as

$$0 \leq \lambda^2(\mathbf{z}_0 - \mathbf{z})(\mathbf{z}_0 - \mathbf{z})^T + 2\lambda\mathbf{k}(\mathbf{z}_0 - \mathbf{z})^T$$

and this holds for all λ, $0 \leq \lambda \leq 1$. Hence we must have $\mathbf{k}(\mathbf{z}_0 - \mathbf{z})^T \geq 0$. But $\mathbf{z}$ was at any point of $\mathbf{C}$, so in fact

$$\mathbf{k}(\mathbf{z}_0 - \mathbf{z})^T \geq 0 \text{ for every } \mathbf{z} \in \mathbf{C} . \tag{3.6}$$

From (3.5) $f(\mathbf{z}_0) = (\mathbf{b} - \mathbf{z}_0)(\mathbf{b} - \mathbf{z}_0)^T > 0$ so that $\mathbf{k}\,\mathbf{b}^T > \mathbf{k}\,\mathbf{z}_0^T$. Combining this with (3.6) we obtain

$$k\,b^T > k\,z_0{}^T \geq k\,z^T \text{ for every } z \in C .$$

Hence the vector $y = k$ satisfies the conclusion of the lemma.

If this proof seems a little obtuse a glance at Figure 3.1 should make matters clear. The lemma amounts to the assertion that a hyperplane such as

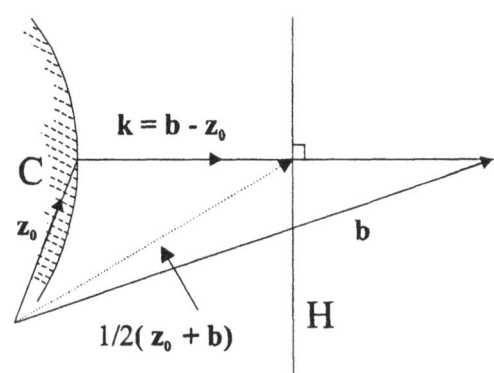

$$H = \{\, x ; k\,x^T = \frac{1}{2} k\,(\,z^0 + b\,)^T \,\}$$

separates C and b.

Figure 3.1 H separates C and b.

Theorem 3.2. Either

$$\text{(i)} \quad \text{the equations } x\,A = b \text{ have a solution } x \geq 0, \tag{3.7}$$

or

$$\text{(ii)} \quad \text{the inequalities } A\,y^T \leq 0,\; y\,b^T > 0 \text{ have a solution,} \tag{3.8}$$

but not both.

Proof. If both (3.7) and (3.8) hold, then

$$0 \geq x\,A\,y^T = b\,y^T = y\,b^T > 0 ,$$

which is impossible.

Suppose (3.7) is false, then

$$b \notin C = \{\, z = x\,A ; x \geq 0 \,\} .$$

Now C is just the image of the first quadrant $x \geq 0$ under the linear transformation $x \to x\,A$ (in fact C is just a polyhedral convex cone). Since the quadrant $x \geq 0$ is closed and convex the linearity of the transformation readily ensures that C is closed and convex. Hence by Lemma 3.1 there exists a y for which

$$y\,b^T > y\,z^T = y\,(\,x\,A\,)^T \text{ for all } x \geq 0 . \tag{3.9}$$

In particular this inequality is satisfied with $x = 0$ so that $y\,b^T > 0$. It remains to show that $A\,y^T \leq 0$. Suppose this is false, then for some i $(A\,y^T)_i = \lambda > 0$, where $(A\,y^T)_i$ denotes the ith component of $A\,y^T$. Put

$$x = \lambda^{-1}(\,y\,b^T\,)\,\sigma_i ,$$

where $\sigma_i = (0, ..., 0, 1, 0, ..., 0)$ the 1 being the ith component. Then

$$y(xA)^T = y(\lambda^{-1}(yb^T)\sigma_i A)^T = \lambda^{-1}(yb^T)(\sigma_i A y^T)^T$$
$$= \lambda^{-1}(yb^T)(A y^T)_i = yb^T ,$$

which contradicts (3.9). Hence the inequalities in (3.8) have a solution, so (3.7) and (3.8) cannot be simultaneously false, which proves the theorem.

We can use Theorem 3.2 to prove

Theorem 3.3 Either

$$\qquad\qquad (i) \quad \text{the inequalities } xA \geq b \text{ have a solution } x \geq 0 , \qquad\qquad (3.10)$$

or

$$\qquad\qquad (ii) \quad \text{the inequalities } A y^T \leq 0, \, yb^T > 0 \text{ have a solution } y \geq 0 , \qquad\qquad (3.11)$$

but not both.

Proof. If both (3.10) and (3.11) hold, then

$$0 \geq xA y^T \geq b y^T = y b^T > 0 ,$$

which is impossible.

Suppose (3.10) is false, then $x A - z = b$ has no solution with $x \geq 0$ and $z \geq 0$. Let

$$I = \begin{pmatrix} 1 & 0 & 0 & \dots & 0 \\ 0 & 1 & 0 & \dots & 0 \\ & & \dots & \\ 0 & 0 & 0 & \dots & 1 \end{pmatrix}$$

Then in an obvious notation we can restate this as

$$(x, z) \begin{pmatrix} A \\ -I \end{pmatrix} = b \text{ has no solution with } (x, z) \geq 0 .$$

Hence by Theorem 3.2 the inequalities

$$\begin{pmatrix} A \\ -I \end{pmatrix} y^T \leq 0, \, y b^T > 0$$

have a solution, and this solution satisfies (3.11).

Now Theorem 3.3 has its own 'dual':

Theorem 3.3[*]. Either

$$(i) \quad \text{the inequalities } A\,\mathbf{y}^{T} \le \mathbf{c}^{T} \text{ have a solution } \mathbf{y} \ge \mathbf{0}, \qquad (3.12)$$

or

$$(ii) \quad \text{the inequalities } \mathbf{x}A \ge \mathbf{0}, \ \mathbf{x}\mathbf{c}^{T} < 0 \text{ have a solution } \mathbf{x} \ge \mathbf{0}. \qquad (3.13)$$

but not both.

Proof. If both (3.12) and (3.13) hold, then

$$0 \le \mathbf{x}A\,\mathbf{y}^{T} \le \mathbf{x}\mathbf{c}^{T} < 0,$$

which is impossible.

Suppose (3.12) is false, then $A\,\mathbf{y}^{T} + \mathbf{z}^{T} = \mathbf{c}^{T}$ has no solution with $\mathbf{y}^{T} \ge \mathbf{0}$ and $\mathbf{z}^{T} \ge \mathbf{0}$. Taking transposes we can restate this as

$$(\mathbf{y}, \mathbf{z}) \begin{pmatrix} A^{T} \\ I \end{pmatrix} = \mathbf{c} \text{ has no solution with } (\mathbf{y}, \mathbf{z}) \ge \mathbf{0} \ .$$

Hence by Theorem 3.2 there exists ξ^{T} such that

$$\begin{pmatrix} A^{T} \\ I \end{pmatrix} \xi^{T} \le \mathbf{0} \text{ and } \xi\,\mathbf{c}^{T} > 0 \ .$$

Putting $\mathbf{x} = -\xi$ and taking transposes this becomes

$$\mathbf{x}\,(A, I) \ge \mathbf{0} \text{ and } \mathbf{x}\,\mathbf{c}^{T} < 0 \ ,$$

and this solution satisfies (3.13).

3.3 DUALITY THEORY.

Theorem 3.4. If one of the pure linear programs is feasible and its dual is infeasible, then the original program is unbounded.

Proof. Suppose first that (3.2) is feasible and (3.1) is infeasible. If (3.2) is feasible there exists $\mathbf{y}_0$ such that

$$A\,\mathbf{y}_0^{T} \le \mathbf{c}^{T} \text{ and } \mathbf{y}_0 \ge \mathbf{0} \ . \qquad (3.14)$$

If (3.1) is infeasible the inequalities $\mathbf{x}\,A \ge \mathbf{b}$ have no solution with $\mathbf{x} \ge \mathbf{0}$, and so by Theorem 3.3 there exists $\mathbf{y}_1 \ge \mathbf{0}$ such that

$$A\,\mathbf{y}_1^{T} \le \mathbf{0} \text{ and } \mathbf{y}_1\,\mathbf{b}^{T} > 0 \ . \qquad (3.15)$$

Now if $\lambda \geq 0$ we have $y_0 + \lambda y_1 \geq 0$ and from (3.14) and (3.15)

$$A(y_0 + \lambda y_1)^T = A y_0^T + \lambda A y_1^T \leq c^T ,$$

so $y_0 + \lambda y_1$ is a feasible vector for (3.2) for all $\lambda \geq 0$. Since $y_1 b^T > 0$, $(y_0 + \lambda y_1) b^T$ can be made arbitrarily large by choosing λ sufficiently large. Hence $y b^T$ is unbounded on the set of feasible vectors of (3.2), that is (3.2) is unbounded.

If (3.1) is feasible and (3.2) is infeasible the proof is similar, except that we use Theorem 3.3*. If x_0 is a feasible vector for (3.1) and x_1 is a vector satisfying (3.13), then $x_0 + \lambda x_1$ is a feasible vector for (3.1) with the property that $(x_0 + \lambda x_1) c^T$ can be made arbitrarily small by choosing λ sufficiently large. Hence (3.1) is unbounded.

Theorem 3.5. Let the dual programs (3.1) and (3.2) both be feasible. Then both will have solutions x^* and y^* respectively and

$$x^* c^T = y^* b^T .$$

Proof. The theorem will be proved if we can show that there exist vectors $x^* \geq 0$, $y^* \geq 0$ which satisfy

$$x^* A \geq b \tag{3.16}$$

$$A(y^*)^T \leq c^T \tag{3.17}$$

$$x^* c^T \leq y^* b^T , \tag{3.18}$$

for then by Theorem 3.1 $x^* c^T = y^* b^T$ and so by Corollary 3.1 x^* and y^* are solutions of (3.1) and (3.2) respectively.

The system of inequalities (3.16) to (3.18) can be written for general x, y as

$$(x, y) \begin{pmatrix} A, & O, & -c^T \\ O, & -A^T, & b^T \end{pmatrix} \geq (b, -c, 0) , \tag{3.19}$$

where the symbols O are appropriately shaped zero matrices.

Suppose (3.19) has no solution with $x \geq 0$ and $y \geq 0$, then by Theorem 3.3 there exists a vector $(z, w, \alpha) \geq 0$, where α is a scalar, such that

$$\begin{pmatrix} A, & 0, & -c^T \\ 0, & -A^T, & b^T \end{pmatrix} \begin{pmatrix} z^T \\ w^T \\ \alpha \end{pmatrix} \le 0 \text{ and } (z, w, \alpha) \begin{pmatrix} b^T \\ -c^T \\ 0 \end{pmatrix} > 0 ,$$

which we can rewrite as

$$A z^T \le \alpha c^T , \tag{3.20}$$

$$w A \ge \alpha b , \tag{3.21}$$

$$z b^T > w c^T . \tag{3.22}$$

We first examine the case $\alpha = 0$. Since (3.1) and (3.2) are feasible there exists $x \ge 0$ with $x A \ge b$ and $y \ge 0$ with $A y^T \le c^T$. Using these inequalities and (3.20), (3.21) with $\alpha = 0$ we obtain

$$z b^T = b z^T \le x A z^T \le 0 ,$$
$$0 \le w A y^T \le w c^T ,$$

thus $z b^T \le w c^T$ which contradicts (3.22). Hence $\alpha > 0$, so that (3.21) and (3.20) imply that $\alpha^{-1} w$, $\alpha^{-1} z$ are feasible vectors for (3.1) and (3.2) respectively. Hence by Theorem 3.1

$$\alpha^{-1} z b^T \le \alpha^{-1} w c^T$$

which again contradicts (3.22). Hence our assumption that (3.19) had no solution with $x \ge 0$ and $y \ge 0$ is false, that is there exist $x^* \ge 0$, $y^* \ge 0$ which satisfy (3.16) to (3.18), and this proves the theorem.

As explained in the concluding paragraph of §3.1 we can put together Corollary 3.2, Theorem 3.4 and Theorem 3.5 to obtain

Theorem 3.6. If one of the dual pair of linear programs (3.1) and (3.2) has a solution then so does the other. If both programs are feasible then they have the same value.

This fundamental property of linear inequalities was proved by Farkas [1] in 1902. An immediate corollary is

Theorem 3.7 (Minimax theorem). For any matrix game A there exist mixed strategies p^*, q^* for players 1 and 2 respectively such that

$$p A (q^*)^T \le p^* A (q^*)^T \le p^* A q^T$$

for all mixed strategies p for player 1 and q for player 2.

Proof. Without loss of generality the entries of A are all positive. By Proposition 3.1 the dual programs (3.3) and (3.4) are feasible. By Theorem 3.6, (3.3) has a solution $x^* \geq 0$, (3.4) a solution $y^* \geq 0$ and

$$x^* J_m^T = y^* J_n^T = \alpha , \qquad (3.23)$$

say. Moreover $\alpha > 0$ since $x^* \geq 0$ must have at least one positive component in order to satisfy the constraint $x A \geq J_n$ of (3.3). Also since $A (y^*)^T \leq J_m^T$ and $x^* A \geq J_n$ we have

$$\alpha = x^* J_m^T \geq x^* A (y^*)^T \geq J_n (y^*)^T = y^* J_n^T = \alpha$$

Putting $p^* = \alpha^{-1} x^*$, $q^* = \alpha^{-1} y^*$ this gives

$$\alpha^{-1} = p^* A (q^*)^T . \qquad (3.24)$$

Moreover $p^* \geq 0$, $q^* \geq 0$; and (3.23) expresses the condition that the components of p^* and q^* each sum to 1. Thus p^* and q^* are strategies. Finally we can rewrite the constraints $A (y^*)^T \leq J_m^T$, $x^* A \geq J_n$ as

$$p^* A \geq \alpha^{-1} J_n , \quad A (q^*)^T \leq \alpha^{-1} J_m^T .$$

Hence for all strategies p, q

$$p^* A q \geq \alpha^{-1} \geq p A (q^*)^T$$

and this together with (3.24) proves the theorem.

Since the dual programs (3.3) and (3.4) are special forms of the programs (3.1) and (3.2) respectively, it is quite reasonable that the duality theorem should imply the minimax theorem. In fact it is possible to give a game-theoretic proof of the duality theorem. The argument involves the minimax theorem and the converse of Theorem 2.9. However, since the proof of this converse involves a lemma substantially equivalent to Theorem 3.2 we shall not go into details.

The next result is also a corollary of Theorem 3.6 and can be used to test whether a pair of feasible vectors provides solutions.

Theorem 3.8 (Equilibrium theorem). Let x be feasible for (3.1) and y for (3.2), then x and y are solutions of their respective programs if and only if

$$x_i > 0 \text{ implies } (A y^T)_i = c_i \qquad (1 \leq i \leq m) , \qquad (3.25)$$

and

$$y_j > 0 \text{ implies } (x A)_j = b_j \qquad (1 \leq j \leq n) . \qquad (3.26)$$

Proof. Suppose x and y satisfy (3.25) and (3.26), then

$$\mathbf{x} A \mathbf{y}^T = \sum_{i=1}^{m} x_i (A \mathbf{y}^T)_i = \sum_{i=1}^{m} x_i c_i = \mathbf{x} \mathbf{c}^T ,$$

and

$$\mathbf{x} A \mathbf{y}^T = \sum_{j=1}^{n} (\mathbf{x} A)_j y_j = \sum_{j=1}^{n} b_j y_j = \mathbf{y} \mathbf{b}^T .$$

Hence $\mathbf{x} \mathbf{c}^T = \mathbf{y} \mathbf{b}^T$, and so by Corollary 3.1 $\mathbf{x}$ and $\mathbf{y}$ are solutions of (3.1) and (3.2) respectively.

Conversely suppose $\mathbf{x}$, $\mathbf{y}$ are solutions. The constraints give

$$\mathbf{y} \mathbf{b}^T = \mathbf{b} \mathbf{y}^T \le \mathbf{x} A \mathbf{y}^T \le \mathbf{x} \mathbf{c}^T ,$$

and by Theorem 3.6 $\mathbf{x} \mathbf{c}^T = \mathbf{y} \mathbf{b}^T$ so that

$$\mathbf{x} \mathbf{c}^T = \mathbf{x} A \mathbf{y}^T = \mathbf{b} \mathbf{y}^T$$

that is

$$\mathbf{x} (\mathbf{c}^T - A \mathbf{y}^T) = 0 \text{ and } (\mathbf{x} A - \mathbf{b}) \mathbf{y}^T = 0 .$$

We can rewrite these as

$$\sum_{i=1}^{m} x_i (c_i - (A \mathbf{y}^T)_i) = 0 \text{ and } \sum_{j=1}^{n} ((\mathbf{x} A)_j - b_j) y_j = 0 .$$

Since $\mathbf{x} \ge 0$, $\mathbf{c}^T \ge A \mathbf{y}^T$ the first equation yields (3.25), and since $\mathbf{y} \ge 0$, $\mathbf{x} A \ge \mathbf{b}$ the second yields (3.26).

We infer from (3.26) that if $(\mathbf{x} A)_j > b_j$ for some solution $\mathbf{x}$ of (3.1) then $y_j = 0$ for every solution $\mathbf{y}$ of (3.2). Interpreted in terms of the diet problem this says that if the nutrient N_j is oversupplied in an optimal diet then the price of one unit of the N_j pills must be zero. Similarly (3.25) says that if in some optimal pricing of the pills the equivalent pill form of the food F_i is cheaper than the food itself then the amount of F_i in every optimal diet is zero.

One can interpret the equilibrium theorem in more general economic terms. The history of the interplay between game theory and linear programming on the one hand and competitive equilibrium models in economics on the other is worthy of a brief digression.

Let us assume two kinds of goods: **commodities** that enter into the demand functions of consumers, and **factors** (raw materials) that are used to produce commodities. Each commodity is produced from factors, and the input-output coefficients are assumed constant. Let a_{ij} be the amount of factor F_i used in the production of commodity C_j, y_j the total output of commodity

C_j, b_j the price of C_j, c_i the total available initial supply of factor F_i and x_i the price of F_i. Then to maximise revenue subject to the limits of available supply amounts to the program

$$\text{Maximise } \mathbf{y}\,\mathbf{b}^T$$
$$\text{subject to } A\,\mathbf{y}^T \le \mathbf{c}^T \text{ and } \mathbf{y} \ge \mathbf{0}.$$

Interpretation of the dual program is somewhat elusive ([2], §3.9) and requires that we think of x_i, the component of some solution, as the maximum price that the producer would be willing to pay for one *extra* (small) unit of F_i. This is called the **marginal value** of F_i. The dual program then demands a minimal accounting cost for the available factors of production.

The equations $A\,\mathbf{y}^T = \mathbf{c}^T$ and $\mathbf{x}\,A = \mathbf{b}$ would reflect a situation in which demand exactly equals supply, and each commodity is produced with zero profit. Early economic models, such as Cassel's 1924 [3] simplification of Walras's system, suffered from the defect that even with perfectly plausible values of the input-output coefficients a_{ij} the prices or quantities satisfying these equations might well be negative.

Zeuthen reconsidered the meaning of the equations. He noted that economists, at least since Menger, had recognised that some factors (for example, air) are so abundant that there would be no price charged for them. These would not enter into the list of factors in Cassel's system. But, Zeuthen argued, the division of factors into free and scarce should not be taken as given *a priori*. Hence all that can be said is that the use of a factor should not exceed its supply, but if it falls short then the factor is free. In symbols

$$(A\,\mathbf{y}^T)_i \le c_i\,; \text{ if the inequality is strict then } x_i = 0.$$

Of course this is just the equilibrium theorem.

Independently of Zeuthen, Schlesinger, a Viennese banker and amateur economist, had come to the same conclusion and grasped intuitively the fact that replacing equalities with inequalities also resolves the problem of negative values of the equilibrium variables. In 1933-34 Schlesinger realised the complexity of a rigorous treatment, and at his request Morgenstern put him in touch with a young mathematician, Abraham Wald. The result of their collaboration was the first rigorous analysis of general competitive equilibrium theory.

Like Cassel, Wald dealt with a static model of production in which a prescribed endowment of the factors of production is used to produce commodities, and the output of each commodity is assumed to be a linear function of the input of factors. A static demand structure, relating the prices of commodities to the quantity produced, is also assumed. The object of the model is to determine the amount of each commodity to be produced and the prices of the commodities and of the factors. in a series of papers over the period 1933-35 Wald demonstrated the existence of competitive equilibrium in various alternative models (see [4,5] for a summary). At equilibrium the cost of each commodity in factors of production is equal to its market price. The endowment of each factor is to be sufficient for the goods produced; any factor present in excess is free.

Wald's papers were of forbidding mathematical depth, but help eventually came from the parallel developments of von Neumann and Morgenstern in game theory and linear programming. In 1956 Kuhn [6] re-examined Wald's theorem in the light of these developments, and was able to show that there is a pair of dual linear programs implicit in the Walrasian equations of equilibrium. By this observation the existence of an equilibrium for a fixed set of prices for the goods becomes a consequence of the duality theorem, and the problem is reduced to that of satisfying the demand relations; a question Kuhn settled by an application of the Kakutani fixed point theorem. Thus, using tools that had not been fashioned when Wald wrote, a great simplification in the analysis was effected. Subsequent developments were rapid and a systematic account of more recent existence conditions can be found in Debreu [7].

Returning to our mathematical development we observe that so far nothing has been said regarding the process of finding the solution of a linear program. With this thought now in mind we introduce the following terminology.

Let $M_1, M_2 \subseteq \{1, 2, ..., m\}$ and $N_1, N_2 \subseteq \{1, 2, ..., n\}$. Then each of

$$\left. \begin{aligned} (\mathbf{x}A)_j &= b_j \text{ for all } j \in N_1 , \\ x_i &= 0 \text{ for all } i \in M_2 , \end{aligned} \right\} \tag{3.27}$$

$$\left. \begin{aligned} (A\mathbf{y}^T)_i &= c_i \text{ for all } i \in M_1 , \\ y_j &= 0 \text{ for all } j \in N_2 , \end{aligned} \right\} \tag{3.28}$$

is called a **system of equated constraints**.

Systems (3.27) and (3.28) are **dual** if

$$\left. \begin{aligned} M_1 \cup M_2 &= \{1, 2, ..., m\} , \quad M_1 \cap M_2 = \phi , \\ N_1 \cup N_2 &= \{1, 2, ..., n\} , \quad N_1 \cap N_2 = \phi . \end{aligned} \right\} \tag{3.29}$$

In the event that (3.27) and (3.28) are dual systems we can re-index the quantities involved so that $M_1 = \{1, 2, ..., p\}$, $N_1 = \{1, 2, ..., q\}$, and the systems become

$$(x_1, \ ... \ x_p) \begin{pmatrix} a_{11} \cdots a_{1q} \\ \cdot \quad \cdot \\ \cdot \quad \cdot \\ \cdot \quad \cdot \\ a_{p1} \cdots a_{pq} \end{pmatrix} = (b_1, \ ... \ b_q) , \tag{3.30}$$

$$x_{p+1} = ... = x_m = 0 ,$$

$$
\begin{pmatrix} a_{11} \dots a_{1q} \\ \cdot \quad \cdot \\ \cdot \quad \cdot \\ \cdot \quad \cdot \\ a_{p1} \dots a_{pq} \end{pmatrix}
\begin{pmatrix} y_1 \\ \cdot \\ \cdot \\ \cdot \\ y_q \end{pmatrix}
=
\begin{pmatrix} c_1 \\ \cdot \\ \cdot \\ \cdot \\ c_q \end{pmatrix},
\tag{3.31}
$$

$$y_{q+1} = \dots = y_m = 0$$

respectively. If the $p \times q$ matrix

$$
A_1 = \begin{pmatrix} a_{11} \dots a_{1q} \\ \cdot \quad \cdot \\ \cdot \quad \cdot \\ \cdot \quad \cdot \\ a_{p1} \dots a_{pq} \end{pmatrix}
$$

is square and non-singular we say that each system is a **non-singular square** system.

The following result which characterises solutions of (3.1) and (3.2) is a corollary of the equilibrium theorem.

Theorem 3.9. Feasible vectors $\mathbf{x}$ and $\mathbf{y}$ are solutions of (3.1) and (3.2) respectively if and only if they satisfy a dual system of equated constraints.

Proof. Suppose $\mathbf{x}$ and $\mathbf{y}$ are solutions and define
$$N_1 = \{ j; (\mathbf{x}A)_j = b_j \}, \ N_2 = \{1, 2, \dots, n \} \setminus N_1 .$$

Now $(\mathbf{x}A)_j \le b_j$ for all j since $\mathbf{x}$ is feasible, hence if $j \in N_2$ we have $(\mathbf{x}A)_j < b_j$, and (3.26) gives $y_j = 0$. We define M_1 and M_2 in an analogous fashion and find that if $i \in M_2$ then $x_i = 0$. Thus (3.27) to (3.29) are satisfied.

Conversely, suppose $\mathbf{x}$ and $\mathbf{y}$ satisfy (3.27) to (3.29). Then by hypothesis, $\mathbf{x}$ and $\mathbf{y}$ are feasible and

$$
\begin{aligned}
\mathbf{x}\mathbf{c}^T &= \sum_{i=1}^m x_i c_i = \sum_{i \in M_1} x_i c_i && (\text{since } x_i = 0 \text{ if } i \in M_2) \\
&= \sum_{i \in M_1} x_i (A\mathbf{y}^T)_i = \sum_{i \in M_1} x_i (\sum_{j=1}^n a_{ij} y_j) && (\text{from } (3.28)) \\
&= \sum_{i \in M_1} x_i (\sum_{j \in N_1} a_{ij} y_j) && (\text{since } y_j = 0 \text{ if } j \in N_2)
\end{aligned}
$$

$$= \sum_{j \in N_1} (\sum_{i \in M_1} x_i a_{ij}) y_j = \sum_{j \in N_1} ((xA)_j) y_j$$

$$= \sum_{j \in N_1} b_j y_j \qquad \text{(from (3.27))}$$

$$= \sum_{j=1}^{n} b_j y_j \qquad \text{(since } y_j = 0 \text{ if } j \in N_2)$$

$$= \mathbf{b}\,\mathbf{y}^T \,.$$

Hence by Corollary (3.1) **x** and **y** are solutions as required.

3.4 THE GEOMETRIC SITUATION.

The object of this section is to sketch the geometric ideas which underlie the theorems we have proved and are about to prove. Indeed it is possible to arrive at the same conclusions by a purely geometric route, but this would involve a lengthy digression. Even so, when such an attractive geometric interpretation is available it is helpful to keep it in mind. However, the theorems of later sections will not be logically dependent upon the contents of this section (although the definition of an extreme point will be used).

We begin by observing that the feasible regions

$$S = \{\, \mathbf{x} \in \mathbb{R}^m; \; \mathbf{x}A \geq \mathbf{b}, \, \mathbf{x} \geq \mathbf{0}\,\} \,, \qquad (3.32)$$

$$T = \{\, \mathbf{y} \in \mathbb{R}^n; \; A\mathbf{y}^T \leq \mathbf{c}^T, \, \mathbf{y} \geq \mathbf{0}\,\} \,. \qquad (3.33)$$

for (3.1) and (3.2) respectively are closed and convex. Of course either may be empty. For example that S is closed follows from the continuity of the linear transformation $\mathbf{x} \rightarrow \mathbf{x}\,A$. To prove S is convex suppose $\mathbf{x}_1, \mathbf{x}_2 \in S$ and $0 \leq \lambda \leq 1$, then

$$\lambda \mathbf{x}_1 + (1 - \lambda)\mathbf{x}_2 \geq \mathbf{0}$$

and

$$(\lambda \mathbf{x}_1 + (1 - \lambda)\mathbf{x}_2)A = \lambda \mathbf{x}_1 A + (1 - \lambda)\mathbf{x}_2 A \geq \lambda \mathbf{b} + (1 - \lambda)\mathbf{b} = \mathbf{b} \,,$$

so that $\lambda \mathbf{x}_1 + (1 - \lambda)\mathbf{x}_2 \in S$. Similarly for T.

Example 3.4 Consider the problem

$$\text{Minimise} \quad 3x_1 + 2x_2$$
subject to
$$4x_1 + x_2 \geq 4$$
$$3x_1 + 2x_2 \geq 6$$
$$x_1 + 3x_2 \geq 3$$
and
$$x_1 \geq 0, \, x_2 \geq 0 \,.$$

Written in our customary form this is

$$
\begin{aligned}
&\text{Minimise} \quad (x_1, x_2)\,(3, 2)^T \\
&\text{subject to} \\
&\qquad\qquad (x_1, x_2)\begin{pmatrix} 4 & 3 & 1 \\ 1 & 2 & 3 \end{pmatrix} \ge (4, 6, 3) \\
&\text{and} \qquad (x_1, x_2) \ge 0 .
\end{aligned} \tag{3.34}
$$

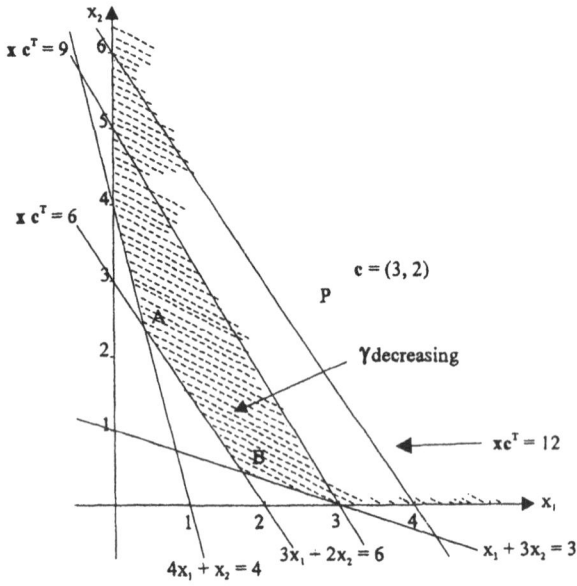

Figure 3.2 Feasible region for the minimum problem (3.34)

Figure 3.2 illustrates this problem graphically.

This figure suggests how to set about a geometrical proof of the existence of a solution to a feasible bounded minimising linear program. Consider a point P in the feasible region and the hyperplane $x\,c^T = \gamma_p$, say through P. There are two possibilities: either there exist parallel translates of this hyperplane $x\,c^T = \gamma$ which intersect the feasible region and for which γ is arbitrarily small, or for sufficiently small γ $x\,c^T = \gamma$ implies x is not in the feasible region. In the first case the program is unbounded. In the second case, since the feasible region is closed, there exists a *least* value of γ such that $x\,c^T = \gamma$ for some x in the feasible region, which provides the required solution (the possible unboundedness of the region causes difficulties here). For this value of γ, by using the convexity of the feasible region, one can further show that $x\,c^T = \gamma$ is a support hyperplane for the region. In Figure 3.2 this support hyperplane is $x\,c^T = 6$, and any point on the line segment AB provides a solution. Of course similar remarks apply to a maximising problem.

The dual problem to (3.34) is

$$
\left.
\begin{array}{l}
\text{Maximise } \left(y_1,\ y_2,\ y_3\right) \left(4,\ 6,\ 3\right)^{\mathrm{T}} \\
\text{Subject to} \\
\begin{pmatrix} 4 & 3 & 1 \\ 1 & 2 & 3 \end{pmatrix} \left(y_1,\ y_2,\ y_3\right)^{\mathrm{T}} \le \left(3,\ 2\right)^{\mathrm{T}} \\
\text{and} \qquad \left(y_1,\ y_2,\ y_3\right) \ge \mathbf{0}
\end{array}
\right\}
\qquad (3.35)
$$

Figure 3.3 illustrates this problem.

Here the support plane $\mathbf{y}\,\mathbf{b}^{\mathrm{T}} = 6$ contains the sole solution $(y_1, y_2, y_3) = (0, 1, 0)$. Notice that if $\mathbf{x} = (x_1, x_2)$ is any point, other than A or B, on AB, the constraints $4x_1 + x_2 \ge 4$, $x_1 + 3x_2 \ge 3$ are satisfied with *strict* inequality, and so the equilibrium theorem predicts that $y_1 = 0$ and $y_3 = 0$ in any solution of (3.35), which is the case.

Definition. Let $\mathbb{C}$ be a convex subset of Euclidean space then $\mathbf{x} \in \mathbb{C}$ is an **extreme point** of $\mathbb{C}$ if for any points $\mathbf{u}, \mathbf{v} \in \mathbb{C}$ and $0 < \lambda < 1$

$$ \mathbf{x} = \lambda\,\mathbf{u} + (1 - \lambda)\,\mathbf{v} \quad \text{implies} \quad \mathbf{x} = \mathbf{u} = \mathbf{v}. $$

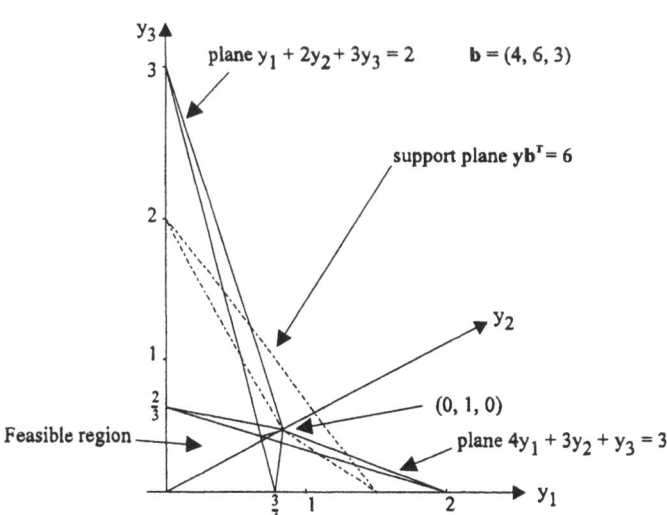

Figure 3.3 Feasible region for the maximum problem (3.35).

Thus $\mathbf{x}$ is an extreme point of $\mathbb{C}$ if it cannot be expressed as a convex linear combination of two points $\mathbf{u}, \mathbf{v} \in \mathbb{C}$ distinct from $\mathbf{x}$. For example the extreme points of a closed disc in the plane are every boundary point. The extreme points of a convex polygon are its vertices.

Consider the region S defined in (3.32). If we denote the columns of A by the vectors $\mathbf{a}_1, \dots, \mathbf{a}_n$

then S is defined by the inequalities

$$\mathbf{x}\,\mathbf{a}_j^{\mathrm{T}} \ge b_j \quad (1 \le j \le n), \quad x_j \ge 0 \quad (1 \le j \le m) \ .$$

Thus S is the intersection of $n + m$ half-spaces associated with $n + m$ hyperplanes

$$\mathbf{x}\,\mathbf{a}_j^{\mathrm{T}} = b_j \quad (1 \le j \le n), \quad x_j = 0 \quad (1 \le j \le m) \ .$$

A convex region defined by the intersection of finitely many half-spaces is known as a **convex polyhedral set**. It can be shown that the extreme points of such a set in $\mathbb{R}^m$ are those points on the boundary where m linearly independent defining hyperplanes intersect. In particular a convex polyhedral set has only finitely many extreme points.

A standard theorem in convexity theory is that a *bounded* convex polyhedral set is a convex polytope.

Definition. A **convex polytope** in $\mathbb{R}^m$ is just the convex hull of a finite number of points $\mathbf{u}_1, ..., \mathbf{u}_k \in \mathbb{R}^m$, that is a set of the type

$$[\mathbf{u}_1, ..., \mathbf{u}_k] = \{\, \mathbf{x} = \sum_{i=1}^{k} \lambda_i \mathbf{u}_i \; ; \; 0 \le \lambda_i \le 1, \; \sum_{i=1}^{k} \lambda_i = 1 \,\} \ .$$

Theorem 3.10. A convex polytope is the convex hull of its extreme points.

Proof. Let $\mathbf{C} = [\mathbf{u}_1, ..., \mathbf{u}_k]$ be a convex polytope and $V = \{\mathbf{v}_1, ..., \mathbf{v}_l\}$ its set of extreme points. We have to prove that

$$\{\mathbf{v}_1, ..., \mathbf{v}_l\} = [\mathbf{u}_1, ..., \mathbf{u}_k].$$

Firstly we show that

$$V \subseteq \{\mathbf{u}_1, ..., \mathbf{u}_k\} \ . \tag{3.36}$$

For supose $\mathbf{v} \in V$ and $\mathbf{v} \notin [\mathbf{u}_1, ..., \mathbf{u}_k]$, then since

$$\mathbf{v} \in V \subseteq \mathbf{C} = \{\mathbf{u}_1, ..., \mathbf{u}_k\} \ ,$$

$\mathbf{v}$ is a convex linear combination of $\mathbf{u}_1, ..., \mathbf{u}_k$. Hence there is such a representation of $\mathbf{v}$ which involves a *minimal* number of the $\mathbf{u}_i$.

Let

$$\mathbf{v} = \sum_{i \in P} \lambda_i \mathbf{u}_i, \quad 0 \le \lambda_i \le 1, \quad \sum_{i \in P} \lambda_i = 1 \tag{3.37}$$

be such a minimal representation. Since $\mathbf{v} \notin [\mathbf{u}_1, ..., \mathbf{u}_k]$, at least two λ_i, $i \in P$, satisfy

$0 < \lambda_i < 1$. Hence we can partition the sum as

$$\mathbf{v} = \sum_{i \in P} \lambda_i \mathbf{u}_i = \sum_{i \in Q} \lambda_i \mathbf{u}_i + \sum_{j \in R} \lambda_j \mathbf{u}_j ,$$

where $P = Q \subset R$, $Q \cap R = \phi$, and there is an $i \in Q$ and a $j \in R$ such that $0 < \lambda_i, \lambda_j < 1$. Write

$$\lambda = \sum_{i \in Q} \lambda_i, \quad 1 - \lambda = \sum_{j \in R} \lambda_j . \qquad (3.38)$$

Then $0 < \lambda < 1$. amd we can define

$$\mathbf{u} = \frac{1}{\lambda} \sum_{i \in Q} \lambda_i \mathbf{u}_i, \quad \mathbf{w} = \frac{1}{1 - \lambda} \sum_{j \in R} \lambda_j \mathbf{u}_j .$$

Note that

$$\mathbf{u} \in \{\mathbf{u}_i ; i \in Q\} \subseteq \mathbb{C}, \quad \mathbf{w} \in \{\mathbf{u}_j ; j \in R\} \subseteq \mathbb{C},$$

and

$$\mathbf{v} = \lambda \mathbf{u} + (1 - \lambda) \mathbf{w} \quad (0 < \lambda < 1) .$$

Hence, as $\mathbf{v}$ is an extreme point of $\mathbb{C}$, $\mathbf{v} = \mathbf{u} = \mathbf{w}$. In particular

$$\mathbf{v} = \sum_{i \in Q} \left(\frac{\lambda_i}{\lambda} \right) \mathbf{u}_i, \quad 0 \le \frac{\lambda_i}{\lambda} \le 1, \quad \sum_{i \in Q} \frac{\lambda_i}{\lambda} = 1 ,$$

which contradicts the minimality of the representation (3.37). This proves (3.36).

Now by definition, no extreme point can be a convex linear combination of other extreme points. Hence if we consider sets S with

$$V \subseteq S \subseteq \{\mathbf{u}_1, \ldots, \mathbf{u}_k\} ,$$

we can partially order them by inclusion and select a *maximal* subset S with the property that no element of S is a convex linear combination of the other elements of S. We do not assume that such a maximal set S is unique, although this turns out to be the case. Since S is maximal, every element of $[\mathbf{u}_1, \ldots, \mathbf{u}_k]$ is a convex linear combination of elements of S, that is if $S = [\mathbf{s}_1, \ldots, \mathbf{s}_p]$

$$[\mathbf{u}_1, \ldots, \mathbf{u}_k] = [\mathbf{s}_1, \ldots, \mathbf{s}_p].$$

Hence

$$\mathbb{C} = \{\mathbf{u}_1, \ldots, \mathbf{u}_k\} \subseteq \{\mathbf{s}_1, \ldots, \mathbf{s}_p\} . \qquad (3.39)$$

If we can prove $S = V$ the theorem will therefore follow.

Let $\mathbf{s}_1 \in S$. If

$$s_1 = \lambda \mathbf{u} + (1 - \lambda) \mathbf{w}; \quad \mathbf{u}, \mathbf{w} \in \mathbb{C}, \quad 0 < \lambda < 1 .$$

by using (3.39) we can express $\mathbf{u}$ and $\mathbf{w}$ as convex linear combinations of $s_1, \ldots, s_p$ to obtain

$$s_1 = \lambda \sum_{i=1}^{p} \lambda_i s_i + (1 - \lambda) \sum_{i=1}^{p} \mu_i s_i ,$$

where $0 \le \lambda_i, \mu_i \le 1$ and $\Sigma \lambda_i = \Sigma \mu_i = 1$. Hence

$$(1 - \alpha_1) s_1 = \sum_{i=2}^{p} \alpha_i s_i , \qquad (3.40)$$

where

$$\alpha_i = \lambda \lambda_i + (1 - \lambda) \mu_i \quad (1 \le i \le p), \quad \sum_{i=1}^{p} \alpha_i = 1 .$$

If $\alpha_1 < 1$, (3.40) expresses s_1 as a convex linear combination of $s_2, \ldots, s_p$ which contradicts the definition of S. Hence $\alpha_1 = 1$, and since $0 < \lambda < 1$ we conclude $\lambda_1 = \mu_1 = 1$. Thus $s_1 = \mathbf{u} = \mathbf{w}$, and so s_1 is an extreme point of $\mathbb{C}$, that is $s_1 \in V$. Similarly $s_i \in V$ for $2 \le i \le p$, and since $V \subsetneq S$ it follows that $V = S$. The proof is complete.

Our previous line of geometric argument tended to suggest that for a feasible bounded linear program the solution will be found on a hyperplane on which the objective function is constant, which is a support hyperplane to the feasible region. Another geometric theorem is that a support hyperplane to a (bounded) convex set always contains extreme points of the set. The conclusion is that we should look for a solution at an extreme point of the feasible region, and since this is a convex polyhedral set there are only finitely many extreme points to examine. In principle this reduces the problem of finding a solution to a finite computation.

For a convex polytope there is no difficulty in rigourising this line of argument.

Theorem 3.11. A real-valued linear function defined on a convex polytope $\mathbb{C}$ attains its maximum or minimum at an extreme point.

Proof. Since $\mathbb{C}$ is a convex polytope it follows from Theorem 3.10 that $\mathbb{C} = [\mathbf{v}_1, \ldots, \mathbf{v}_l]$ where $\mathbf{v}_i$ $(1 \le i \le l)$ are the extreme points of $\mathbb{C}$. Let the linear function be $f(\mathbf{x})$ and define

$$m = \min_{1 \le i \le l} f(\mathbf{v}_i), \quad M = \max_{1 \le i \le l} f(\mathbf{v}_i) .$$

For any $\mathbf{x} \in \mathbb{C}$

$$\mathbf{x} = \sum_{i=1}^{l} \lambda_i \mathbf{v}_i, \quad 0 \le \lambda_i \le 1, \quad \sum_{i=1}^{l} \lambda_i = 1$$

and by the linearity of f

$$f(\mathbf{x}) = f(\sum_{i=1}^{l} \lambda_i \mathbf{v}_i) = \sum_{i=1}^{l} \lambda_i f(\mathbf{v}_i) \le M \sum_{i=1}^{l} \lambda_i = M .$$

Similarly $f(\mathbf{x}) \ge m$.

However, this argument is inadequate for an unbounded polyhedral set. Fortunately it is unnecessary to pursue the geometric approach further because, as we shall see, all the information we require is already contained in the duality theorem, in particular in its corollary Theorem 3.9.

We have seen that the feasible region S is a possibly unbounded convex polyhedral set; the system (3.27) of equated constraints determines a linear subspace of $\mathbb{R}^m$ whose intersection with this polyhedral set, if non-empty, is a closed face of the latter. Similarly (3.28) defines a face of the convex polyhedral set T of feasible vectors for the maximising problem. If systems (3.27) and (3.28) are dual, that is if (3.29) is satisfied, we say they define **dual faces** of the two polyhedral sets S, T. Theorem 3.9, when translated into this geometric language, asserts that feasible vectors $\mathbf{x}$, $\mathbf{y}$ are solutions if and only if they lie on dual faces. In Figure 3.2 and Figure 3.3 the line segment AB and the point $(0, 1, 0)$ are dual faces.

3.5 EXTREME POINTS OF THE FEASIBLE REGION.

We can characterise algebraically the extreme points of the feasible region by

Theorem 3.12. A feasible non-zero $\mathbf{x}$ (or $\mathbf{y}$) is an extreme feasible vector if and only if it satisfies a non-singular square system of equated constraints.

Proof. Suppose first that $\mathbf{x}$ satisfies the non-singular square system (3.30) with associated matrix A_1. For any $\mathbf{u} \in \mathbb{R}^m$ we write $\tilde{\mathbf{u}} = (u_1, ..., u_p)$. Suppose

$$\mathbf{x} = \lambda \mathbf{u} + (1 - \lambda) \mathbf{v}, \quad 0 < \lambda < 1 ,$$

where $\mathbf{u}$ and $\mathbf{v}$ are feasible. For $i > p$ we have $x_i = \lambda u_i + (1-\lambda)v_i = 0$ and so, since $u_i \geq 0$, $v_i \geq 0$ and $0 < \lambda < 1$, we have $u_i = v_i = 0$.

It is also true that

$$\tilde{\mathbf{u}} A_1 \geq \mathbf{b}_1, \; \tilde{\mathbf{v}} A_1 \geq \mathbf{b}_1, \; (\lambda \tilde{\mathbf{u}} + (1 - \lambda)\tilde{\mathbf{v}}) A_1 = \mathbf{b}_1$$

where $\mathbf{b}_1 = (b_1, ..., b_q)$. From these relations it follows that $\tilde{\mathbf{u}} A_1 = \tilde{\mathbf{v}} A_1 = \mathbf{b}_1$, and so, since A_1 is non-singular, $\tilde{\mathbf{u}} = \tilde{\mathbf{v}}$. Thus

$$\mathbf{x} = \mathbf{u} = \mathbf{v},$$

and from this we conclude $\mathbf{x}$ is an extreme point of the feasible region.

Suppose now $\mathbf{x} \neq \mathbf{0}$ is an extreme feasible vector. Let

$$N_1 = \{ j; \; (\mathbf{x} A)_j = b_j \} .$$

We first show that $N_1 \neq \phi$. Since $\mathbf{x} \neq \mathbf{0}$, $x_i > 0$ for some i. Let $\mathbf{u}$, $\mathbf{v}$ be the vectors obtained from $\mathbf{x}$ by replacing x_i by $x_i + \varepsilon$, $x_i - \varepsilon$ respectively. Then

$$\mathbf{x} = \frac{1}{2}\mathbf{u} + \frac{1}{2}\mathbf{v}$$

and if $N_1 = \phi$, so that $(\mathbf{x}A)_j > b_j$ for all j, then $\mathbf{u}$ and $\mathbf{v}$ would be feasible for some sufficiently small $\varepsilon > 0$. This contradicts the fact that $\mathbf{x}$ is extreme by taking $\lambda = \frac{1}{2}$ in the definition of an extreme point.

Now define

$$M_2 = \{ i ; x_i = 0 \} .$$

Delete from A the columns which are *not* 'in N_1' and the rows which *are* 'in M_2'. Since $N_1 \neq \phi$ and $M_1 = \{1, 2, ..., m\} \backslash M_2 \neq \phi$ (because $\mathbf{x} \neq \mathbf{0}$), we actually have a submatrix $\tilde{A}$ left, and it is clear that

$$\tilde{\mathbf{x}} \tilde{A} = \tilde{\mathbf{b}} ,$$

where the components of $\tilde{\mathbf{x}}$ are the positive components of $\mathbf{x}$ and the components of $\tilde{\mathbf{b}}$ are the appropriate components of $\mathbf{b}$.

We next show that the rows of $\tilde{A}$ are linearly independent. If this is not so there exists $\tilde{\mathbf{x}}' \neq \mathbf{0}$ such that $\tilde{\mathbf{x}}'\tilde{A} = \mathbf{0}$. Put $\tilde{\mathbf{u}} = \tilde{\mathbf{x}} + \varepsilon\tilde{\mathbf{x}}'$, $\tilde{\mathbf{v}} = \tilde{\mathbf{x}} - \varepsilon\tilde{\mathbf{x}}'$ and adjoin appropriate zero components to $\tilde{\mathbf{u}}$ and $\tilde{\mathbf{v}}$ to obtain m-dimensional vectors $\mathbf{u}$ and $\mathbf{v}$. Then

$$\mathbf{x} = \frac{1}{2}\mathbf{u} + \frac{1}{2}\mathbf{v} ,$$

and for sufficiently small $\varepsilon > 0$, $\mathbf{u}$ and $\mathbf{v}$ are feasible but $\mathbf{x} \neq \mathbf{u}$ and $\mathbf{x} \neq \mathbf{v}$, which contradicts the fact that $\mathbf{x}$ is extreme.

Since the rows of $\tilde{A}$ are linearly independent, we can delete suitable columns of $\tilde{A}$ to obtain a non-singular square submatrix A_1, which completes the proof.

The restriction to non-zero vectors $\mathbf{x}$, $\mathbf{y}$ in this theorem is no real limitation, since plainly the zero vector is an extreme feasible vector if it is feasible.

Following accepted tradition we shall call a solution vector of (3.1) or (3.2) an **optimal** vector. The next theorem characterises points, other than the origin, which are extreme points of the set of optimal vectors.

Theorem 3.13. Feasible non-zero $\mathbf{x}$ and $\mathbf{y}$ are extreme optimal vectors if and only if they satisfy dual non-singular square systems of equated constraints.

Proof. If non-zero vectors $\mathbf{x}$ and $\mathbf{y}$ satisfy non-singular square systems of equated constraints they are extreme feasible vectors by Theorem 3.12. If in addition the systems of equated constraints are dual then $\mathbf{x}$ and $\mathbf{y}$ are optimal by Theorem 3.9. Any extreme feasible vector which is also optimal is *a fortiori* an extreme optimal vector.

Suppose now $\mathbf{x}$ and $\mathbf{y}$ are non-zero extreme optimal vectors. By Theorem 3.9 $\mathbf{x}$ and $\mathbf{y}$ satisfy dual systems of equated constraints, and we can re-index the quantities involved to obtain (3.30) and (3.31), where $x_i = 0$ $(p + 1 \leq i \leq m)$, $y_j = 0$ $(q + 1 \leq j \leq n)$. Now some x_i with $i \leq p$ or some y_j with $j \leq q$ can also vanish, so we further re-index, using $r \leq p$ and $s \leq q$ so that $x_i > 0$

for $1 \leq i \leq r$ and $y_j > 0$ for $1 \leq j \leq s$. Since $\mathbf{x} \neq \mathbf{0}$ and $\mathbf{y} \neq \mathbf{0}$ we have $r \geq 1$, $s \geq 1$, hence $p \geq 1$, $q \geq 1$. Let $\tilde{A}$ be the matrix obtained as the intersection of the first p rows and first q columns of A.

The first r rows of $\tilde{A}$ are linearly independent. The proof of this fact closely parallels the corresponding part of the proof of Theorem 3.12, for the choice of r places us in the same situation. The only difference is that here $\mathbf{x}$ is an extreme *optimal* vector. Hence in order to obtain the contradiction as before, if we put $\tilde{\mathbf{u}} = \tilde{\mathbf{x}} + \varepsilon\tilde{\mathbf{x}}'$, $\tilde{\mathbf{v}} = \tilde{\mathbf{x}} - \varepsilon\tilde{\mathbf{x}}'$ and derive $\mathbf{u}$, $\mathbf{v}$ in the same way we need to show not only that for suitable $\varepsilon > 0$, $\mathbf{u}$ and $\mathbf{v}$ are feasible but also that they are optimal. Now since $\mathbf{u}$, $\mathbf{v}$ are feasible and $\mathbf{x} = \frac{1}{2}\mathbf{u} + \frac{1}{2}\mathbf{v}$, $\mathbf{y}$ are optimal we have from Theorem 3.1 and Theorem 3.5

$$\mathbf{y}\,\mathbf{b}^T \leq \mathbf{u}\,\mathbf{c}^T, \quad \mathbf{y}\,\mathbf{b}^T \leq \mathbf{v}\,\mathbf{c}^T, \quad \mathbf{y}\,\mathbf{b}^T = (\frac{1}{2}\mathbf{u} + \frac{1}{2}\mathbf{v})\,\mathbf{c}^T,$$

and so

$$\mathbf{y}\,\mathbf{b}^T = \mathbf{u}\,\mathbf{c}^T = \mathbf{v}\,\mathbf{c}^T.$$

Corollary 3.1 now tells us that $\mathbf{u}$ and $\mathbf{v}$ are optimal as required.

Similarly the first s columns of $\tilde{A}$ are linearly independent. We now re-index in the intervals $r \leq i \leq p$, $s \leq j \leq q$ so that the first p' rows of $\tilde{A}$, $p' \geq r$, are a maximally linearly independent subset of the rows of $\tilde{A}$, and the first q' columns of $\tilde{A}$, $q' \geq s$, are a maximally linearly independent subset of the columns of $\tilde{A}$.

The intersection of the first p' rows and q' columns of $\tilde{A}$ is a matrix A_1. Because the last $q - q'$ columns of $\tilde{A}$ are linearly dependent on the first q' columns one can readily verify that any linear relationship between the p' rows of A_1 can be extended to a linear relationship between the first p' rows of $\tilde{A}$. The latter, however, are linearly independent, and so the rows of A_1 must be linearly independent. Similarly the columns of A_1 are linearly independent. Hence A_1 is a non-singular square matrix. Clearly $\mathbf{x}$ and $\mathbf{y}$ satisfy the dual non-singular square systems of equated constraints associated with A_1, which completes the proof.

Corollary 3.3. The sets of feasible and optimal vectors of the linear programs (3.1) and (3.2) have finitely many extreme vectors.

Proof. This is an immediate consequence of Theorems 3.12 and 3.13 since a vector which satisfies a non-singular square system of equated constraints is uniquely determined, and there are at most

$$\sum_{r=1}^{\min\,\{m,\,n\}} \binom{m}{r}\binom{n}{r}$$

square submatrices of A.

Theorem 3.13 provides a systematic method for finding all extreme optimal vectors; one examines all square systems of equated constraints, discarding those whith singular associated matrices, and then lists those solutions of the remaining systems which are feasible vectors. This method is known as the *Shapley-Snow procedure*. It is of great theoretical importance since it provides the simplest known way of determining all extreme vectors. However, it is not frequently applied in practice because the algorithm usually requires a prohibitive amount of computation. In §3.7 to §3.9 we shall discuss the simplex algorithm which finds an extreme optimal vector relatively easily.

3.6 THE SHAPLEY-SNOW PROCEDURE FOR GAMES.

A system of equated constraints for a game with matrix $A = (a_{ij})$ has, by definition, one of the forms

$$\left.\begin{array}{r}(\mathbf{p}A)_j = v \text{ for all } j \in N_1 , \\ p_i = 0 \text{ for all } i \in M_2 , \\ p_1 + \ldots + p_m = 1 , \end{array}\right\} \tag{3.41}$$

$$\left.\begin{array}{r}(A\mathbf{q}^T)_i = u \text{ for all } i \in M_1 , \\ q_j = 0 \text{ for all } j \in N_2 , \\ q_1 + \ldots + q_n = 1 . \end{array}\right\} \tag{3.42}$$

Dual systems and **square** systems are defined as for linear programs; a square system is **non-singular** if it has a unique solution $(\mathbf{p}, v)$ or $(\mathbf{q}, u)$. An **extreme** optimal strategy is an optimal strategy is an optimal strategy which is not a convex linear combination of two other optimal strategies.

Before deriving the matrix game analog of Theorem 3.13 we remark that the process of assigning the dual programs (3.3) and (3.4) to the game with matrix A really only depends on the fact that the value of the game is strictly positive. The condition $a_{ij} > 0$ all i, j was used only to show that (3.3) is feasible and, provided $v > 0$, we can now infer this fact from Theorem 3.7! Thus for any matrix game A with positive value we may associate the feasible pair of dual linear programs (3.3) and (3.4). The optimal strategies for the game are put into one-one correspondence with the optimal vectors of the dual linear programs by the relations

$$\mathbf{p}^0 = v\mathbf{x}^0 , \quad \mathbf{q}^0 = v\mathbf{y}^0 . \tag{3.43}$$

It is clear that (3.43) also gives a one-one correspondence between the extreme optimal strategies of the matrix game A and the extreme optimal vectors of the dual programs. We can further set up a one-one correspondence between the systems of equated constraints; (3.27) and (3.41) correspond if and only if they have the same N_1 and M_2, whilst (3.28) and (3.42) correspond if and only if they have the same M_1 and N_2. Here and subsequently it is understood we are taking $\mathbf{b} = \mathbf{J}_n$, $\mathbf{c} = \mathbf{J}_m$ in (3.27), (3.28) respectively.

Lemma 3.2. Let A be a matrix game with positive value. If $\mathbf{p}^0$ is an optimal strategy and $(\mathbf{p}^0, v)$ is the (unique) solution of a non-singular square system (3.41), then the corresponding square system (3.27) for the associated program (3.3) is also non-singular.

Conversely, if an optimal vector x^0 is the (unique) solution of a non-singular square system (3.27), then the corresponding square system (3.41) is also non-singular.

The dual statement for systems (3.28) and (3.42) also holds.

Proof. Suppose first that p^0 is an optimal strategy and (p^0, v) is a solution of a non-singular square system (3.41), then $x^0 = p^0/v$ is a solution of (3.27), and $\Sigma x_i^0 = 1/v \neq 0$. If x is any solution of (3.27) with $\Sigma x_i \neq 0$ then $(x/\Sigma x_i, 1/\Sigma x_i)$ is a solution of (3.41); thus, since any such solution is obviously unique,

$$\frac{x}{\sum x_i} = p^0, \qquad \frac{1}{\sum x_i} = v,$$

so $x = p^0/v = x^0$. If x is any solution of (3.27) with $\Sigma x_i = 0$, then

$$x' = \frac{1}{2}(x^0 + x)$$

is a solution of (3.27) with $\Sigma x_i' \neq 0$, and so $x' = x^0$ as above. But this implies $x = x^0$, contradicting $\Sigma x_i \neq 0$. So (3.27) has x^0 as its unique solution and is therefore non-singular.

Now suppose x^0 is the solution of a non-singular square system (3.27). If (p, v) is any solution of (3.41) with $v \neq 0$, tahen p/v is a solution of (3.27); clearly different choices of (p, v) yield different solutions of (3.27) in this way. Since (3.27) is non-singular, only one such (p, v) can exist, and it must be (p^0, v) where $p^0 = vx^0$. If (p, s) is any solution of (3.27) other than (p^0, v), then $s = 0$ and

$$(p', v') = (\frac{1}{2}(p^0 + p), \frac{1}{2}(v + s))$$

is a solution with $\frac{1}{2}(v+s) \neq 0$, and so $(p', v) = (p^0, v)$ as above. But this implies $(p, s) = (p^0, v)$, a contradiction since $s = 0$. Thus (3.41) has (p^0, v) as its unique solution and is therefore non-singular.

Theorem 3.14. Let A be a matrix game with value $v \neq 0$. Optimal strategies p^0 and q^0 are extreme optimal strategies if and only if (p^0, v) and (q^0, v) satisfy dual non-singular square systems of equated constraints.

Proof. If $v > 0$ then the theorem follows from Theorem 3.13, Lemma 3.2 and the fact that the relations (3.43) make extreme optimal strategies of the matrix game A correspond to extreme optimal vectors of the associated dual programs (3.3) and (3.4).

The case $v < 0$ can be reduced to the previous one by noting that players 1 and 2 for A can be considered to be the second and first players, respectively, in the game $-A^T$ with value $-v > 0$. This completes the proof.

We are now able to prove a theorem which gives us a method of finding all extreme optimal strategies for a matrix game. It transpires there are only finitely many. Since the set of optimal strategies for a player is a closed *bounded* convex set (Theorem 2.8) we infer that it is in fact a convex polytope, and according to Theorem 3.10 we can find *all* optimal strategies by forming all convex combinations of the extreme optimal strategies.

We recall that given an $r \times r$ square matrix game $M = (a_{ij})$, the **minor** m_{ij} is formed by deleting the ith row and j^{th} column from M and taking the determinant of the remaining $(r-1) \times (r-1)$ matrix. The **cofactor** A_{ij} is defined as

$$A_{ij} = (-1)^{i+j} m_{ij} .$$

If we define the **adjoint** of M by

$$\text{adj } M = (A_{ij})^T \text{ if } r > 1, \quad \text{adj } M = 1 \text{ if } r = 1 ,$$

then for all square matrices M we have

$$M \text{ adj} M = (\det M) I ,$$

where I is the identity matrix. This identity is proved in linear algebra and provides a formula for the inverse of a non-singular matrix M.

Lemma 3.3. Let k be a real number, M an $r \times r$ matrix, and U the $r \times r$ matrix with all entries 1. Then

$$(i) \quad \det(M + kU) = \det M + k \mathbf{J}_r \text{adj} M \mathbf{J}_r^T$$

$$(ii) \quad \mathbf{J}_r \text{adj}(M + kU) = \mathbf{J}_r \text{adj} M$$

$$(iii) \quad \text{adj}(M + kU) \mathbf{J}_r^T = \text{adj} M \mathbf{J}_r^T .$$

Proof. If we regard $\det M$ as a real valued function of the r^2 entries (a_{ij}) and expand by the ith row we have

$$\det M = \sum_{k=1}^{r} a_{ik} A_{ik} ,$$

and no A_{ik} $(1 \le i \le r)$ involves a_{ij} $(1 \le j \le r)$.

Hence

$$\frac{\partial}{\partial a_{ij}} (\det M) = A_{ij} .$$

We apply this formula to the matrix $M + kU$ with entries $a_{ij}' + k$ and cofactors A_{ij}' to obtain

$$\frac{d}{dk} \det(M + kU) = \sum_{i,j} \frac{\partial}{\partial a'_{ij}} \det(M + kU) \frac{d}{dk}(a'_{ij}) = \sum_{i,j} A'_{ij}.$$

But since

$$\sum_{j=1}^{r} A'_{ij} . 1$$

is evidently the expansion about the ith row of a matrix obtained from $M + kU$ by replacing the ith row by $(1, 1, ..., 1)$, we can subtract k times the ith row from all the others in this matrix, which leaves the determinant unchanged, to obtain

$$\sum_{j=1}^{r} A'_{ij} = \sum_{j=1}^{r} A_{ij}. \tag{3.44}$$

Hence summing over i we obtain

$$\frac{d}{dk} \det(M + kU) = \sum_{i,j} A_{ij} = \mathbf{J}_r \, \mathrm{adj} M \, \mathbf{J}_r^T$$

which is independent of k. It follows that $\det(m + kU)$ is the linear function

$$\det M + k \, \mathbf{J}_r \, \mathrm{adj} M \, \mathbf{J}_r^T$$

which proves (i). Moreover (3.44) is exactly (iii) written in a less obvious fashion. We can prove (ii) similarly by the same argument but using a column expansion instead of a row expansion.

Theorem 3.15 (Shapley-Snow). Let A be an $m \times n$ matrix game with value v. Optimal strategies $\mathbf{p}$ and $\mathbf{q}$ are extreme optimal strategies if and only if there is a square $r \times r$ submatrix M of A with $\mathbf{J}_r \, \mathrm{adj} M \, \mathbf{J}_r^T \neq 0$ and

$$v = \frac{\det M}{\mathbf{J}_r \, \mathrm{adj} M \, \mathbf{J}_r^T}. \tag{3.45}$$

$$\tilde{\mathbf{p}} = \frac{\mathbf{J}_r \, \mathrm{adj} M}{\mathbf{J}_r \, \mathrm{adj} M \, \mathbf{J}_r^T}, \tag{3.46}$$

$$\tilde{\mathbf{q}}^T = \frac{\mathrm{adj} M \, \mathbf{J}_r^T}{\mathbf{J}_r \, \mathrm{adj} M \, \mathbf{J}_r^T}, \tag{3.47}$$

where $\tilde{\mathbf{p}}$ and $\tilde{\mathbf{q}}$ are r-dimensional vectors obtained by deleting from $\mathbf{p}$ and $\mathbf{q}$ respectively the components corresponding to the rows and columns of A which must be deleted to obtain M.

Remark. We observe that even if A is the matrix with all entries zero there will be at least mn submatrices M of A such that $\mathbf{J}_r \, \mathrm{adj} M \, \mathbf{J}_r^T \neq 0$, namely the 1×1 submatrices, since according to our convention regarding adjoints the adjoint of any 1×1 matrix is always 1. Thus in this special case any pure strategy is an extreme optimal strategy.

Proof. Suppose (3.45) to (3.47) hold and $v \neq 0$. We have the identity

$$M \text{ adj} M = (\det M) \, I \, ,$$

where I is the identity matrix. Since $v \neq 0$ we have $\det M \neq 0$ from (3.45). Hence M is non-singular and M^{-1} exists and is given by

$$M^{-1} = \frac{1}{\det M} \text{ adj} M \, . \tag{3.48}$$

Thus (3.45) to (3.47) may be rewritten as

$$v = \frac{1}{\mathbf{J}_r \, M^{-1} \, \mathbf{J}_r^T} \, ,$$
$$\tilde{\mathbf{p}} = v \, \mathbf{J}_r \, M^{-1} \, ,$$
$$\tilde{\mathbf{q}}^T = v \, M^{-1} \, \mathbf{J}_r^T \, .$$

Hence $\tilde{\mathbf{p}} \, M = v \, \mathbf{J}_r$ and $M \, \tilde{\mathbf{q}}^T = v \, \mathbf{J}_r^T$; furthermore $\tilde{\mathbf{p}} \, \mathbf{J}^T = \mathbf{q} \, \mathbf{J}_r \, \tilde{\mathbf{q}}^T = 1$, so that $\tilde{\mathbf{p}}$ and $\tilde{\mathbf{q}}$ are probability vectors (recall $\mathbf{p} \geq 0$, $\mathbf{q} \geq 0$), and the components of $\tilde{\mathbf{p}}$ and $\tilde{\mathbf{q}}$ not in $\mathbf{p}$ and $\mathbf{q}$ must all vanish. We have now shown that $\mathbf{p}$ and $\mathbf{q}$ satisfy the dual non-singular square system of equated constraints associated with the matrix M, hence by Theorem 3.14 $\mathbf{p}$ and $\mathbf{q}$ are extreme optimal strategies.

Suppose next that $\mathbf{p}$ and $\mathbf{q}$ are extreme optimal strategies and $v \neq 0$. By Theorem 3.14 $\mathbf{p}$ and $\mathbf{q}$ satisfy dual non-singular square systems of equated constraints associated with some $r \times r$ submatrix M of A. Thus we have

$$\tilde{\mathbf{p}} \, M = v \, \mathbf{J}_r \, , \quad M \, \tilde{\mathbf{q}}^T = v \, \mathbf{J}_r^T \, , \quad \tilde{\mathbf{p}} \, \mathbf{J}_r^T = \mathbf{J}_r \, \tilde{\mathbf{q}}^T = 1 \, ,$$

or

$$\tilde{\mathbf{p}} = v \, \mathbf{J}_r \, M^{-1} \, , \quad \tilde{\mathbf{q}}^T = v \, M^{-1} \, \mathbf{J}_r^T \, , \quad \tilde{\mathbf{p}} \, \mathbf{J}_r^T = \mathbf{J}_r \, \tilde{\mathbf{q}}^T = 1 \, ,$$

which implies

$$v \, \mathbf{J}_r \, M^{-1} \, \mathbf{J}_r^T = 1$$

so that

$$v = \frac{1}{\mathbf{J}_r \, M^{-1} \, \mathbf{J}_r^T} \, .$$

These relations together with (3.48) give (3.45) to (3.47).

It remains to deal with the case $v = 0$. Suppose $v = 0$; add a positive quantity k to each entry of A, obtaining a strategically equivalent game $A(k)$ with positive value. By the first two parts of the proof $\mathbf{p}$ and $\mathbf{q}$ are extreme optimal strategies for $A(k)$ (and thus for A) if and only if there is a square submatrix $M(k) = M + kU$ of $A(k)$ with $\mathbf{J}_r\,\mathrm{adj}M(k)\,\mathbf{J}_r^T \neq 0$ and

$$k = \frac{\det M(k)}{\mathbf{J}_r\,\mathrm{adj}M(k)\,\mathbf{J}_r^T}\,, \tag{3.49}$$

$$\tilde{\mathbf{p}} = \frac{\mathbf{J}_r\,\mathrm{adj}M(k)}{\mathbf{J}_r\,\mathrm{adj}M(k)\,\mathbf{J}_r^T}\,, \tag{3.50}$$

$$\tilde{\mathbf{q}}^T = \frac{\mathrm{adj}M(k)\,\mathbf{J}_r^T}{\mathbf{J}_r\,\mathrm{adj}M(k)\,\mathbf{J}_r^T}\,. \tag{3.51}$$

By Lemma 3.3(ii) or (iii), $\mathbf{J}_r\,\mathrm{adj}M(k)\,\mathbf{J}_r^T \neq 0$ if and only if $\mathbf{J}_r\,\mathrm{adj}M\,\mathbf{J}_r^T \neq 0$. Hence $\mathbf{J}_r\,\mathrm{adj}M\,\mathbf{J}_r^T \neq 0$, and from the identities (i) to (iii) of Lemma 3.3 it is clear that $M(k)$ has the properties (3.49) to (3.51) if and only if M has the properties (3.45) to (3.47) with $v = 0$ (which incidentally means $\det M = 0$), and the proof is complete.

Thus a systematic procedure to find all extreme optimal strategies is to consider all $r \times r$ submatrices $1 \leq r \leq \min\{m, n\}$. For each such matrix M first check that $\mathbf{J}_r\,\mathrm{adj}M\,\mathbf{J}_r^T \neq 0$. If this does not hold, discard M. Next apply the formulae (3.45) to (3.47). The relations $\Sigma\tilde{p}_i = \Sigma\tilde{q}_j = 1$ automatically hold by virtue of (3.46) and (3.47), but it might happen that $\tilde{p}_i < 0$ or $\tilde{q}_j < 0$ for some i or j. In this event reject M. If all the components of $\tilde{\mathbf{p}}$ and $\tilde{\mathbf{q}}$ are non-negative, form $\mathbf{p}$ and $\mathbf{q}$ by inserting zeros for the components corresponding to deleted rows and columns respectively. Finally, check if $\mathbf{p}A \geq v\mathbf{J}_n$ and $A\mathbf{q}^T \leq v\mathbf{J}_m^T$. If this condition is satisfied, $\mathbf{p}$ and $\mathbf{q}$ are extreme optimal strategies; if not, M must be rejected.

Example 3.5 Consider the game with matrix

$$A = \begin{pmatrix} 0 & 1 & -1 \\ -1 & 0 & 1 \\ 1 & -1 & 0 \end{pmatrix}$$

This is a 3×3 symmetric game of the type considered in 1.6. Being symmetric it has value zero (Theorem 2.12), and since there is no saddle point the analysis in §1.6 shows that $(1/3, 1/3, 1/3)$ is the unique optimal strategy for each player. Let us now apply the Shapley-Snow procedure.

There are nine 1×1 submatrices each of the form $M = (0)$, $M = (1)$, or $M = (-1)$.

Type $M = (0)$. These have

$$\mathrm{adj}M = 1,\quad \mathbf{J}_1\,\mathrm{adj}M\,\mathbf{J}_1^T = 1 \neq 0,\quad \det M = 0\,.$$

Thus $v = 0$, $\tilde{\mathbf{p}} = (1)$, $\tilde{\mathbf{q}} = (1)$. This corresponds to taking $\mathbf{p}$ and $\mathbf{q}$ to be pure strategies, in which case one can readily verify the conditions $\mathbf{p}A \geq 0$, $A\mathbf{q}^T \leq 0$ are violated.

Type M = (±1). These have

$$\text{adj} M = 1, \quad J_1 \text{ adj} M \, J_1^T = 1 \neq 0, \quad \det M = \pm 1 .$$

Thus $v = \pm 1$, $\tilde{p} = (1)$, $\tilde{q} = (1)$. Again **p** and **q** are pure strategies, and the conditions $\mathbf{p} \, A \geq v \, J_3$, $A \, \mathbf{q}^T \leq v \, J_3^T$ are violated.

So far we have merely verified the obvious, namely that there is no solution in pure strategies, that is, no saddle point.

There are nine 2×2 submatrices obtained by deleting the ith row and j^{th} column from A. These are listed below, each in the i, j position.

$$\begin{pmatrix} 0 & 1 \\ -1 & 0 \end{pmatrix} \begin{pmatrix} -1 & 1 \\ 1 & 0 \end{pmatrix} \begin{pmatrix} -1 & 0 \\ 1 & -1 \end{pmatrix}$$

$$\begin{pmatrix} 1 & -1 \\ -1 & 0 \end{pmatrix} \begin{pmatrix} 0 & -1 \\ 1 & 0 \end{pmatrix} \begin{pmatrix} 0 & 1 \\ 1 & -1 \end{pmatrix}$$

$$\begin{pmatrix} 1 & -1 \\ 0 & 1 \end{pmatrix} \begin{pmatrix} 0 & -1 \\ -1 & 0 \end{pmatrix} \begin{pmatrix} 0 & 1 \\ -1 & 0 \end{pmatrix}$$

The corresponding adjoints are

$$\begin{pmatrix} 0 & -1 \\ 1 & 0 \end{pmatrix} \begin{pmatrix} 0 & -1 \\ -1 & -1 \end{pmatrix} \begin{pmatrix} -1 & 0 \\ -1 & -1 \end{pmatrix}$$

$$\begin{pmatrix} 0 & 1 \\ 1 & 1 \end{pmatrix} \begin{pmatrix} 0 & 1 \\ -1 & 0 \end{pmatrix} \begin{pmatrix} -1 & -1 \\ -1 & 0 \end{pmatrix}$$

$$\begin{pmatrix} 1 & 1 \\ 0 & 1 \end{pmatrix} \begin{pmatrix} 1 & 1 \\ 1 & 0 \end{pmatrix} \begin{pmatrix} 0 & -1 \\ 1 & 0 \end{pmatrix}$$

The diagonal matrices fail to satisfy $J_2 \text{ adj} M \, J_2^T \neq 0$. For the remaining M we take the one in the $(1, 2)$ position as typical.

$$J_2 \text{ adj} M \, J_2^T = -3, \quad \det M = -1, \quad \text{giving } v = \frac{1}{3} .$$

$$\tilde{p} = \frac{1}{(-3)} (-1, -2) = (\tfrac{1}{3}, \tfrac{2}{3}), \quad \tilde{q} = \frac{1}{(-3)} (-1, -2) = (\tfrac{1}{3}, \tfrac{2}{3}) .$$

Thus $\mathbf{p} = (0, \, {}^1/_3, \, {}^2/_3)$, $\mathbf{q} = ({}^1/_3, \, 0, \, {}^2/_3)$. But although

$$A \, \mathbf{q}^T = (-\tfrac{2}{3}, \tfrac{1}{3}, \tfrac{1}{3})^T \leq \tfrac{1}{3} (1, 1, 1)^T$$

we have

$$\mathbf{p} \, A = (\tfrac{1}{3}, \, -\tfrac{2}{3}, \, \tfrac{1}{3}) < \tfrac{1}{3} \, (1, \, 1, \, 1) \ .$$

Hence this M must be rejected. Similarly the other remaining 2×2 matrices are rejected.

There is now only one possibility remaining, that is the original 3×3 matrix A. We find

$$\text{adj} A = \begin{pmatrix} 1 & 1 & 1 \\ 1 & 1 & 1 \\ 1 & 1 & 1 \end{pmatrix}, \quad J_3 \, \text{adj} A \, J_3^T = 9, \quad \det A = 0 \ .$$

Hence $v = 0$ and $\mathbf{p} = \frac{1}{9}(3, 3, 3) = (\frac{1}{3}, \frac{1}{3}, \frac{1}{3}) = \mathbf{q}$, which of course is the required solution.

In this example the game was **completely mixed**, that is every optimal strategy for both players involves each of thier respective pure strategies with positive probability. The example suggests the following corollaries of Theorem 3.15.

Corollary 3.4. The solution to a completely mixed matrix game is unique.

Proof. If the game is completely mixed, no optimal strategy can have a zero component. Hence in Theorem 3.15 the only possibility for the matrix M is that it be the full matrix A, but then (3.46) and (3.47) determine $\mathbf{p}$ and $\mathbf{q}$ uniquely.

Corollary 3.5. A completely mixed game with matrix A has value zero if and only if A is singular.

Proof. From (3.45) with $M = A$ we have $v = 0$ if and only if $\det A = 0$.

Example 3.6 Find all solutions of the matrix game

$$A = \begin{pmatrix} 0 & 4 & -2 \\ 3 & 1 & 4 \\ 3 & 2 & 1 \end{pmatrix} .$$

Beginning with the full 3×3 matrix we find

$$\text{adj} A = \begin{pmatrix} -7 & -8 & 18 \\ 9 & 6 & -6 \\ 3 & 12 & -12 \end{pmatrix}, \quad J_3 \, \text{adj} A \, J_3^T = 15, \quad \det A = 30 \ .$$

Hence $v = 2$ and

$$\mathbf{p} = \frac{1}{15} \, (5, \, 10, \, 0) = (\tfrac{1}{3}, \tfrac{2}{3}, 0), \quad \mathbf{q} = \frac{1}{15} \, (3, \, 9, \, 3) = (\tfrac{1}{5}, \tfrac{3}{5}, \tfrac{1}{5})$$

Hence

$$\mathbf{p} \, A = (\tfrac{1}{3}, \tfrac{2}{3}, 0) \, A = (2, 2, 2) \geq 2 \, \mathbf{J}_3$$

$$\text{and } A \, \mathbf{q}^{\,T} = A \, (\tfrac{1}{5}, \tfrac{3}{5}, \tfrac{1}{5})^{T} = (2, 2, 2)^{T} \leq 2 \, \mathbf{J}_3^{\,T} \, ,$$

which shows that we have found a solution.

We may obtain the 2×2 matrices by deleting the ith row and j^{th} column of A. Doing this and writing the resulting matrix in the (i, j) position we get

$$\begin{pmatrix} 1 & 4 \\ 2 & 1 \end{pmatrix} \begin{pmatrix} 3 & 4 \\ 3 & 1 \end{pmatrix} \begin{pmatrix} 3 & 1 \\ 3 & 2 \end{pmatrix}$$
$$\begin{pmatrix} 4 & -2 \\ 2 & 1 \end{pmatrix} \begin{pmatrix} 0 & -2 \\ 3 & 1 \end{pmatrix} \begin{pmatrix} 0 & 4 \\ 3 & 2 \end{pmatrix}$$
$$\begin{pmatrix} 4 & -2 \\ 1 & 4 \end{pmatrix} \begin{pmatrix} 0 & -2 \\ 3 & 4 \end{pmatrix} \begin{pmatrix} 0 & 4 \\ 3 & 1 \end{pmatrix}$$

The corresponding adjoints are

$$\begin{pmatrix} 1 & -4 \\ -2 & 1 \end{pmatrix} \begin{pmatrix} 1 & -4 \\ -3 & 3 \end{pmatrix} \begin{pmatrix} 2 & -1 \\ -3 & 3 \end{pmatrix}$$
$$\begin{pmatrix} 1 & 2 \\ -2 & 4 \end{pmatrix} \begin{pmatrix} 1 & 2 \\ -3 & 0 \end{pmatrix} \begin{pmatrix} 2 & -4 \\ -3 & 0 \end{pmatrix}$$
$$\begin{pmatrix} 4 & 2 \\ -1 & 4 \end{pmatrix} \begin{pmatrix} 4 & 2 \\ -3 & 0 \end{pmatrix} \begin{pmatrix} 1 & -4 \\ -3 & 0 \end{pmatrix}$$

Only the $(2, 2)$ matrix is eliminated at this stage, because $\mathbf{J}_2 \, \text{adj} M \, \mathbf{J}_2^{\,T} = 0$. We systematically check the remaining 2×2 matrices;

$$M = \begin{pmatrix} 1 & 4 \\ 2 & 1 \end{pmatrix}, \quad \mathbf{J}_2 \, \text{adj} M \, \mathbf{J}_2^{T} = -4, \quad \det M = -7 \, .$$

This gives $v = \,^{7}/_{4}$, so we can discard M since we already know $v = 2$.

$$M = \begin{pmatrix} 3 & 4 \\ 3 & 1 \end{pmatrix}, \quad \mathbf{J}_2 \, \text{adj} M \, \mathbf{J}_2^{T} = -3, \quad \det M = -9 \, ,$$

$$v = \frac{9}{3} = 3 \neq 2, \quad \text{discard } M.$$

$$M = \begin{pmatrix} 3 & 1 \\ 3 & 2 \end{pmatrix}, \quad \mathbf{J_2} \text{ adj} M \ \mathbf{J_2^T} = 1, \ \det M = 3,$$

$$v = \frac{3}{1}, \neq 2, \text{ discard } M.$$

$$M = \begin{pmatrix} 4 & -2 \\ 2 & 1 \end{pmatrix}, \quad \mathbf{J_2} \text{ adj} M \ \mathbf{J_2^T} = 5, \ \det M = 8,$$

$$v = \frac{8}{5} \neq 2, \text{ discard } M.$$

$$M = \begin{pmatrix} 0 & 4 \\ 3 & 2 \end{pmatrix}, \quad \mathbf{J_2} \text{ adj} M \ \mathbf{J_2^T} = -5, \ \det M = -12,$$

$$v = \frac{12}{5} \neq 2, \text{ discard } M.$$

$$M = \begin{pmatrix} 4 & -2 \\ 1 & 4 \end{pmatrix}, \quad \mathbf{J_2} \text{ adj} M \ \mathbf{J_2^T} = 9, \ \det M = 18$$

$$v = \frac{18}{9} = 2. \text{ Continuing}$$

$$\tilde{\mathbf{p}} = \frac{1}{9}(3, 6) = (\tfrac{1}{3}, \tfrac{2}{3}), \quad \tilde{\mathbf{q}} = \frac{1}{9}(6, 3) = (\tfrac{2}{3}, \tfrac{1}{3}),$$

which give $\mathbf{p} = (^1/_3, \ ^2/_3, \ 0)$, $\mathbf{q} = (0, \ ^2/_3, \ ^1/_3)$. Hence

$$\mathbf{p} A = (2, 2, 2) \geq 2 \mathbf{J_3}, \quad A \mathbf{q^T} = (2, 2, \tfrac{5}{3})^\mathrm{T} \leq 2 \mathbf{J_3^T},$$

and we have found another extreme optimal strategy for player 2.

$$M = \begin{pmatrix} 0 & -2 \\ 3 & 4 \end{pmatrix}, \quad \mathbf{J_2} \text{ adj} M \ \mathbf{J_2^T} = 3, \ \det M = 6$$

$$v = \frac{6}{3} = 2. \text{ Continuing}$$

$$\tilde{\mathbf{p}} = \frac{1}{3}(1, 2) = (\tfrac{1}{3}, \tfrac{2}{3}), \quad \tilde{\mathbf{q}} = \frac{1}{3}(6, -3) = (2, -1),$$

we already know $\mathbf{p} = (^1/_3, \ ^2/_3, \ 0)$ is an extreme optimal strategy but $\mathbf{q} = (2, 0, -1)$ is not a strategy.

$$M = \begin{pmatrix} 0 & 4 \\ 3 & 1 \end{pmatrix}, \quad \mathbf{J}_2 \, \mathrm{adj} M \, \mathbf{J}_2^{\mathrm{T}} = -6, \quad \det M = -12$$

$v = \dfrac{12}{6} = 2$. Continuing

$$\tilde{\mathbf{p}} = -\frac{1}{6}(-2, -4) = (\frac{1}{3}, \frac{2}{3}), \quad \tilde{\mathbf{q}} = -\frac{1}{6}(-3, -3) = (\frac{1}{2}, \frac{1}{2})$$

which give $\mathbf{p} = (^1/_3, \, ^2/_3, \, 0)$, $\mathbf{q} = (^1/_2, \, ^1/_2, \, 0)$. Hence

$$\mathbf{p} A = (2, 2, 2) \geq 2\mathbf{J}_3, \quad A \mathbf{q}^{\mathrm{T}} = (2, 2, \frac{5}{2})^{\mathrm{T}} > 2\mathbf{J}_3^{\mathrm{T}},$$

so we discard M.

This exhausts the 2×2 submatrices. As we saw in the last example, looking at 1×1 submatrices amounts to searching for saddle points: there are no saddle points , so we are done. The complete solution of the game is

$$\mathbf{p} = (\frac{1}{3}, \frac{2}{3}, 0) \, ,$$
$$\mathbf{q} = \lambda(\frac{1}{5}, \frac{3}{5}, \frac{1}{5}) + (1 - \lambda)(0, \frac{2}{3}, \frac{1}{3}), \quad (0 \leq \lambda \leq 1) \, ,$$
$$v = 2 \, .$$

3.7 MIXED CONSTRAINTS, SLACK VARIABLES, AND THE TABLEAU.

Although the Shapley-Snow procedure will find *all* extreme optimal vectors of a linear program, our experience in applying it to games shows that for a large program it is not a practical method to use, especially if *only* one solution is required. In the remainder of this chapter we shall describe a surprisingly efficient procedure, which finds an extreme optimal vector, known as the *simplex algorithm*. Since whole books have been written on this subject (for example [2]) our object is merely to describe the computational procedure rather than to give a detailed theoretical treatment.

We begin by observing that the requirement that the variables of a linear program be non-negative actually involves no real loss of generality, for by writing $x_j = u_j - v_j$, where $u_j \geq 0$, $v_j \geq 0$, any value of x_j is then permitted. Thus a method which solves a program for non-negative variables can be used to solve the same program with unrestricted variables (assuming it remains bounded) at a costs of doubling the number of variables. Henceforth we shall therefore continue as before to assume that the variables are non-negative.

In general a linear program need not be presented in one of the pure forms (3.1) or 3.2). There may be inequality constraints in both directions and some equalities as well. We note in passing that such a mixed problem can always be rewritten in pure form. For example w can rewrite the program

$$\left.\begin{array}{l}
\text{Minimise } x_2 - 3x_3 + 2x_4 \\
\text{Subject to} \\
\quad x_1 + 3x_2 - x_3 + 2x_4 = 7, \\
\quad\quad\quad - 2x_2 + 4x_3 \geq -12, \\
\quad\quad\quad - 4x_2 + 3x_3 + 8x_4 \leq 10, \\
\text{and } x_j \geq 0 \quad (1 \leq j \leq 4).
\end{array}\right\} \tag{3.52}$$

as

$$\begin{array}{l}
\text{Minimise } x_2 - 3x_3 + 2x_4 \\
\text{Subject to} \\
\quad x_1 + 3x_2 - x_3 + 2x_4 \leq 7 \\
\quad -x_1 - 3x_2 + x_3 - 2x_4 \leq -7 \\
\quad\quad\quad - 2x_2 + 4x_3 \leq 12 \\
\quad\quad\quad - 4x_2 + 3x_3 + 8x_4 \leq 10, \\
\text{and } x_j \geq 0 \quad (1 \leq j \leq 4).
\end{array}$$

Thus once again our earlier adherence to (3.1) or (3.2) involved no loss of generality. In practice, however, one never bothers to rewrite a problem such as (3.52) in pure form, since the first step in preparing a program for the simplex algorithm is to *eliminate all inequalities*. This is accomplished by introducing one new variable for each inequality. The new variables are called **slack variables**.

Thus (3.52) becomes

$$\begin{array}{l}
\text{Minimise } x_2 - 3x_3 + 2x_4 \\
\text{Subject to} \\
\quad x_1 + 3x_2 - x_3 + 2x_4 \qquad\qquad = 7, \\
\quad\quad\quad 2x_2 - 4x_3 \qquad - x_5 \quad = -12, \\
\quad\quad\quad - 4x_2 + 3x_3 + 8x_4 \qquad + x_6 = 10, \\
\text{and } x_j \geq 0 \; (1 \leq j \leq 6).
\end{array}$$

Notice that since the first inequality in (3.52) is $\geq$, we *subtract* $x_5 \geq 0$, but since the second inequality is $\leq$ we *add* $x_6 \geq 0$. Computationally it is convenient to *always work with the numbers on the right positive*, so we rewrite the program as

$$\left.\begin{array}{l}
\text{Minimise } x_2 - 3x_3 + 2x_4 \\
\text{Subject to} \\
\quad x_1 + 3x_2 - x_3 + 2x_4 \qquad\qquad = 7, \\
\quad\quad\quad - 2x_2 + 4x_3 \qquad + x_5 \quad = 12, \\
\quad\quad\quad - 4x_2 + 3x_3 + 8x_4 \qquad + x_6 = 10,
\end{array}\right\} \tag{3.53}$$

Our problem is now in the form

$$\left. \begin{array}{l} \text{Minimise } x\,c^T \\ \text{subject to } A\,x^T = b^T \text{ and } x \geq 0, \end{array} \right\} \qquad (3.54)$$

where

$$A = \begin{pmatrix} 1 & 3 & -1 & 2 & 0 & 0 \\ 0 & -2 & 4 & 0 & 1 & 0 \\ 0 & -4 & 3 & 8 & 0 & 1 \end{pmatrix}.$$

The relationship with our earlier notation is simply that we have replaced A^T by A because this is the natural and universally accepted thing to do in setting up the simplex method.

Let $a_1, ..., a_n$ denote the column vectors of the matrix A. Then $A\,x^T = b^T$ can be rewritten as

$$\sum_{i=1}^{n} x_i\,a_i = b. \qquad (3.55)$$

Definition. A solution $x = (x_1, ..., x_n)$ of (3.55) is called **basic** if the set of vectors $\{\ a_i;\ x_i \neq 0\}$ is linearly independent.

It is not hard to show that basic feasible vectors correspond to extreme points of the feasible region for the problem (3.54) with $A\,x^T = b^T$. The situation is analogous to that in Theorem 3.12 where we were dealing with a region defined by inequality constraints.

The first requirement of the simplex algorithm is a rather special kind of basic feasible vector. Geometrically, the fact that we begin with any kind of basic feasible vector means we start at an extreme point of the feasible region. For convenience of calculation we wish to begin with a basic feasible vector $x = (x_1, ..., x_n)$ for which the a_i associated with $x_i \neq 0$ are not only linearly independent but are actually standard basis vectors, that is vectors with one component equal to 1 and the remainder equal to 0. In general there will not be such a convenient basic feasible vector available, and the procedure for coping with this eventuality is dealt with in §3.9. For (3.53) a vector of the required type is $x = (7, 0, 0, 0, 12, 10) \geq 0$, for the columns of A associated with 7, 12 and 10 respectively are just a standard basis in $\mathbb{R}^3$.

Having found a basic feasible vector of the required type we can begin to draw up the first simplex tableau for (3.53)

| | | c_1 | c_2 | c_3 | c_4 | c_5 | c_6 | | |
| | | 0 | 1 | -3 | 2 | 0 | 0 | | |
Basis	c_j	$\mathbf{a}_1$	$\mathbf{a}_2$	$\mathbf{a}_3$	$\mathbf{a}_4$	$\mathbf{a}_5$	$\mathbf{a}_6$	b	Ratio
$\mathbf{a}_1$	0	1	3	-1	2	0	0	7	
$\mathbf{a}_5$	0	0	-2	4	0	1	0	12	
$\mathbf{a}_6$	0	0	-4	3	8	0	1	10	
	z								
	z-c								

In the first two rows we list the vector $\mathbf{c}$. Underneath the components of $\mathbf{c}$ we list the column vectors $\mathbf{a}_1$, ..., $\mathbf{a}_6$ of A and write the first column under $\mathbf{a}_1$, the second column under $\mathbf{a}_2$, etc. Similarly for $\mathbf{b}$. In the column headed 'Basis' we list the $\mathbf{a}_i$ for which $x_i \neq 0$ in our initial basic feasible vector $\mathbf{x} = (7, 0, 0, 0, 12, 10)$. The second column, headed 'c_j', lists the components of $\mathbf{c}$ which correspond to each vector listed under 'Basis', thus $c_1 = 0$ corresponds to $\mathbf{a}_1$, $c_5 = 0$ corresponds to $\mathbf{a}_5$ and $c_6 = 0$ to $\mathbf{a}_6$.

We will now fill in the remaining gaps in the tableau. This will prepare the problem for the initial step of the simplex algorithm. We first compute the entries for the row headed '$\mathbf{z}$'. Suppose the contents of this row from left to right are z_1, z_2, ..., z_6, s, then z_1 is the inner product of the column headed 'c_j' and the column headed '$\mathbf{a}_1$'. Thus

$$z_1 = (0)(1) + (0)(0) + (0)(0) = 0 .$$

Similarly

$$z_2 = (0)(3) + (0)(-2) + (0)(-4) = 0 ,$$
$$z_3 = (0)(-1) + (0)(4) + (0)(3) = 0 ,$$
$$z_4 = (0)(2) + (0)(0) + (0)(8) = 0 ,$$
$$z_5 = (0)(0) + (0)(1) + (0)(0) = 0 ,$$
$$z_6 = (0)(0) + (0)(0) + (0)(1) = 0 ,$$
$$s = (0)(7) + (0)(12) + (0)(10) = 0 .$$

Next form the row $\mathbf{z}$-$\mathbf{c}$ by taking the difference between the entry in the row $\mathbf{z}$ and the corresponding entry in the second row, that is $z_j - c_j$ ($1 \leq j \leq 6$). We next look for the *largest positive* quantity in the row $\mathbf{z}$ - $\mathbf{c}$ which is *not* in a basis column and arrow the corresponding column. In this case we exclude columns $\mathbf{a}_1$, $\mathbf{a}_5$, $\mathbf{a}_6$ and the larges positive value of $z_j - c_j$ in the remaining columns is 3. In a *maximising* problem we should look for the *smallest negative* quantity in the row $\mathbf{z}$ - $\mathbf{c}$ which is not in a basis column; in other respects the procedure is the same for a maximising problem at all stages. Geometrically, this chooses an edge from the current extreme point of the feasible region along which the change in the objective function is most favourable. If there is no positive (or negative, for a maximising problem) entry among the non-basis columns of the row z - c, it means that our initial basic feasible vector was actually an optimal vector, and we are done.

Finally, take the ratio of each entry in the column headed 'b' with the corresponding entry in the arrowed column, in this case the column headed 'a_3'. Enter these ratios in the 'Ratio' column, choose the *smallest positive* ratio, and arrow the corresponding row. Geometrically, this tells us how far we can go along the chosen edge before a constraint becomes critical; we then have reached the next extreme point of the feasible region. If every ratio were indeterminate or negative *it would mean that the problem was unbounded*.

3.8 THE PIVOT OPERATION.

We have now reached the tableau

			c_1	c_2	c_3	c_4	c_5	c_6		
			0	1	-3	2	0	0		
	Basis	c_j	a_1	a_2	a_3	a_4	a_5	a_6	b	Ratio
R_1	a_1	0	1	3	-1	2	0	0	7	$-7/_1$
R_2	a_5	0	0	-2	4	0	1	0	12	$12/_4$ ←
R_3	a_6	0	0	-4	3	8	0	1	10	$10/_3$
		z	0	0	0	0	0	0	0	
		z-c	0	-1	3	-2	0	0		

$$\uparrow$$

First simplex tableau for (3.53).

The entry, 4, in both the arrowed row and column is of special interest and is known as the **pivot**. Our current basic feasible vector (7, 0, 0, 0, 12, 10) involves basis vector a_5 in the arrowed row, and our next operation, the pivot operation, will compute a new basic feasible vector which involves a_3 (in the arrowed column) *instead of* a_5. Because of this x_5 is called the **departing** variable and x_3 the **entering** variable.

The first step in forming the second tableau is to divide the entries in the pivot row which correspond to columns a_j ($1 \le j \le 6$) and b by the pivot. Using R_i ($1 \le i \le 3$) to denote a row in the first tableau (an 'old' row) and R_i' a row in the second (a 'new' row) we have $R_2' = ¼ R_2$. We also replace a_5 by a_3 in the Basis column and $c_5 = 0$ by $c_3 = -3$ in the c_j column. The result is

$R_2' = \tfrac{1}{4}R_2$

	c_1	c_2	c_3	c_4	c_5	c_6		
	0	1	-3	2	0	0		
Basis	c_j	a_1	a_2	a_3	a_4	a_5	a_6	b	Ratio
a_1	0								
a_3	-3	0	$-\tfrac{1}{2}$	1	0	$\tfrac{1}{4}$	0	3	
a_6	0								
z									
z-c									

To produce R_1' and R_3' we now perform elementary row operations on R_1 and R_3 in such a way that the new entries in column a_3 are zero. Thus

$$R_1' = R_1 + R_2' = R_1 + \tfrac{1}{4}R_2, \quad R_3' = R_3 - 3R_2' = R_3 - \tfrac{3}{4}R_2,$$

and this gives

		c_1	c_2	c_3	c_4	c_5	c_6		
		0	1	-3	2	0	0		
Basis	c_j	a_1	a_2	a_3	a_4	a_5	a_6	b	Ratio
a_1	0	1	$\tfrac{5}{2}$	0	2	$\tfrac{1}{4}$	0	10	
a_3	-3	0	$-\tfrac{1}{2}$	1	0	$\tfrac{1}{4}$	0	3	
a_6	0	0	$-\tfrac{5}{2}$	0	8	$-\tfrac{3}{4}$	1	1	
z									
z-c									

$R_1' = R_1 + R_2'$

R_2'

$R_3' = R_3 - 3R_2'$

Finally we fill in rows z and z - c and the Ratio column as described in the previous section to arrive at

			c_1	c_2	c_3	c_4	c_5	c_6		
			0	1	-3	2	0	0		
	Basis	c_j	a_1	a_2	a_3	a_4	a_5	a_6	b	Ratio
R_1'	a_1	0	1	$5/2$	0	2	$1/4$	0	10	4 ←
R_2'	a_3	-3	0	$-1/2$	1	0	$1/4$	0	3	-6
R_3'	a_6	0	0	$-5/2$	0	8	$-3/4$	1	1	$-2/5$
	z		0	$3/2$	-3	0	$-3/4$	0	-9	
	z-c		0	$1/2$	0	-2	$-3/4$	0		

$$\uparrow$$

Second simplex tableau for (3.53).

We have now moved to the basic feasible vector $(10, 0, 3, 0, 0, 1)$, which can be read off from the tableau, and the corresponding value of the objective function is -9, the entry in the bottom right corner. The new pivot is $5/2$, which tells us that at the next step x_1 is the departing variable and x_2 the entering variable.

If we repeat the whole sequence of operations once more we obtain

			c_1	c_2	c_3	c_4	c_5	c_6		
			0	1	-3	2	0	0		
	Basis	c_j	a_1	a_2	a_3	a_4	a_5	a_6	b	Ratio
$R_1'=2/5\,R_1$	a_2	1	$2/5$	1	0	$4/5$	$1/10$	0	4	
$R_2'=R_2+\frac{1}{2}R_1'$	a_3	-3	$1/5$	0	1	$2/5$	$3/10$	0	5	
$R_3'=R_3+\frac{5}{2}R_1'$	a_6	0	1	0	0	10	$-1/2$	1	11	
	z		$-1/5$	1	-3	$-2/5$	$-4/5$	0	-11	
	z-c		$-1/5$	0	0	$-12/5$	$-4/5$	0		

Final simplex tableau for (3.53).

Since all entries in row $z - c$ are zero or negative the calculation is ended. We can read off the optimal vector as $(0, 4, 5, 0, 0, 11)$ and the minimised value of the objective function as -11.

Assuming the problem is feasible only two things can go wrong with the simplex algorithm described above. Firstly the problem may be unbounded, a situation we can recognise by being unable to choose a smallest positive ratio at some stage. The second difficulty is that it is theoretically possible for the calculation to cycle through a sequence of non-optimal basic feasible vectors. This may occur if at some stage the minimum positive ratio is not unique, a phenomenon known as **degeneracy**. Algebraically, degeneracy amounts to the fact that one or more of the constraints is redundant, and will occur if the rank of the matrix A in (3.54) is less than the dimension of **b**, that is less than the number of equations. Geometrically, degeneracy occurs when the n-dimensional convex set of feasible vectors has more than n bounding hyperplanes passibg through an extreme point. In this case one extreme point corresponds to more than one basic feasible vector.

Were it not for degeneracy it would be easy to prove that the simplex algorithm must converge; for assuming non-degeneracy it is not hard to show that the value of the objective function is *strictly* improved by each pivot operation, and that a basic feasible vector once visited will therefore not reappear in any subsequent iteration. Since the number of basic feasible vectors is finite, the process must terminate after a finite number of steps.

But degeneracy does not necessarily imply that cycling occurs. It is an interesting fact that *cycling has never yet occurred in a practical problem*! Degeneracy has persisted for several successive iterations but has not led to cycling, although several examples have been constructed to show that cycling is possible. Because of this experience, anticycling precautions which would involve extra computer time and storage are simply not used in practice. From a statistical viewpoint these time-consuming tests are too high a price to pay for insurance against the remote risk of a costly cycling disaster.

As an extremely crude rule of thumb, a linear program with m constraints can be expected to require something like $2m$ pivot operations for solution, the number of variables n being relatively unimportant. This insensitivity to n holds over a surprising wide range which includes most problems arising in the real world.

In our next application of the simplex algorithm we use it to solve a 3×3 game.

Example 3.7 Find a solution of the matrix game

$$\begin{pmatrix} -2 & 1 & 0 \\ 2 & -3 & -1 \\ 0 & 2 & -3 \end{pmatrix}.$$

We first convert the matrix into one with all positive entries by adding 4 to every entry to obtain

$$\begin{pmatrix} 2 & 5 & 4 \\ 6 & 1 & 3 \\ 4 & 6 & 1 \end{pmatrix};$$

this new game has positive value.

We now have a choice. We can either set up player 1's problem as in (3.3) or player 2's problem as in (3.4). It turns out that player 1's problem has no basic feasible vector of the type required to start the simplex algorithm. This will become apparent in §3.9, where we shall also see what to do about it. However, this difficulty does not occur with player 2. To solve a matrix game by the simplex method it is usually easier to use player 2's problem, precisely because there is always a suitable initial basic feasible vector.

Let $\mathbf{q} = (q_1, q_2, q_3)$ be a strategy for player 2. Writing $y_j = q_j/v$ $(1 \leq j \leq 3)$ we can state player 2's problem as

$$\text{Maximise } y_1 + y_2 + y_3 \ \left(= \frac{1}{v} \right)$$

subject to
$$2y_1 + 5y_2 + 4y_3 \leq 1,$$
$$6y_1 + y_2 + 3y_3 \leq 1,$$
$$4y_1 + 6y_2 + y_3 \leq 1,$$
$$\text{and} \quad y_j \geq 0 \ (1 \leq j \leq 3)$$

Introducing slack variables this becomes

$$\text{Maximise } y_1 + y_2 + y_3$$
subject to
$$2y_1 + 5y_2 + 4y_3 + y_4 \qquad\qquad = 1,$$
$$6y_1 + y_2 + 3y_3 \qquad + y_5 \qquad = 1,$$
$$4y_1 + 6y_2 + y_3 \qquad\qquad + y_6 = 1,$$
$$\text{and} \quad y_j \geq 0 \ (1 \leq j \leq 6)$$

A basic feasible vector of the required type is $\mathbf{y} = (0, 0, 0, 1, 1, 1)$, and the first simplex tableau is

| | | c_1 | c_2 | c_3 | c_4 | c_5 | c_6 | | |
		1	1	1	0	0	0		
Basis	c_j	$\mathbf{a}_1$	$\mathbf{a}_2$	$\mathbf{a}_3$	$\mathbf{a}_4$	$\mathbf{a}_5$	$\mathbf{a}_6$	b	Ratio
$\mathbf{a}_4$	0	2	5	4	1	0	0	1	$1/2$
$\mathbf{a}_5$	0	6	1	3	0	1	0	1	$1/6$ ←
$\mathbf{a}_6$	0	4	6	1	0	0	1	1	$1/4$
z		0	0	0	0	0	0	0	
z-c		-1	-1	-1	0	0	0		

                ↑

Remember this is a maximising problem, so we must choose the smallest negative entry in $z - c$ which is not in a basis column. Since every non-zero entry in $z - c$ is -1 we choose the first arbitrarily. The subsequent tableaux are

	Basis	c_j	a_1	a_2	a_3	a_4	a_5	a_6	b	Ratio
			c_1	c_2	c_3	c_4	c_5	c_6		
			1	1	1	0	0	0		
$R_1'=R_1-2R_2'$	a_4	0	0	$14/3$	3	1	$-1/3$	0	$2/3$	$1/7$
$R_2'=1/6\ R_2$	a_1	1	1	$1/6$	$1/2$	0	$1/6$	0	$1/6$	1
$R_3'=R_3-4R_1'$	a_6	0	0	$16/3$	-1	0	$-2/3$	1	$1/3$	$1/16$ ←
	z		1	$1/6$	$1/2$	0	$1/6$	0	$1/6$	
	z-c		0	$-5/6$	$-1/2$	0	$1/6$	0		

$\uparrow$

	Basis	c_j	a_1	a_2	a_3	a_4	a_5	a_6	b	Ratio
			c_1	c_2	c_3	c_4	c_5	c_6		
			1	1	1	0	0	0		
$R_1'=R_1-\frac{14}{3}R_3'$	a_4	0	0	0	$31/8$	1	$1/4$	$-7/8$	$3/8$	$3/31$ ←
$R_2'=R_2-\frac{1}{6}R_3'$	a_1	1	1	0	$17/32$	0	$3/16$	$-1/32$	$5/32$	$5/17$
$R_3'=\frac{3}{16}R_3$	a_2	1	0	1	$-3/16$	0	$-1/8$	$3/16$	$1/16$	$-1/3$
	z		1	1	$11/32$	0	$1/16$	$5/32$	$7/32$	
	z-c		0	0	$-21/32$	0	$1/16$	$5/32$		

$\uparrow$

		c_1	c_2	c_3	c_4	c_5	c_6		
		1	1	1	0	0	0		
Basis	c_j	a_1	a_2	a_3	a_4	a_5	a_6	b	Ratio
a_3	1	0	0	1	$8/31$	$2/31$	$-7/31$	$3/31$	
a_1	1	1	0	0	$-17/124$	$19/124$	$11/124$	$13/124$	
a_2	1	0	1	0	$3/62$	$-7/62$	$9/62$	$10/124$	
z		1	1	1	$21/124$	$13/124$	$1/124$	$35/124$	
$z-c$		0	0	0	$21/124$	$13/124$	$1/124$		

Rows at left of table:
$R_1' = 8/31 R_1$
$R_2' = R_2 - 17/32 R_1'$
$R_3' = R_3 + 3/16 R_1'$

Since all entries in $z - c$ are non-negative the calculation is ended. We conclude that an optimal vector is

$$y = (y_1, y_2, y_3) = \left(\frac{13}{124}, \frac{10}{124}, \frac{3}{31} \right), \text{ and } \frac{1}{v} = \frac{35}{124}.$$

Hence the corresponding optimal strategy for player 2 is

$$q = (q_1, q_2, q_3) = \frac{124}{35} \left(\frac{13}{124}, \frac{10}{124}, \frac{3}{31} \right) = \left(\frac{13}{35}, \frac{10}{35}, \frac{12}{35} \right),$$

and the value of the original game is

$$\frac{124}{35} - 4 = -\frac{16}{35}.$$

There is an unexpected bonus contained within our solution of player 2's problem. If the initial $m \times n$ simplex tableau contains an $m \times m$ unit matrix, then the solution of the problem yields an explicit solution to the dual. We read off the solution to the dual from the non-basis columns of the z row (for a minimising problem it is necessary to change the sign throughout). In this case the solution to player 1's problem is

$$(x_1, x_2, x_3) = \left(\frac{21}{124}, \frac{13}{124}, \frac{1}{124} \right), \quad \frac{1}{v} = \frac{35}{124}.$$

which gives player 1's optimal strategy as

$$(p_1, p_2, p_3) = \frac{124}{35} \left(\frac{21}{124}, \frac{13}{124}, \frac{1}{124} \right) = \left(\frac{21}{35}, \frac{13}{35}, \frac{1}{35} \right).$$

Thus a complete solution to the original game is

$$p = (\frac{21}{35}, \frac{13}{35}, \frac{1}{35})$$

$$q = (\frac{13}{35}, \frac{10}{35}, \frac{12}{35})$$

$$v = -\frac{16}{35}.$$

3.9 ARTIFICIAL VARIABLES.

The technique described in §3.7 - 3.8 is often called the **phase-2 algorithm**. It can proceed if the initial simplex tableau contains an $m \times m$ unit matrix. We now consider the **phase-1 algorithm** which will find a suitable initial basic feasible vector provided the program is feasible.

The idea behind the phase-1 procedure is to introduce new non-negative variables $w_1, ..., w_m$, called **artificial** variables, which *are* associated with an $m \times m$ unit matrix. In many instances it is not necessary to introduce as many as m artificial variables, but m is the worst possibility. In a minimising (maximising) program we modify the objective function to give the new variables a positive (negative) 'cost', the idea being that if we now judiciously apply the phase-2 simplex algorithm to the enlarged problem we can drive Σw_i to zero and thereby obtain a suitable basic feasible vector for the original problem.

Before we discuss possible complications the following example should serve to clarify the procedure.

Example 3.8 Solve the linear program

$$\text{Minimise } - x_1 - 2x_2 - 3x_3 + x_4$$
subject to
$$x_1 + 2x_2 + 3x_3 \qquad = 15$$
$$2x_1 + x_2 + 5x_3 \qquad = 20$$
$$x_1 + 2x_2 + x_3 + x_4 = 10$$
and $x_j \geq 0 \quad (1 \leq j \leq 4)$

Here the last column of the matrix is a standard basis vector, so we need only introduce two artificial variables w_1, w_2. We associate with each of these a positive cost $w > 0$. The enlarged problem is then

$$\text{Minimise } - x_1 - 2x_2 - 3x_3 + x_4 + ww_1 + ww_2$$
subject to
$$x_1 + 2x_2 + 3x_3 \qquad + w_1 \qquad = 15$$
$$2x_1 + x_2 + 5x_3 \qquad + w_2 = 20$$
$$x_1 + 2x_2 + x_3 + x_4 \qquad = 10$$
and $x_j \geq 0 \quad (1 \leq j \leq 4) , \; w_j \geq 0 \; (1 \leq j \leq 2)$.

The first simplex tableau for the enlarged problem is

Basis	c_j	$\begin{array}{c}c_1\\-1\\ \mathbf{a}_1\end{array}$	$\begin{array}{c}c_2\\-2\\ \mathbf{a}_2\end{array}$	$\begin{array}{c}c_3\\-3\\ \mathbf{a}_3\end{array}$	$\begin{array}{c}c_4\\1\\ \mathbf{a}_4\end{array}$	$\begin{array}{c}c_5\\w\\ \mathbf{a}_5\end{array}$	$\begin{array}{c}c_6\\w\\ \mathbf{a}_6\end{array}$	b	Ratio
$\mathbf{a}_5$	w	1	2	3	0	1	0	15	5
$\mathbf{a}_6$	w	2	1	5	0	0	1	20	4 ←
$\mathbf{a}_4$	1	1	2	1	1	0	0	10	10
z		$3w+1$	$3w+2$	$8w+1$	1	w	w	$35w+10$	
z-c		$3w+2$	$3w+4$	$8w+4$	0	0	0		

↑

We treat w as if it were very large so that in practice it is only necessary to enter the coefficient of w in the z and z - c rows. Thus we write 3 for $3w+1$, and such nonsense as 0-1=0 meaning $0w-1$ is effectively 0 for the purpose of filling in the tableau. Moreover once an artificial variable is eliminated from the basis *it is never selected to re-enter the basis*. The subsequent tableaux are

	Basis	c_j	$\begin{array}{c}c_1\\-1\\ \mathbf{a}_1\end{array}$	$\begin{array}{c}c_2\\-2\\ \mathbf{a}_2\end{array}$	$\begin{array}{c}c_3\\-3\\ \mathbf{a}_3\end{array}$	$\begin{array}{c}c_4\\1\\ \mathbf{a}_4\end{array}$	$\begin{array}{c}c_5\\w\\ \mathbf{a}_5\end{array}$	$\begin{array}{c}c_6\\w\\ \mathbf{a}_6\end{array}$	b	Ratio
$R_1'=R_1-3R_2'$	$\mathbf{a}_5$	w	$-\tfrac{1}{5}$	$\tfrac{7}{5}$	0	0	1	$-\tfrac{3}{5}$	3	$\tfrac{15}{7}$ ←
$R_2'=\tfrac{1}{5}R_2$	$\mathbf{a}_3$	-3	$\tfrac{2}{5}$	$\tfrac{1}{5}$	1	0	0	$\tfrac{1}{5}$	4	20
$R_3'=R_3-R_2'$	$\mathbf{a}_4$	1	$\tfrac{2}{5}$	$\tfrac{9}{5}$	0	1	0	$-\tfrac{1}{5}$	6	$\tfrac{10}{3}$
	z		$-\tfrac{1}{5}$	$\tfrac{7}{5}$	0	0	w	$-\tfrac{3}{5}w$	3	
	z-c		$-\tfrac{1}{5}$	$-\tfrac{7}{5}$	0	0	0	$-\tfrac{8}{5}w$		

↑

		c_1	c_2	c_3	c_4	c_5	c_6		
		-1	-2	-3	1	w	w		
Basis	c_j	a_1	a_2	a_3	a_4	a_5	a_6	b	Ratio
a_4	-2	$-1/7$	1	0	0	$5/7$	$-3/7$	$15/7$	
a_1	-3	$3/7$	0	1	0	$-1/7$	$2/7$	$25/7$	
a_6	1	$6/7$	0	0	1	$-9/7$	$4/7$	$15/7$	
	z	0	0	0	0	0	0	0‡	
	z-c	0	0	0	0	$-w$	$-w$		

$R_1'=5/7 R_1$
$R_2'=R_2-1/5 R_1'$
$R_3'=R_3-9/5 R_1'$

‡ N.B The artificial part of the objective function is zero.

At this stage we have eliminated $\mathbf{a}_5$ and $\mathbf{a}_6$ from the basis and found an initial basic feasible vector $\mathbf{x} = (0,\ 15/7,\ 25/7,\ 15/7)$ for the original problem with the property that column vectors associated with the non-zero components of $\mathbf{x}$ form a 3×3 unit matrix in the new tableau. The artificial variables have performed their function, and we can now drop the columns of the tableau associated with $\mathbf{a}_5$ and $\mathbf{a}_6$. In fact we could have dropped these columns as soon as the associated vector was pivoted out of the basis. We therefore contract the tableau, fill in the rows z, z - c in the usual way and complete the solution by using phase-2.

		c_1	c_2	c_3	c_4		
		-1	-2	-3	1		
Basis	c_j	a_1	a_2	a_3	a_4	b	Ratio
a_2	-2	$-1/7$	1	0	0	$15/7$	-15
a_3	-3	$3/7$	0	1	0	$25/7$	$25/3$
a_4	1	$6/7$	0	0	1	$15/7$	$15/6$ ←
	z	$-1/7$	-2	-3	1	$-90/7$	
	z-c	$-6/7$	0	0	0		

↑

			c_1	c_2	c_3	c_4		
			-1	-2	-3	1		
	Basis	c_j	$\mathbf{a}_1$	$\mathbf{a}_2$	$\mathbf{a}_3$	$\mathbf{a}_4$	$\mathbf{b}$	Ratio
$R_1'=R_1+\frac{1}{7}R_3'$	$\mathbf{a}_2$	-2	0	1	0	$\frac{1}{6}$	$\frac{5}{2}$	
$R_2'=R_2-\frac{3}{7}R_3'$	$\mathbf{a}_3$	-3	0	0	1	$-\frac{1}{2}$	$\frac{5}{2}$	
$R_3'=\frac{7}{6}R_3$	$\mathbf{a}_1$	-1	1	0	0	$\frac{7}{6}$	$\frac{5}{2}$	
	$\mathbf{z}$		-1	-2	-3	0	-15	
	$\mathbf{z}$-$\mathbf{c}$		0	0	0	-1		

This gives the solution $\mathbf{x} = (\frac{5}{2}, \frac{5}{2}, \frac{5}{2}, 0)$ and the value of the objective function as -15.

Thus phase-1 amounts to using the phase-2 simplex algorithm to minimise Σw_i, which corresponds to the artificial part of the objective function, subject to the extended constraints. In general one must not expect to see one artificial variable pivot out of the basis at each step; as with phase-2, a crude estimate is that $2m$ pivot operations may be required. However, it is possible that a stage may be reached when there are no positive (negative, for a maximising problem) entries in the non-artificial part of the $\mathbf{z}$ - $\mathbf{c}$ row but $\Sigma w_i \neq 0$. In this case phase-1 is forced to terminate, and it means that *the original program was infeasible.*

If pivoting continues until $\Sigma w_i = 0$ there are two possibilities. Either, as in Example 3.8, the resulting basic feasible vector does not involve any artificial basis vector, or it involves artificial basis vectors for which necessarily the corresponding values of the artificial variables are zero. Both cases provide a suitable basic feasible vector for the original problem, but the second case is rather interesting. It is an indication of degeneracy in the original problem. In fact if $\Sigma w_i = 0$ in a phase-1 simplex algorithm but an artificial basis remains in the 'Basis' column of the tableau, it means that the corresponding constraint in the original problem was redundant (for example [2], Theorem 5.1, p. 137).

Example 3.9 Solve the linear program

$$\text{Maximise } x_1 + x_2 - 2x_3$$
subject to
$$x_1 + x_2 + x_3 = 12$$
$$2x_1 + 5x_2 - 6x_3 = 10$$
$$7x_1 + 10x_2 - x_3 = 70$$
$$\text{and } x_j \geq 0 \quad (1 \leq j \leq 3)$$

We introduce artificial variables w_1, w_2, w_3 and because this is a maximising problem we associate with each a negative cost $-w$ ($w > 0$). The enlarged problem is then

Maximise $x_1 + x_2 - 2x_3 - ww_1 - ww_2 - ww_3$
subject to
$$x_1 + x_2 + x_3 + w_1 \qquad\qquad = 12$$
$$2x_1 + 5x_2 - 6x_3 \qquad + w_2 \qquad = 10$$
$$7x_1 + 10x_2 - x_3 \qquad\qquad + w_3 = 70$$
and $x_j \geq 0 \;\; (1 \leq j \leq 3)$, $w_j \geq 0 \; (1 \leq j \leq 3)$.

The first two tableaux are

		c_1	c_2	c_3	c_4	c_5	c_6		
		1	1	-2	$-w$	$-w$	$-w$		
Basis	c_j	$\mathbf{a}_1$	$\mathbf{a}_2$	$\mathbf{a}_3$	$\mathbf{a}_4$	$\mathbf{a}_5$	$\mathbf{a}_6$	b	Ratio
$\mathbf{a}_4$	$-w$	1	1	1	1	0	0	12	$^{12}/_1$
$\mathbf{a}_5$	$-w$	2	5	-6	0	1	0	10	2 ←
$\mathbf{a}_6$	$-w$	7	10	-1	0	0	1	70	7
z		-10	-16	6	$-w$	$-w$	$-w$	-92	
z-c		-10	-16	6	0	0	0		

$\uparrow$

			c_1	c_2	c_3	c_4	c_5	c_6		
			1	1	-2	$-w$	$-w$	$-w$		
	Basis	c_j	$\mathbf{a}_1$	$\mathbf{a}_2$	$\mathbf{a}_3$	$\mathbf{a}_4$	$\mathbf{a}_5$	$\mathbf{a}_6$	b	Ratio
$R_1'=R_1-R_2'$	$\mathbf{a}_4$	$-w$	$^3/_5$	0	$^{11}/_5$	1	$-^1/_5$	0	10	$^{50}/_{11}$ ←
$R_2'=^1/_5 R_2$	$\mathbf{a}_2$	1	$^2/_5$	1	$-^6/_5$	0	$^1/_5$	0	2	$-^{10}/_6$
$R_3'=R_3-10R_1'$	$\mathbf{a}_6$	$-w$	3	0	11	0	-2	1	50	$^{50}/_{11}$ ←
	z		$-^{18}/_5$	0	$-^{66}/_5$	$-w$	$^{11}/_5 w$	$-w$	-60	
	z-c		$-^{18}/_5$	0	$-^{66}/_5$	0	$^{16}/_5 w$	0		

$\uparrow$

At this stage a_3 is set to enter the basis, but the duplicate ratio $^{50}/_{11}$ tells us that either a_4 or a_6 can exit, and degeneracy has occurred. To break the tie we arbitrarily exit a_4. Pivoting about $^{11}/_5$ yields

			c_1	c_2	c_3	c_4	c_5	c_6		
			1	1	-2	-w	-w	-w		
	Basis	c_j	a_1	a_2	a_3	a_4	a_5	a_6	b	Ratio
$R_1'=\,^5/_{11}R_1$	a_3	-2	$^3/_{11}$	0	1	$^5/_{11}$	$-^1/_{11}$	0	$^{50}/_{11}$	
$R_2'=R_2+^6/_5R_1'$	a_2	1	$^8/_{11}$	1	0	$^6/_{11}$	$^1/_{11}$	0	$^{82}/_{11}$	
$R_3'=R_3-11R_1'$	a_6	-w	0	0	0	-5	-1	1	0	
	z		0	0	0	5w	w	-w	0	
	z-c		0	0	0	6w	2w	0		

Phase-1 has ended by finding a basic feasible vector dependent upon the artificial basis a_6, but of course with $w_3 = 0$, Since the third component of $a_1 = (^3/_{11},\,^8/_{11},\,0)$ is zero we are unable to pivot a_6 out of the basis and replace it by a_1. We conclude that the third constraint of the original problem was redundant. The initial phase-2 tableau is therefore

		c_1	c_2	c_3		
		1	1	-2		
Basis	c_j	a_1	a_2	a_3	b	Ratio
a_3	-2	$^3/_{11}$	0	1	$^{50}/_{11}$	$^{50}/_3$
a_2	1	$^8/_{11}$	1	0	$^{82}/_{11}$	$^{41}/_4$ ←
z		$^2/_{11}$	1	-2	$-^{18}/_{11}$	
z-c		$-^9/_{11}$	0	0		

                            ↑

Pivoting about $^8/_{11}$ yields the final tableau

	c_j	c_1 1	c_2 1	c_3 -2		
Basis		a_1	a_2	a_3	b	Ratio
a_3	-2	0	$-\tfrac{3}{8}$	1	$\tfrac{7}{4}$	
a_1	1	1	$\tfrac{11}{8}$	0	$\tfrac{41}{4}$	
z	1		$\tfrac{17}{8}$	-2	$\tfrac{27}{4}$	
z-c	0		$\tfrac{9}{8}$	0		

$R_1'=R_1-\tfrac{3}{11}R_2'$

$R_2'=\tfrac{11}{6}R_2'$

Hence an optimal vector is $\mathbf{x} = (^{41}/_4, 0, ^7/_4,)$ and the value of the objective function is $^{27}/_4$.

In the normal course of events such exotica as degeneracy cannot be expected to occur! For a fairly routine illustration of the use of artificial variables we return to Example 3.7 and solve player 1's problem.

The game matrix with 4 added throughout is

$$\begin{pmatrix} 2 & 5 & 4 \\ 6 & 1 & 3 \\ 4 & 6 & 1 \end{pmatrix}.$$

Let $\mathbf{p} = (p_1, p_2, p_3)$ be a strategy for player 1. Putting $x_i = p_i/v$ $(1 \leq i \leq 3)$ we can state player 1's problem as

$$\text{Minimise } x_1 + x_2 + x_3 \quad (=\tfrac{1}{v})$$

subject to
$$2x_1 + 6x_2 + 4x_3 \geq 1,$$
$$5x_1 + x_2 + 6x_3 \geq 1,$$
$$4x_1 + 3x_2 + x_3 \geq 1,$$
and $x_i \geq 0$ $(1 \leq i \leq 3)$.

Inserting the slack variables this becomes

$$\text{Minimise } x_1 + x_2 + x_3$$
subject to
$$2x_1 + 6x_2 + 4x_3 - x_4 \qquad = 1,$$
$$5x_1 + x_2 + 6x_3 \qquad - x_5 \qquad = 1,$$
$$4x_1 + 3x_2 + x_3 \qquad - x_6 = 1,$$
and $x_i \geq 0$ $(1 \leq i \leq 6)$.

No standard basis vectors are available, so we introduce three artificial variables to obtain

Minimise $x_1 + x_2 + x_3 + ww_1 + ww_2 + ww_3$ $(w > 0)$
subject to

$$
\begin{aligned}
2x_1 + 6x_2 + 4x_3 - x_4 \qquad\qquad + w_1 \qquad\qquad &= 1, \\
5x_1 + \ x_2 + 6x_3 \qquad - x_5 \qquad\qquad + w_2 \qquad &= 1, \\
4x_1 + 3x_2 + \ x_3 \qquad\qquad - x_6 \qquad\qquad + w_3 &= 1, \\
\end{aligned}
$$

and $x_i \geq 0$ $(1 \leq i \leq 6)$, $w_i \geq 0$ $(1 \leq i \leq 3)$.

The phase-1 tableaux are continued in the following pages:

Basis	c_j	c_1 1	c_2 1	c_3 1	c_4 0	c_5 0	c_6 0	c_7 w	c_8 w	c_9 w	b	Ratio
		a_1	a_2	a_3	a_4	a_5	a_6	a_7	a_8	a_9		
a_7	w	2	6	4	-1	0	0	1	0	0	1	$1/2$
a_8	w	5	1	6	0	-1	0	0	1	0	1	$1/5$
a_9	w	4	3	1	0	0	-1	0	0	1	1	$1/4$
	z	11	10	11	-1	-1	-1	w	w	w	3	
	z-c	11	10	11	-1	-1	-1	0	0	0		

↑ (under a_1 column) ↓ (above Ratio column)

Basis	c_j	c_1 1	c_2 1	c_3 1	c_4 0	c_5 0	c_6 0	c_7 w	c_8 w	c_9 w	b	Ratio
		a_1	a_2	a_3	a_4	a_5	a_6	a_7	a_8	a_9		
a_7	w	0	$28/5$	$8/5$	-1	$2/5$	0	1	$-2/5$	0	$3/5$	$3/28$
a_1	1	1	$1/5$	$6/5$	0	$-1/5$	0	0	$1/5$	0	$1/5$	1
a_8	w	0	$11/5$	$-19/5$	0	$4/5$	-1	0	$-4/5$	1	$1/5$	$1/11$
	z	0	$39/5$	$-11/5$	-1	$6/5$	-1	w	$-6/5 w$	w	$4/5$	
	z-c	0	$39/5$	$-11/5$	-1	$6/5$	-1	0	$-11/5 w$	0		

↑ (under a_2 column) ↓ (above Ratio column)

$R_1' = R_1 - 2R_2'$

$R_2' = {}^1\!/_5 R_2$

$R_3' = R_3 - 4R_2'$

First tableau

| | | c_1 | c_2 | c_3 | c_4 | c_5 | c_6 | c_7 | c_8 | c_9 | | |
| | | 1 | 1 | 1 | 0 | 0 | 0 | w | w | w | | |
Basis	c_j	a_1	a_2	a_3	a_4	a_5	a_6	a_7	a_8	a_9	b	Ratio
a_7	w	0	0	$124/11$	-1	$-18/11$	$28/11$	1	$18/11$	$-28/11$	$1/11$	$1/124$
a_1	1	1	0	$17/11$	0	$-3/11$	$1/11$	0	$3/11$	$-1/11$	$2/11$	$2/17$
a_2	1	0	1	$-19/11$	0	$4/11$	$-5/11$	0	$-4/11$	$5/11$	$1/11$	$-1/19$
z		0	0	$124/11$	-1	$-18/11$	$28/11$	w	$18/11\,w$	$-28/11\,w$	$1/11$	
$z-c$		0	0	$124/11$	-1	$-18/11$	$28/11$	0	$7/11\,w$	$-39/11\,w$		

$R_1' = R_1 - \frac{28}{5}R_3'$

$R_2' = R_2 - \frac{1}{5}R_3'$

$R_3' = -\frac{5}{11}R_3$

Second tableau

| | | c_1 | c_2 | c_3 | c_4 | c_5 | c_6 | c_7 | c_8 | c_9 | | |
| | | 1 | 1 | 1 | 0 | 0 | 0 | w | w | w | | |
Basis	c_j	a_1	a_2	a_3	a_4	a_5	a_6	a_7	a_8	a_9	b	Ratio
a_3	1	0	0	1	$-11/124$	$-9/62$	$7/31$	$11/124$	$-2/5$	0	$1/124$	
a_1	1	1	0	0	$17/124$	$-3/62$	$-8/31$	$-17/124$	$1/5$	0	$21/124$	
a_2	1	0	1	0	$-19/124$	$7/62$	$2/31$	$19/124$	$-4/5$	1	$13/124$	
z		0	0	0	0	0	0	0	0	0	0	
$z-c$		0	0	0	0	0	0					

$R_1 = \frac{11}{124}R_1'$

$R_2 = R_2 - \frac{17}{11}R_1'$

$R_3 = R_3 + \frac{19}{11}R_1'$

The phase-1 simplex algorithm is successfully completed. We contract the tableau and proceed to phase-2.

		c_1	c_2	c_3	c_4	c_5	c_6		
		1	1	1	0	0	0		
Basis	c_j	a_1	a_2	a_3	a_4	a_5	a_6	b	Ratio
a_3	1	0	0	1	$-11/124$	$-9/62$	$7/31$	$1/124$	
a_1	1	1	0	0	$17/124$	$-3/62$	$-8/31$	$21/124$	
a_2	1	0	1	0	$-19/124$	$7/62$	$-2/31$	$13/124$	
z	1	1	1	$-13/124$	$-5/62$	$-3/31$	$35/124$	
z-c	0	0	0	$-13/124$	$-5/62$	$-3/31$		

But, lo and behold, the phase-2 simplex algorithm tableau indicates we have reached optimality, and we can read off the solution as

$$x = \left(\frac{21}{124}, \frac{13}{124}, \frac{1}{124} \right), \quad \frac{1}{v} = \frac{35}{124}.$$

Hence an optimal strategy for player 1 is

$$p = \frac{124}{35} \left(\frac{21}{124}, \frac{13}{124}, \frac{1}{124} \right) = \left(\frac{21}{35}, \frac{13}{35}, \frac{1}{35} \right)$$

and the value of the original game is

$$\frac{124}{35} - 4 = -\frac{16}{35}.$$

Finally we can read off player 2's solution from the non-basic columns of the z row, remembering to change the sign since the problem we have solved was a minimising one.

PROBLEMS FOR CHAPTER 3.

1. Use Lemma1 3.1 to prove that precisely one of the sets

$$\left\{ x \in \mathbb{R}^m; \ xA = b \right\}, \quad \left\{ y \in \mathbb{R}^n; \ Ay^T = 0 \text{ and } yb^T = 1 \right\}$$

is empty. [Hint: use the ideas in the proof of Theorem 3.2]

2. Use Theorem 3.2 to show that the equations

$$2x_1 + 3x_2 - 5x_3 = -4,$$
$$x_1 + 2x_2 - 6x_3 = -3,$$

have no non-negative solution. Interpret this geometrically.

3. Apply Theorem 3.8 to verify that $x = (1, 1, 2, 0)$ is a solution of the linear program

$$\text{Minimise } 8x_1 + 6x_2 + 3x_3 + 6x_4$$

subject to
$$x_1 + 2x_2 \qquad + x_4 \geq 3,$$
$$3x_1 + x_2 + x_3 + x_4 \geq 6,$$
$$x_3 + x_4 \geq 2,$$
$$x_1 \qquad + x_3 \qquad \geq 2,$$
$$\text{and } x_j \geq 0 \quad (1 \leq j \leq 4)$$

4. Verify that $y = (2, 2, 4, 0)$ is a solution of the linear program

$$\text{Maximise } 2y_1 + 4y_2 + y_3 + y_4$$

subject to
$$y_1 + 3y_2 \qquad + y_4 \leq 8,$$
$$2y_1 + y_2 \qquad \leq 6,$$
$$y_2 + y_3 + y_4 \leq 6,$$
$$y_1 + y_2 + y_3 \qquad \leq 9,$$
$$\text{and } y_j \geq 0 \quad (1 \leq j \leq 4).$$

5. The manager of a downtown 24-hour Deli has divided an average weekday into four-hour periods and figured out how many assistants he needs serving in each four-hour period. His conclusions are given below.

3.01 - 7.00	7.01 - 11.00	11.01 - 3.00	3.01 - 7.00	7.01 - 11.00	11.01 - 3.00
2	10	14	8	10	3

The assistants report for duty at 3 a.m., 7 a.m., etc. and their shifts last for eight hours. The manager's problem is to determine how the given numbers can be supplied for each period, using the minimum number of assistants each day. If $x_1, x_2, ..., x_6$ are the numbers starting at 3 a.m., 7 a.m., ..., 11 p.m. respectively, verify that $x = (0, 14, 0, 8, 2, 2)$ is a solution.

6. A matrix A is called a Minkowski-Leontief matrix if (i) A is square, $m \times m$, say, (ii) $a_{ii} > q \geq 0$, (iii) $a_{ij} \leq q$ for $i \neq j$, and (iv)

$$\sum_{i=1}^{m} a_{ij} > mq \geq 0 .$$

Prove that a Minkowski-Leontief matrix genertes a completely mixed game. [Hint: compare with Chapter 2, Problem 7.] Hence deduce that every Minkowski-Leontief matrix is non-singular.

7. Find the complete solution to the game

$$\begin{pmatrix} 1 & 1 & 3 \\ 1 & 1 & 0 \\ 0 & 2 & -5 \end{pmatrix}$$

8. Solve by the phase-2 simplex algorithm

Minimise $x_1 + 6x_2 + 2x_3 - x_4 + x_5 - 3x_6$,
subject to
$$x_1 + 2x_2 + x_3 \qquad\qquad + 5x_6 = 3,$$
$$- 3x_2 + 2x_3 + x_4 \qquad + x_6 = 1,$$
$$5x_2 + 3x_3 \qquad + x_5 - 2x_6 = 2,$$
and $x_j \geq 0 \;\; (1 \leq j \leq 6)$.

9. Solve using phase-1 and phase-2 simplex algorithms

Minimise $x_1 + x_2 + 3x_3 + x_4$
subject to
$$2x_1 + 3x_2 \qquad\quad + 6x_4 \geq 8,$$
$$3x_1 + x_2 + 2x_3 - 7x_4 = -4,$$
and $x_j \geq 0 \;\; (1 \leq j \leq 4)$.

10. Solve by the simplex method

Maximise $x_1 + 6x_2 + 4x_3$
subject to
$$-x_1 + 2x_2 + 2x_3 \leq 13,$$
$$4x_1 - 4x_2 + x_3 \leq 20,$$
$$x_1 + 2x_2 + x_3 \leq 17,$$
and $x_1 \leq 7, x_2 \geq 2, x_3 \geq 3$.

CHAPTER REFERENCES

[1] Farkas, J., 'Uber die Theorie der einfachen Ungleichungen'. *Journal für reine und die angewandte Mathematik,* **124**, 1-27 (1902)/

[2] Simmons, D., *Linear Programming for Operations Research.* Holden-Day Inc. (1972), San Francisco.

[3] Cassel, G., *The Theory of Social Economy.* Harcourt Brace (1924), New York.

[4] Wald, A., 'Uber einige Gleichungssysteme der mathematischen ökonomie'. *Zeitschrift für Nationalökonomie,* **7**, 637-670 (1936).

[5] Wald, A., 'On some systems of equations of mathematical economics'. *Econometrica,* **19**, 368-403 (1951) (English translation of [4]).

[6] Kuhn, H.W. 'On a theorem of Wald', *Linear inequalities and related systems.* Ed by Kuhn and Tucker. Princeton University Press (1956), Princeton.

[7] Debreu, G., 'New concepts and techniques for equilibrium analysis'. *International Economic Review,* **3**, 257-273 (1962).

4

Cooperative games

... but one can't please all of the people all of the time...

A **cooperative** game is a game in which the players have complete freedom of preplay communication to make joint *binding* agreements. These agreements may be of two kinds: to coordinate strategies or to share payoffs.

One might venture to guess that this element of agreement between players in a cooperative game would simplify the analysis. (Certainly if there is perfect agreement bewteen all players the analysis is trivial!) However, in general it is not the case, and partial agreement complicates the issue to such an extent that n-person cooperative game theory is neither as elegant nor as cohesive as the non-cooperative case.

To share payoffs is not always possible since these may be in non-transferable units such as 'years in prison' or 'early parole'. An alternative in this situation is to include the possibility of side payments in some transferable unit such as money. Considerations of this kind prompt us to take up a question which, in our hurry to begin solving games, we neglected to examine at the start.

4.1 UTILITIES AND SCALES OF MEASUREMENT.

We have assumed that all payoffs can be interpreted as numbers, and that players are seeking to maximise their payoffs. *How* these numbers are to be assigned raises some rather delicate problems. For example a chess player may rather risk losing a game than settle for a draw. A woman playing checkers (draughts) with her daughter may rather lose than win. If payoffs are in money it is even more likely to introduce differences between players, depending on the psychological 'worth' of the amount of money paid for each outcome.

Suppose on the toss of a fair coin one can win \$20 on heads and lose \$10 on tails. The expected value of the game is \$5, and it seems, therefore, that we should accept the bet. Would we be ready to say, however, that refusing the bet is an indication of irrationality? To win \$5 may be 'worth' much more to a student than to a banker, but the student may not be able to afford the loss of \$10. On the other hand it is well known that people gamble when the odds are not fair; that they play games whose expected value is negative. Recognition of the fact that such actions need not be irrational has led to the formulation of the concept of **individual utility**, which reflects numerically that individual's preferences over a set of gambles with given probabilities. The way in which an individual utility may vary against, for example, money is illustrated in Figure 4.1.

Suppose in some gamble you win $20 with probability $^4/_{10}$ and lose $10 with probability $^6/_{10}$. If you assign a utility of -50 to the loss of $10, then it will not pay you to engage in this gamble if your utility for a gain of $20 is less than 75, since $0.4(75) + 0.6(-50) = 0$. The objection might be raised at this point that the assignment of utilities to outcomes was completely arbitrary, but interestingly enough this objection can, at least to some extent, be met, as we shall see. The importance of this to game theory is clear. Since the payoffs form an integral part of the definition of the game, the game remains undefined until the payoff magnitudes in terms of personal utilities are assigned by the players to the outcomes.

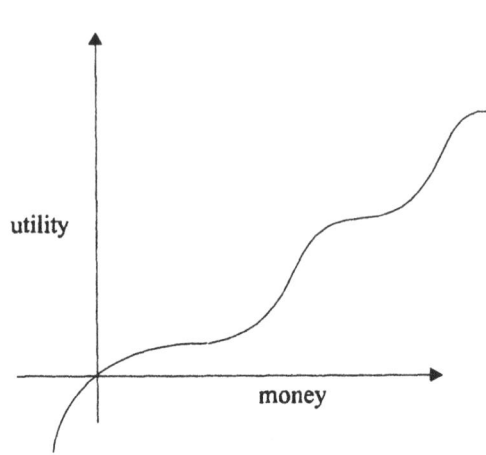

Figure 4.1 Utility for money need not be linear and may vary from one individual to another.

To fix our ideas let us suppose that $A_1,..., A_r$ are prizes. We can define a lottery L over the basic prizes $A_1, ..., A_r$ to be an r-tuple $(p_1A_1, ..., p_rA_r)$ where $p_i \geq 0$ and $\Sigma p_i = 1$. We interpret this to mean that one and only one prize will be won, and the probability that it will be A_i is p_i. We consider an individual's preference between pairs of lotteries L, L' over the same basic set of prizes. If L is preferred or viewed equal to L' we write $L \geq L'$. We seek to assign a numerical utility $u(L)$ to each lottery so that $u(L) \geq u(L')$ if and only if $L \geq L'$. If only a finite number of alternative lotteries were involved there would be far too many different ways to assign a utility function, but if we insist that utility accurately reflect the individuals preference of lotteries over *all possible* probability r-tuples $(p_1, ..., p_r)$ we are then dealing with an infinite set, and by its very size there will be many more constraints on the utility function. Roughly speaking, von Neumann and Morgenstern were able to show that if a person is able to express preferences between every possible pair of gambles (lotteries), where the gambles are taken over some basic set of alternatives (prizes), then provided that there is an element of consistency in these preferences, one *can* assign numerical utilities to the basic alternatives in such a way that if the person is guided solely by the expected utility value of a gamble then that person will be acting in accord with his or her true tastes. The element of consistency required is that the preference relation over lotteries be transitive, namely if $L \geq L'$ and $L' \geq L''$, then $L \geq L''$.

The pivotal step in this construction is the idea of combining two lotteries. Given two lotteries L, L' over a basic set of prizes $A_1, ..., A_r$ we can define a new lottery $qL \oplus (1 - q)L'$, where $0 < q < 1$, over the same set of prizes, in the following way. We interpret $qL \oplus (1 - q)L'$ to mean that one and only one of the lotteries L, L' will be the prize, and the probability it will be L is q. If utility is to behave as we would like, then one of our requirements should be that

$$u(q L \oplus (1 - q)L') = qu(L) + (1 - q)u(L') . \tag{4.1}$$

This condition is said to characterise a **linear** utility function. (One should not confuse this notion with the idea that the graph in Figure 4.1 is a straight line, for the two are altogether distinct.) It is not our purpose here to give an axiomatic development of utility theory, for despite its significance to our theme it is not strictly part of game theory. Suffice it to say that on the basis of a few simple axioms one can establish the existence of a function $u(L)$ which satisfies (4.1) and is in fact given by

$$u(L) = p_1 u_1 + \ldots + p_r u_r , \tag{4.2}$$

where $L = (p_1 A_1, \ldots, p_r A_r)$ and $u_1, \ldots, u_r$ are numbers which can be interpreted as the numerical utilities of the prizes $A_1, \ldots, A_r$. Moreover $u(L) \geq u(L')$ if and only if $L \geq L'$ as required. Sometimes (4.1) is referred to as the *expected utility hypothesis* since it amounts to the assertion (4.2) that the utility of a lottery is equal to the expected utility of its component prizes. A more detailed discussion of axiomatic utility from a game theoretic viewpoint can be found in Luce and Raiffa [1].

Given that a subjects preferences can be represented by a linear utility function, then the subject behaves *as if he or she were a maximiser of expected utility.*

It is also true in risky situations that any order-preserving transformation of a utility function is again a utility function, but such a transformation of a *linear* utility function does not generally result in a *linear* utility function. If u is a linear utility function and a, b are real numbers with $a > 0$, then $u' = au + b$ is again a linear utility function over the same set. Conversely, if u and u' are two linear utility functions for a preference relation over the same set of alternatives, then there exist constants a, b with $a > 0$ such that $u' = au + b$.

The fact that a linear utility function is defined only up to a linear transformation leads to the problem of interpersonal comparisons of utility when there is more than one person in the situation. (As one writer put it "we must not equate a childs pleasure in childishness with an adults pleasure in adultery".) At present this problem is unresolved, which leaves us with the options of assuming that such comparisons are possible, which creates a possible flaw in the theory, or attempting to devise theories in which comparisons are not made.

One should remark that the general theory of utility is not confined to a finite basic set of alternatives as in the simplified description given above. Moreover it sohould be stressed that a certain caution must be maintained in interpreting the concept of utility: one alternative possesses a larger utility than another because it is preferred, not the other way round.

We next turn our attention to the *scale* by which measurements are made. Ther are three basic scales possible, an *ordinal* scale, an *interval* scale and a *ratio* scale. Thus temperature was originally a number only up to a positive monotone transformation and so determined on an ordinal scale. With the development of thermometry the transformations were restricted to the linear ones, that is, only the absolute zero and the absolute unit were missing, which determined temperature on an interval scale. Subsequent developments in thermodynamics even fixed the absolute zero, so that the admissible transformations are only multiplying by a positive constant, and this determines temperature on a ratio scale.

The (consistent) preference relation $\geq$ discussed above is determined only on an ordinal scale. This might suffice for a player to decide the best pure strategy if the game in question happened to have a saddle point, but if the concept of mixed strategy is to be meaningfully applied more is needed. In fact, we need individual utility to be determined at least on an interval scale. Fortunately, however, we have seen that such a numerical description of individual utility is indeed possible, on the basis of certain axioms, so that with these observations made we may reasonably press ahead.

To summarise: utility is that quantity whose expected gain in a risky situation is attempted to be maximised by a decision maker all of whose choices among risky outcomes are consistent. If payoffs in a game are in individual utilities then no conclusion of the theory should be affected if each outcome utility P_i, to the ith player, is replaced by $aP_i + b$, where a, b are real numbers with $a > 0$. Moreover we can choose a and b different for each player in the game. The constant a can be interpretetd as changing the unit of utility, and b as a fixed bonus or fee which the player receives or pays regardless of the outcome of the game.

We single out two special cases of the general problem:

(i) Payoffs are in utility terms, no interpersonal comparisons of utility are permitted, and no side payments are allowed.
(ii) Payoffs are in monetary terms, utility is linear in money, interpersonal comparisons are meaningful and monetary side payments are allowed.

The general theory of non-cooperative games described in Chapter 2 applies even to the very restrictive first case if payoffs are interpreted as personal utilities. However, in this situation one is no longer able to draw a distinction between zero sum and non-zero sum games, since the very definition of a zero sum game requires that payoffs are in some comparable unit.

In this chapter we propose to deal with cooperative games in the context of case (ii) and follow to a large extent the treatment of n-person cooperative game theory developed by von Neumann and Morgenstern. Such a game is called a **constant sum** game if there exists some constant c such that

$$\sum_{i=1}^{n} P_i = c$$

at every terminal state. If we charge each player a fee of $-c/n$ to enter the game we can replace the original game by a strategically equivalent zero sum game. Therefore in the discussion which follows 'constant sum' and 'zero sum' can be regarded as equivalent terms.

4.2 CHARACTERISTIC FUNCTIONS.

Just as in n-person non-cooperative game theory, we can consider a cooperative game as presented in extensive or normal forms. The practical problems of either of these are obvious;

to present a finite n-person game in normal form requires an n-dimensional matrix. Moreover, in a cooperative game new considerations arise. We are especially interested in which coalitions are likely to form. Since payoffs are assumed to be in monetary form we can take it that coalitions will by and large act, by coordinating strategies, to maximise their joint payoff. Because agreements are binding we assume that coalitions once formed, by whatever bargaining process, remain stable for the duration of the game.

Although the information concerning which coalitions are likely to form can be recovered from the extensive or normal form, it is obviously more desirable to have it available explicitly. The stage is now set for us to move to the next level of abstraction in game theory: the *characteristic function form of a game*. Let Γ be an n-person cooperative game with a set of players $I = \{1, 2, ..., n\}$. Any subset $S \subseteq I$ will be called a **coalition**. We allow the possibility of one person coalitions $\{i\}$, $i \in I$, and for convenience, even the no-person coalition ϕ is permitted. The set of all subsets of I, that is, the set of all possible coalitions, is denoted by $\wp(I)$. The **characteristic function** v of the game Γ is a function

$$v : \wp(I) \to \mathbb{R}$$

such that $v(S)$ represents the largest joint payoff which the coalition S is *guaranteed* to obtain if the members coordinate their strategies by preplay agreement. We define $v(\phi) = 0$.

How should we interpret the word 'guaranteed'? The following interpretation is offered by von Neumann and Morgenstern.

> The worst possible situation which a coalition S might face is that the remaining players form a counter-coalition $I \setminus S$ which then proceeds to minimise the joint gain of S.   (4.3)

The technical advantages of this interpretation are clear. It reduces the problem of finding $v(S)$, which is supposed to be a measure of the strength of the coalition S, to finding the value of a *zero sum* non-cooperative game between S and ΛS, a relatively simple precedure. Moreover, if the original game is constant sum then for ΛS to minimise the joint gain of S is precisely equivalent to ΛS maximising its own joint gain.

However, if the game is not constant sum there are serious objections to this interpretation. We have characterised players as maximisers of utility, in which case surely it is inconsistent to suppose that ΛS will seek to minimise the gain of S at *whatever cost to itself*? Although this supposition unquestionably leads to the most conservative estimate of $v(S)$ it will inevitably be a wildly pessimistic estimate of the strength of S in many cases. The following example (McKinsey [2]) illustrates that problems of this kind can arise even in the case $n = 2$. In this game player 1 has but one strategy and player 2 has two pure strategies and the payoff bimatrix is

$$\big((\; 0, \; -1000) \; (\; 10, \; 0) \big)$$

Even though player 1 has no strategy choice it is intuitively clear that player 1 is actually in a stronger position than player 2; for if player 2 seeks to maximise his or her own gain the result will be (10, 0), in which case player 1 gets 10. On the other hand, considering the cost of 1000, is player 2 likely to play so as to minimise player 1s gain? Moreover if, in preplay negotiation, player 2 threatens to implement this strategy unless player 1 agrees to form the coalition {1, 2} and share some of the joint gain of 10, is player 1 likely to take this threat seriously? Given a choice would you prefer to take the role of player 1 or player 2?

If we accept that player 1 is in a better position than player 2, then any 'solution' of the game should reflect this asymmetry. This is not true, however, of the solution in the sense of von Neumann and Morgenstern for, as we shall see, the solution is defined entirely in terms of the characteristic function, and it happens that the characteristic function of this game is

$$v(\phi) = 0, \quad v(\{1\}) = v(\{2\}) = 0, \quad v(\{1, 2\}) = 10 ,$$

which is symmetric in 1 and 2. For these reasons it is difficult to accept that n-person cooperative game theory as it now stands, at least in the non-constant sum case, is a reasonable model of coalition formation between rational players.

Perhaps an alternative procedure for evaluating the strength of a coalition S might be:

> Consider all possible coalition structures involving the coalition S. For each structure evaluate the minimum expectation of S in the non-cooperative game involving $S = S_1, ..., S_k$ (say). Then take the minimum of these numbers over all the coalition structures involving S.

This suggestion has the merit that each of the 'players' $S_1, ..., S_k$, in each subgame considered, will be regarded as a maximiser of utility. In the light of the existing state of the theory of non-cooperative games, however, we must note the disadvantages. For the non-cooperative non-zero sum game our only solution concept is the equilbrium point, which is not satisfactory in every respect. Moreover, there is at present no convenient algorithm for computing equilibria if $k \geq 3$.

We shall, therefore, return to the conservative assumption (4.3) as a means of calculating $v(S)$. This leads to the formula

$$v(S) = \max_{x \in X_S} \min_{y \in X_{I\setminus S}} \sum_{i \in S} P_i(x, y) , \qquad (4.4)$$

where X_S, $X_{\wedge S}$ denote the sets of coordinated mixed strategies for coalitions S, $\wedge S$ respectively and $P_i(x, y)$ denotes the expected payoff to player i when the mixed strategies $x \in X_S, y \in X_\wedge$ are employed.

The following examples demonstrate the process of calculating the characteristic function of a given game.

Example 4.1 (Left-Right). This is a 3-person game of perfect information in which each player chooses 'Left' (= L) or 'Right' (= R). The extensive form is given below.

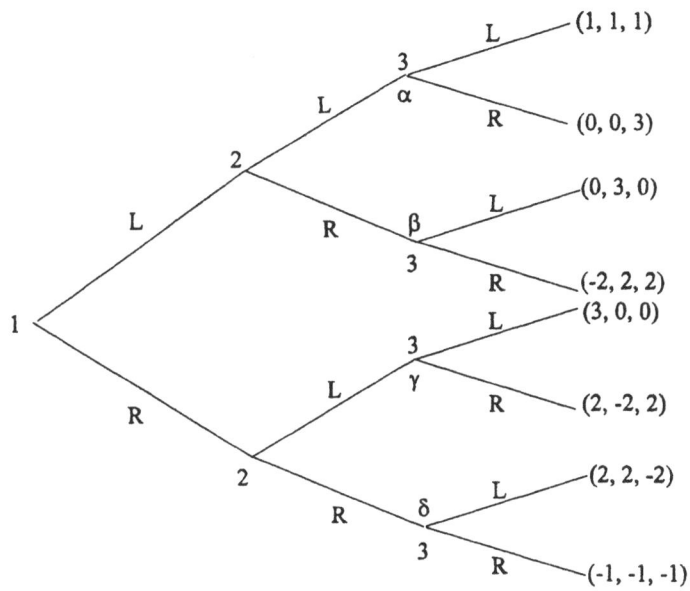

Thus the players have 2, 4 and 16 pure strategies respectively. Note that the game is *not* constant sum. Treated as a non-cooperative game of perfect information Left-Right has a pure strategy equilibrium point corresponding to payoffs (-1, -1, -1).

To calculate the characteristic function we observe that each of the subgames S versus ΛS is a zero sum game of perfect information and hence is soluble in pure strategies. For example if $S = \{1, 2\}$, $\Lambda S = \{3\}$ we enter the *total* payoff to $\{1, 2\}$ into the matrix *and assume* $\{3\}$ *will seek to minimise this.*

Now $\{3\}$ has 16 pure strategies each of the form 'If at α choose L, if at β choose L, if at γ choose L, if at δ choose L.', where any L can be replaced by an R. If we label these strategies by LLLL etc. and those of $\{1, 2\}$ by LL, meaning 1 chooses L and 2 chooses L etc., we obtain Table 4.1. Hence $v(\{1, 2\}) = 0$.

In the 16×4 subgame $\{3\}$ v $\{1, 2\}$ in which $\{1, 2\}$ seeks to minimise $\{3\}$s gains we find $v(\{3\}) = -1$. The remaining subgames are

$$\{1\} \text{ v } \{2, 3\} \text{ which is } 2 \times 16 \text{ and gives } v(\{1\}) = -1,$$
$$\{2, 3\} \text{ v } \{1\} \text{ which is } 16 \times 2 \text{ and gives } v(\{2, 3\}) = 0,$$
$$\{2\} \text{ v } \{1, 3\} \text{ which is } 4 \times 8 \text{ and gives } v(\{2\}) = -1,$$
$$\{1, 3\} \text{ v } \{2\} \text{ which is } 8 \times 4 \text{ and gives } v(\{1, 3\}) = 0.$$

Table 4.1

		LLLL	LLLR	LLRL	LLRR	LRLL	LRLR	LRRL	LRRR	RLLL	RLLR	RLRL	RLRR	RRLL	RRLR	RRRL	RRRR
	LL	2	2	2	2	2	2	2	2	0	0	0	0	0	0	0	0‡
{1,2}	LR	3	3	3	3	3	0	0	0	3	3	3	3	0	0	0	0‡
	RL	3	3	0	0	3	3	0	0	3	3	0	0	3	3	0	0‡
	RR	4	-2	4	-2	4	-2	4	-2	4	-2	4	-2	4	-2	4	-2

‡ = Saddle point.

Table 4.2

	LLLL	LLLR	LLRL	LLRR	LRLL	LRLR	LRRL	LRRR	RLLL	RLLR	RLRL	RLRR	RRLL	RRLR	RRRL	RRRR
LL	(2,1)	(2,1)	(2,1)	(2,1)	(2,1)	(2,1)	(2,1)	(2,1)	(0,3)	(0,3)	(0,3)	(0,3)	(0,3)	(0,3)	(0,3)	(0,3)‡
LR	(3,0)	(3,0)	(3,0)	(3,0)	(0,2)	(0,2)	(0,2)	(0,2)	(3,0)	(3,0)	(3,0)	(3,0)	(0,2)	(0,2)	(0,2)	(0,2)‡
RL	(3,0)	(3,0)	(0,2)	(0,2)	(3,0)	(3,0)	(0,2)	(0,2)	(3,0)	(3,0)	(0,2)	(0,2)	(3,0)	(3,0)	(0,2)	(0,2)‡
RR	(4,-2)	(4,-2)	(4,-2)	(4,-2)	(4,-2)	(4,-2)	(4,-2)	(-2,-1)	(4,-2)	(-2,-1)	(4,-2)	(-2,-1)	(4,-2)	(-2,-1)	(4,-2)	(-2,-1)

‡ = Equilibrium point.

Thus

$$v(\phi) = 0, \quad v(\{1, 2, 3\}) = 3,$$
$$v(\{1\}) = v(\{2\}) = v(\{3\}) = -1,$$
$$v(\{1, 2\}) = v(\{1, 3\}) = v(\{2, 3\}) = 0. \tag{4.5}$$

As a matter of interest let us calculate the alternative evaluation of the 'worth' of S, the function $w(S)$, based on the assumption that each coalition will seek to maximise its own joint gain. In this case the above game $\{1, 2\}$ versus $\{3\}$ becomes the non-constant sum game exhibited in Table 4.2 and we find that $w(\{1, 2\}) = 0$ as before but that $w(\{3\}) = 2$, as distinct from -1 in the previous case. (Considerations of mixed strategy equilibria do not in fact change these values.) Using this procedure it is only necessary to consider three subgames (although if n is large it will be necessary to consider vastly *more* subgames to compute w, which takes account of all possible coalition structures, than to compute v). We find that

$$w(\phi) = 0, \quad w(\{1, 2, 3\}) = 3,$$
$$w(\{1\}) = w(\{2\}) = w(\{3\}) = 2,$$
$$w(\{1, 2\}) = w(\{1, 3\}) = w(\{2, 3\}) = 0. \tag{4.6}$$

The analysis of Left-Right is particularly simple because the game has perfect information. In general one must bear in mind that although members of S are supposed to have formed a coalition by preplay negotiation, nevertheless, once play begins they are bound by the rules of the non-cooperative game and so cannot share information except as allowed by these rules.

A more usual procedure is to calculate the characteristic function from the normal form of the game.

Example 4.2 Consider the 3-person zero sum game in which each player chooses A or B. The pay-off 'matrix' is given below.

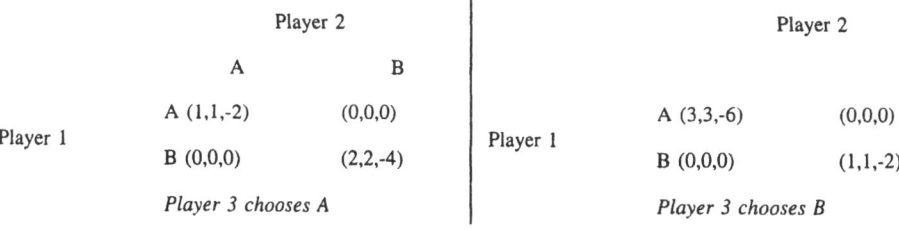

	Player 2			Player 2	
	A	B		A	B
Player 1	A (1,1,-2)	(0,0,0)	Player 1	A (3,3,-6)	(0,0,0)
	B (0,0,0)	(2,2,-4)		B (0,0,0)	(1,1,-2)
	Player 3 chooses A			*Player 3 chooses B*	

Since this game is zero sum there are only three subgames to consider, in order to calculate the characteristic function, as opposed to six.

			{3}		
			$2/3$	$1/3$	
			A	B	{3}'s losses
	$1/3$	AA	2	6	$10/3$
{1, 2}	0	AB	0	0	0
	0	BA	0	0	0
	$2/3$	BB	4	2	$10/3$
	{1, 2}'s gains		$10/3$	$10/3$	value=$10/3$

Thus $v(\{1, 2\}) = {}^{10}/_3$, $v(\{3\}) = -{}^{10}/_3$.

			{2}		
			$1/2$	$1/2$	
			A	B	{2}'s losses
	$1/2$	AA	-1	0	$-1/2$
{1, 3}	0	AB	-3	0	$-3/2$
	0	BA	0	-2	$-2/2$
	$1/2$	BB	0	-1	$-1/2$
	{1, 3}'s gains		$-1/2$	$-1/2$	value=$-1/2$

Thus $v(\{1, 3\}) = -{}^{1}/_2$, $v(\{2\}) = {}^{1}/_2$.

			{1}		
			$1/2$	$1/2$	
			A	B	{1}'s losses
	$1/2$	AA	-1	0	$-1/2$
{2, 3}	0	AB	-3	0	$-3/2$
	0	BA	0	-2	$-2/2$
	$1/2$	BB	0	-1	$-1/2$
	{2, 3}'s gains		$-1/2$	$-1/2$	value=$-1/2$

Thus $v(\{2, 3\}) = -{}^{1}/_{2}$, $v(\{1\}) = {}^{1}/_{2}$. Consequently the characteristic function is

$$
\begin{aligned}
v(\phi) &= v(\{1, 2, 3\}) = 0 \\
v(\{1\}) &= v(\{2\}) = \frac{1}{2}, \quad v(\{3\}) = -\frac{10}{3}, \\
v(\{1, 3\}) &= v(\{2, 3\}) = -\frac{1}{2}, \quad v(\{1, 2\}) = \frac{10}{3}
\end{aligned}
\qquad (4.7)
$$

It should be clear that the set of *coordinated* mixed strategies available to a coalition T is a much larger set than the product of the members mixed strategy sets. In the notation of Chapter 2, if S_T denotes the set of coordinated pure strategies available to T then

$$
S_T = \prod_{i \in T} S_{\{i\}} \;.
$$

If $[S_T]$ denotes the convex hull of S_T viewed as a subset of Euclidean space, and X_T denotes the set of mixed strategies for T, then $X_T = [S_T]$ and

$$
\prod_{i \in T} X_{\{i\}} = \prod_{i \in T} [S_{\{i\}}] \subseteq \left[\prod_{i \in T} S_{\{i\}} \right] = [S_T] = X_T \;.
$$

As an example of how much larger X_T is than $\Pi X_{\{i\}}$ let T consist of three players each of whom has four pure strategies. Denote a pure strategy for T by a triple (i, j, k), $1 \le i, j, k \le 4$, where i denotes the pure strategy used by the first player etc. They could choose to put probability ${}^{1}/_{4}$ on each of (i, i, i) $(1 \le i \le 4)$ and zero on all the other sixty possibilities. Such a joint strategy is impossible without coordination; however, the choice is easily within the grasp of the coalition.

The following property, possessed by any characteristic function, is called **superadditivity**.

Theorem 4.1. For any finite cooperative game Γ

$$
v(S \cup T) \ge v(S) + v(T) \quad \text{for } S, T \subseteq I \text{ and } S \cap T = \phi \qquad (4.8)
$$

Proof. From (4.4) we have

$$
v(S \cup T) = \max_{x \in X_{S \cup T}} \;\; \min_{y \in X_{I \setminus (S \cup T)}} \;\; \sum_{i \in S \cup T} P_i(x, y) \;,
$$

where $X_{S \cup T}$ denotes the set of coordinated mixed strategies for the coalition $S \cup T$ etc. If we restrict our attention to *independent* mixed strategies $\alpha \in X_S$, $\beta \in X_T$ the range of maximisation will decrease and so the value of the maximum above can only decrease. Hence

$$
v(S \cup T) \ge \max_{\alpha \in X_S} \;\; \max_{\beta \in X_T} \;\; \min_{y \in X_{I \setminus (S \cup T)}} \;\; \sum_{i \in S \cup T} P_i(\alpha, \beta, y) \;.
$$

Hence for each $\alpha \in X_S, \beta \in X_T$

$$v(S \cup T) \geq \min_{y \in X_{I\backslash(S\cup T)}} \sum_{i \in S\cup T} P_i(\alpha, \beta, y)$$

$$\geq \min_{y \in X_{I\backslash(S\cup T)}} \left(\sum_{i \in S} P_i(\alpha, \beta, y) + \sum_{i \in T} P_i(\alpha, \beta, y) \right),$$

since $S \cap T = \phi$. Hence

$$v(S \cup T) \geq \min_{y \in X_{I\backslash(S\cup T)}} \sum_{i \in S} P_i(\alpha, \beta, y) + \min_{y \in X_{I\backslash(S\cup T)}} \sum_{i \in T} P_i(\alpha, \beta, y)$$

for all $\alpha \in X_S, \beta \in X_T$. Thus

$$v(S \cup T) \geq \min_{\beta \in X_T} \min_{y \in X_{I\backslash(S\cup T)}} \sum_{i \in S} P_i(\alpha, \beta, y) + \min_{\alpha \in X_S} \min_{y \in X_{I\backslash(S\cup T)}} \sum_{i \in T} P_i(\alpha, \beta, y).$$

In the first term on the right the minimum is taken with respect to mixed strategies $\beta \in X_T$, $y \in X_{I\backslash(S\cup T)}$. The pair $(\beta, y) \in X_T \times X_{I\backslash(S\cup T)}$ defines a mixed strategy in $X_{I\backslash S}$, since

$$X_T \times X_{I\backslash(S\cup T)} = [S_T] \times [S_{I\backslash(S\cup T)}] \subseteq [S_{I\backslash S}] = X_{I\backslash S}.$$

If instead we minimise with respect to an arbitrary mixed strategy in $X_{I\backslash S}$, the range of minimisation will increase and so the value of this minimum can only decrease. A similar remark applies to the second term on the right if the minimum is taken with respect to an arbitrary mixed strategy in $X_{I\backslash T}$. Thus

$$v(S \cup T) \geq \min_{\gamma \in X_{I\backslash S}} \sum_{i \in S} P_i(\alpha, \gamma) + \min_{\delta \in x_{i\backslash T}} \sum_{i \in T} P_i(\beta, \delta)$$

for all $\alpha \in X_S, \beta \in X_T$. Hence

$$v(S \cup T) \geq \max_{\alpha \in X_S} \min_{\gamma \in X_{i\backslash S}} \sum_{i \in S} P_i(\alpha, \gamma) + \max_{\beta \in X_T} \min_{\delta \in X_{i\backslash T}} P_i(\beta, \delta),$$

whence from (4.4)

$$v(S \cup T) \geq v(S) + v(T)$$

as required.

Note that the functions in (4.5) and (4.7) are both superadditive. It is important to realise that (4.8) is a direct consequence of the pessimistic assumption (4.3) regarding the behaviour of counter-coalitions. The superadditivity condition reflects the fact that in a *hostile* universe 'unity is strength'. However, observe that the function w in (4.6), evaluated on the optimistic assumption that coalitions once formed will pursue their own interests, is *not* superadditive. The difference $w(S) - v(S)$ is a measure of the threat potential of the remaining players against S. Lastly the characteristic function of (4.7), a zero sum game, demonstrates that superadditivity does not necessarily imply the inevitable formation of the grand coalition I. For example it would be difficult to convince the coalition $\{1, 2\}$ of the advantages of admitting player 3 to membership!

The weakest form of superadditivity is **additivity**, that is

$$v(S \cup T) = v(S) + v(T) \text{ for } S, T \subseteq I \text{ and } S \cap T = \phi . \tag{4.9}$$

In such a game there is plainly no positive incentive for coalitions involving more than one player to form.

Definition. A cooperative game with an additive characteristic function is called **inessential**. Other cooperative games are called **essential**.

Theorem 4.2. A finite n-person cooperative game Γ is inessential if and only if

$$\sum_{i \in I} v(\{i\}) = v(I) \tag{4.10}$$

Proof. Plainly (4.9) implies (4.10). It remains to show that (4.10) is sufficient to prove (4.9). From Theorem 4.1 we have

$$v(S \cup T) + v(I \backslash (S \cup T)) \le v(I) \tag{4.11}$$

From (4.10)

$$v(I) = \sum_{i \in I} v(\{i\}) = \sum_{i \in S} v(\{i\}) + \sum_{i \in T} v(\{i\}) + \sum_{i \in I \backslash (S \cup T)} v(\{i\})$$

$$V(I) \le v(S) + v(T) + v(I \backslash (S \cup T)) \tag{4.12}$$

again by Theorem 4.1. If we now combine the inequalities (4.11) and (4.12) and again use Theorem 4.1 we obtain (4.9).

For *constant sum* games in addition to the superadditivity of the characteristic function we have the further property known as **complementarity**.

Theorem 4.3. For any finite constant sum cooperative game Γ

$$v(S) + v(I \setminus S) = v(I) \text{ for every } S \subseteq I. \tag{4.13}$$

The converse is false; that is, there are non-constant sum games which also satisfy (4.13) (see Problem 1(b)). The theorem is false if the game Γ is not required to be finite.

Proof. For any constant sum cooperative game

$$v(I) = \sum_{i \in I} P_i(x_1, \ldots, x_n) = c$$

for every mixed strategy n-tuple $(x_1, ..., x_n)$. Hence from (4.4)

$$
\begin{aligned}
v(S) &= \max_{x \in X_S} \min_{y \in X_{I \setminus S}} \sum_{i \in S} P_i(x, y) \\
&= \max_{x \in X_S} \min_{y \in X_{I \setminus S}} \left(c - \sum_{i \in I \setminus S} P_i(x, y) \right) \\
&= c - \min_{x \in X_S} \max_{y \in X_{I \setminus S}} \sum_{i \in I \setminus S} P_i(x, y) \\
&= c - \max_{y \in X_{I \setminus S}} \min_{x \in X_S} \sum_{i \in I \setminus S} P_i(x, y) \quad \text{(by Theorem 2.6)} \\
&= c - v(I \setminus S),
\end{aligned}
$$

as required.

4.3 IMPUTATIONS.

The single aspect of cooperative games which we have so far considered is the problem of estimating the relative strength of coalitions. Distinct from this, though presumably influenced by it, is the question of the payments that the individual players finally receive. Non-cooperative games are strategic games in the sense that an outcome arises as a result of the actions of the players, and this determines their payoffs. On the other hand an outcome of a cooperative game is a distribution of the total available payoff amongst the players rather than as a predetermined consequence of their strategic chioces. A possible distribution of available payoff is called an **imputation**.

In an n-person cooperative game Γ we can represent an imputation by a vector $\mathbf{x} = (x_1, ..., x_n)$, where x_i denotes the amount received by player i. The imputations which can arise as a result of coalition formation are obviously not arbitrary, so we should consider what conditions it is reasonable to impose on any imputation.

Firstly no member of a coalition will consent to receive less than he or she can obtain individually by acting independently of the other players. We therefore require, on grounds of *individual rationality*, that

$$x_i \geq v(\{i\}) \quad \text{for all } i \in I \ . \tag{4.14}$$

Secondly it is plausible that

$$\sum_{i \in I} x_i = v(I) \tag{4.15}$$

For if

$$\sum_{i \in I} x_i < v(I)$$

then each player can be made to gain without loss to the others. For example, if the difference were equidistributed each can be made to gain the amount

$$\frac{1}{n} \left(v(I) - \sum_{i \in I} x_i \right) .$$

On the other hand, because $v(I)$ represents the most that the players can get from the game by forming the grand coalition I, it is impossible that

$$\sum_{i \in I} x_i > v(I) \ .$$

Hence on the grounds of *collective rationality* we require that (4.15) holds for any imputation.

Note that (4.15) is automatically satisfied for constant sum games. However, in a non-constant sum game it would seem that (4.15) should only necessarily hold if the grand coalition I is formed; that is the players reach complete agreement in preplay negotiation. Fortunately, it is quite possible to develop solution theories which do not assume (4.15) [3] but for the time being we shall accept this condition and develop the classical theory. Nevertheless we are on dangerous ground for, as we now see, the assumption (4.15) tempts us to make an even stronger assumption which leads to serious trouble.

The argument leading to collective rationality is an attempt to extend the assumption of individual rationality to groups of players; however, the notion of group rationality is neither a postulate of the model nor does it sem to be a logical consequence of individual rationality. If we accept the reasoning which leads to (4.15) why should the argument apply only to the coalition I and not equally to every coalition S? That is, should we not also impose the condition

$$\sum_{i \in S} x_i \geq v(S) \quad \text{for every } S \subseteq I \ . \tag{4.16}$$

It turns out that there are good reasons for not doing this, basically because (4.16) is far too strong a condition. The imputations which satisfy (4.16) are those which no coalition whatsoever can object to. Certainly any coalition which *actually forms* is in a position to ensure itself at least $v(S)$, but is it reasonable to require the inequality (4.16) to hold for *all* coalitions potential as well as actual? For if two coalitions S and ΛS are bargaining over possible imputations it is quite possible they will at least consider an imputation x with the property

$$\sum_{i \in T} x_i < v(T)$$

for some other coalition T.

Another more pragmatic reason for rejecting (4.16) is that, as we prove in Theorem 4.11, if Γ is an essential constant sum game there are *no* imputations which satisfy (4.16).

We are now in a position to define a cooperative game, or rather a restricted class of cooperative games.

Definition. A **classical cooperative** game Γ is a collection $\langle I, v, X \rangle$ in which I denotes the set of players, v is a function $v: \wp(I) \to \mathbb{R}$ which satisfies $v(\phi) = 0$ and is superadditive, and X denotes a set of imputations which satisfy (4.14) and (4.15).

The classical cooperative game of von Neumann and Morgenstern can be generalised in various directions, for example by dropping the requirement that v be superadditive (that is, weakening (4.3)) and omitting condition (4.15) on the imputations.

Theorem 4.4. A vector $\mathbf{x} \in \mathbb{R}^n$ is an imputation in a classical n-person cooperative game $\langle I, v, X \rangle$ if and only if there exist numbers a_i, $i \in I$, such that

$$x_i = v(\{i\}) + a_i \qquad (a_i \geq 0) \qquad (i \in I) , \tag{4.17}$$

and

$$\sum_{i \in I} a_i = v(I) - \sum_{i \in I} v(\{i\}) . \tag{4.18}$$

Proof. Assume first that such a_i exist. Then since $a_i \geq 0$, (4.17) gives (4.18). From (4.17) and (4.18) we have

$$\sum_{i \in I} (x_i - v(\{i\})) = v(I) - \sum_{i \in I} v(\{i\}) ,$$

that is

$$\sum_{i \in I} x_i = v(I) ,$$

which is (4.15). Since x satisfies both (4.14) and (4.15) it is an imputation.

Suppose on the other hand that **x** is an imputation, and put

$$a_i = x_i - v(\{i\}) . \tag{4.19}$$

From (4.14) we have $a_i \geq 0$ ($i \in I$), so it remains to verify (4.18). Summing (4.19) over $i \in I$ we obtain

$$\sum_{i \in I} a_i = \sum_{i \in I} x_i - \sum_{i \in I} v(\{i\}) ,$$

which on using (4.15) yields (4.18).

Theorem 4.5. Any inessential game has the unique imputation

$$\mathbf{x} = (v(\{1\}), v(\{2\}), \ldots, v(\{n\})) . \tag{4.20}$$

On the other hand any essential game with at least two players possesses infinitely many imputations.

Proof. Using the previous theorem, write the imputations in the form

$$(v(\{1\}) + a_1 , \ldots, v(\{n\}) + a_n) .$$

If the game is inessential, (4.10) gives

$$\sum_{i \in I} v(\{i\}) = v(I)$$

and substituting this into (4.18) we obtain

$$\sum_{i \in I} a_i = 0 .$$

However, since $a_i \geq 0$ this last equation implies $a_i = 0$ for every $i \in I$ as required. Alternatively, if the game is essential then

$$\sum_{i \in I} a_i = v(I) - \sum_{i \in I} v(\{i\}) > 0 ,$$

and this positive number can be written as a sum of $a_i \geq 0$ in infinitely many ways.

4.4 STRATEGIC EQUIVALENCE.

Bearing in mind that any conclusions of our theory should be invariant under a change of scale of utility or if each player is paid a bonus or charged a fee independent of the outcome of the game, we make the following

Definition. Two *n*-person cooperative games with characteristic functions v and v' defined over the same set of players are **strategically equivalent** if there exists $k > 0$ and numbers c_i, $i \in I$, such that

$$v'(S) = kv(S) + \sum_{i \in S} c_i \text{ for every } S \subseteq I . \qquad (4.21)$$

Theorem 4.6. Strategic equivalence is an equivalence relation.

Proof. We write $v \sim v'$ for the relation of strategic equivalence. To prove this theorem it is only necessary to verify the three defining properties of an equivalence relation.

(i) *Reflexivity*: $v \sim v$. Put $k = 1$ and $c_i = 0$ for $i \in I$ in (4.21).

(ii) *Symmetry*: If $v \sim v'$ then $v' \sim v$. Suppose (4.21) holds, then solving for $v(S)$ we obtain

$$v(S) = \frac{1}{k} v'(S) + \sum_{i \in S} \left(-\frac{c_i}{k} \right) .$$

If we now set $k' = 1/k$, $c_i = -c_i/k$ we obtain

$$v(S) = k' v'(S) + \sum_{i \in S} c_i' \qquad (k' > 0) .$$

that is, $v' \sim v$.

(iii) *Transitivity* :If $v \sim v'$ and $v' \sim v''$, then $v \sim v''$. Since $v' \sim v''$ we have

$$v''(S) = k' v'(S) + \sum_{i \in S} c_i' \qquad (k' > 0) .$$

Substituting for $v'(S)$ from (4.21) we obtain

$$v''(S) = kk' v(S) + \sum_{i \in S} (c_i' + k' c_i) ,$$

and setting $k'' = kk'$, $c_i'' = c_i' + k'c_i$ we obtain

$$v''(S) = k'' v(S) + \sum_{i \in S} c_i'' \qquad (k'' > 0) ,$$

that is $v \sim v''$.

A basic result on equivalence relations is that any equivalence relation on some set S induces a partition of the set into disjoint classes, whose union is S, in such a way that any two equivalent elements of S are in the same class. These classes are called **equivalence classes**.

Thus Theorem 4.6 implies that the set of characteristic functions, for a fixed set of players I, can be uniquely subdivided into mutually disjoint classes of strategic equivalence.

Moreover the relation of strategic equivalence between the games (and their characteristic functions) induces an equivalence between imputations, for if $v \sim v'$ then

$$v'(S) = kv(S) + \sum_{I \in S} c_i \quad (k > 0) ,$$

and if $x = (x_1, ..., x_n)$ is an imputation under the characteristic function v, then consider the vector $x' = (x_1', ..., x_n')$, where

$$x_i' = kx_i + c_i .$$

We have, by (4.14),

$$x_i' = kx_i + c_i \geq kv(\{i\}) + c_i = v'(\{i\}) ,$$

that is, the condition of individual rationality is satisfied. Also, by (4.15),

$$\sum_{i \in I} x_i' = \sum_{i \in I} (kx_i + c_i) = k \sum_{i \in I} x_i + \sum_{i \in I} c_i = kv(I) + \sum_{i \in I} c_i = v'(I) .$$

Thus the condition of collective rationality is also satisfied. Hence x' is an imputation under the characteristic function v'. We say x' **corresponds** to x under the strategic equivalence $v' \sim v$.

The idea of strategic equivalence greatly simplifies our study of cooperative games since, for each fixed I, it is now only necessary to study *one* game from each class.

Definition. A cooperative game with a set of players I is called a **zero game** if its characteristic function is identically zero.

Zero games are rather uninteresting!

Theorem 4.7. Any inessential cooperative game is strategically equivalent to a zero game.

Proof. A game is inessential if its characteristic function is additive, that is

$$v(S) = \sum_{i \in S} v(\{i\}) \text{ for every } S \subseteq I.$$

Define v' by

$$v'(S) = v(S) - \sum_{i \in S} v(\{i\}) \text{ for every } S \subseteq I .$$

Then $v \sim v'$ since we can take $k = 1$ and $c_i = -v(\{i\})$ in (4.21), but v' is identically zero.

Definition. A cooperative game with the characteristic function v is in 0, 1 **reduced form** if

$$v(\{i\}) = 0 \ (i \in I), \quad v(I) = 1 .$$

Theorem 4.8. Every essential cooperative game is strategically equivalent to exactly one game in 0, 1 reduced form.

Proof. Let v be the characteristic function of an essential cooperative game. Consider the system of $n + 1$ equations

$$v'(\{i\}) = kv(\{i\}) + c_i = 0 \quad (1 \le i \le n) , \tag{4.22}$$

$$v'(I) = kv(I) + \sum_{i \in I} c_i = 1, \tag{4.23}$$

in the $n+1$ unknowns $c_1, ..., c_n$, k. Adding the equations (4.22) we obtain

$$k \sum_{i \in I} v(\{i\}) + \sum_{i \in I} c_i = 0.$$

Subtracting from (4.23) we obtain

$$k\left(v(I) - \sum_{i \in I} v(\{i\}) \right) = 1 .$$

Since the game is essential we have

$$v(I) - \sum_{i \in I} v(\{i\}) > 0 .$$

Hence

$$k = \frac{1}{v(I) - \sum_{i \in I} v(\{i\})} > 0 .$$

Substituting into the equations (4.22) we obtain

$$c_i = \frac{- v(\{i\})}{v(I) - \displaystyle\sum_{i\in I} v(\{i\})} \ .$$

Thus the unknowns $c_1,..., c_n$, k are determined *uniquely* by equations (4.22) and (4.23), $v \sim v'$ and v' is in 0, 1 reduced form.

It follows from superadditivity that the characteristic function of a 0,1 reduced game is non-negative and a monotonic increasing function with respect to set theoretic inclusion.

For the 0, 1 reduced form conditions (4.14) and (4.15) on imputations become

$$x_i \geq 0 \quad (i \in I), \quad \sum_{i\in I} x_i = 1 \ ,$$

respectively.

For $n = 2$ any constant sum game is inessential and so equivalent to a zero game; for suppose there is an essential constant sum game for two players. By Theorem 4.8 we may reduce this game to 0, 1 reduced form. The characteristic function is then

$$v(\phi) = v(\{1\}) = v(\{2\}) = 0, \quad v(\{1, 2\}) = 1 \ .$$

Now it is easy to see that if a game is constant sum then any strategically equivalent game is also constant sum. Hence v has the property of complementarity established in Theorem 4.3, that is,

$$v(\{2\}) = v(\{1, 2\}) - v(\{1\}) = 1 - 0 = 1 \ ,$$

which is a contradiction. Hence there can be no essential 2-person constant sum game.

Any non-constant sum 2-person cooperative game is strategically equivalent to the 0, 1 reduced game with

$$v(\phi) = v(\{1\}) = v(\{2\}) = 0, \quad v(\{1, 2\}) = 1 \ .$$

Hence up to equivalence there is but one such game and it can be summarised as follows: the players have one unit to share between them provided that they can reach agreement, otherwise each receives nothing.

In the case of 3-person constant sum games we either have an inessential game, equivalent to a zero game, or the unique 0, 1 reduced form

$$\left.\begin{array}{l} v(\phi) = v(\{1\}) = v(\{2\}) = v(\{3\}) = 0 \\ v(\{1, 2\}) = v(\{1, 3\}) = v(\{2, 3\}) = 1 \\ v(\{1, 2, 3\}) = 1 \end{array}\right\} \qquad (4.24)$$

for an essential game. For $v(\{i\}) = 0$, $i \in I$, and $v(I) = 1$ follow from the definition of the 0, 1 reduced form. The remaining values of v are determined by complementarity, for example

$$v(\{1, 2\}) = v(\{1, 2, 3\}) - v(\{3\}) = 1 - 0 = 1 \ .$$

For 3-person non-constant sum games in 0,1 reduced form we have

$$\begin{array}{l} v(\phi) = v(\{1\}) = v(\{2\}) = v(\{3\}) = 0 \\ v(\{1, 2\}) = \alpha, \ v(\{1, 3\}) = \beta, \ v(\{2, 3\}) = \gamma \\ v(\{1, 2, 3\}) = 1 \ , \end{array}$$

where α, β, γ are arbitrary numbers in $[0, 1]$. Thus the strategic equivalence classes of such games are in 1-1 correspondence with the points (α, β, γ) of the unit cube. It turns out that the same is true of 4-person constant sum games.

4.5 DOMINANCE OF IMPUTATIONS.

Ideally the final outcome of a cooperative game would be a unique imputation arrived at through a bargaining procedure conducted by supremely rational players totally motivated by a desire to maximise their individual utility. But consider the 3-person constant sum game whose characteristic function is given in (4.24). We can summarise this game as follows: if two or more players can reach agreement they share one unit, otherwise they each get nothing. Suppose $\{1, 2\}$ form a coalition by agreeing to split the proceeds equally. This leads to the imputation $(\frac{1}{2}, \frac{1}{2}, 0)$. But the game is symmetric; there is no particular reason to single out the coalition $\{1, 2\}$. From the information at our disposal the coalitions $\{1, 3\}$, $\{2, 3\}$ are just as likely to form in the same way. This leads to the set of imputations

$$\{ \ (\tfrac{1}{2}, \tfrac{1}{2}, 0), \ (\tfrac{1}{2}, 0, \tfrac{1}{2}), \ (0, \tfrac{1}{2}, \tfrac{1}{2}) \ \},$$

any one of which is a possible outcome as far as we are concerned. Moreover, it is clear that this problem of non-uniqueness is quite independent of any criticism we may make against the characteristic function as a representation of the game.

Since in general we cannot single out a unique imputation as the obvious solution of a cooperative game, we begin by asking the simpler question: *which imputations can be clearly excluded?*

Let $\mathbf{x} = (x_1, ..., x_n)$, $\mathbf{y} = y_1, ..., y_n)$ be two imputations in a classical cooperative game $\langle I, v, X \rangle$. We first make precise the idea that coalition S prefers $\mathbf{x}$ to $\mathbf{y}$.

Definition. **x dominates y through a coalition S**, written $\mathbf{x} \succ_S \mathbf{y}$, if

$$\sum_{i \in S} x_i \le v(S) \tag{4.25}$$

and

$$x_i > y_i \text{ for every } i \in S . \tag{4.26}$$

Condition (4.25) is called the **effectiveness** condition. It means that the coalition S is actually capable of obtaining what the imputation $\mathbf{x}$ gives it collectively. The second condtion, (4.26), is a plain statement that the members of S unanimously prefer $\mathbf{x}$ to $\mathbf{y}$ and hence is called the **preferability** condition.

Definition. We say $\mathbf{x}$ **dominates** $\mathbf{y}$ if there exists a non-empty coalition S such that $\mathbf{x} \succ_S \mathbf{y}$.

We denote the relation of dominance by $\mathbf{x} \succ \mathbf{y}$. Dominance of $\mathbf{x}$ over $\mathbf{y}$ means that there are "forces" (*viz. S*) in "society" (*viz. I*) which prefer $\mathbf{x}$ to $\mathbf{y}$. Notice that $\succ_S$ is transitive but $\succ$ is not. For example if

$$v(\{1, 2\}) = v(\{2, 3\}) = v(\{1, 3\}) = v(\{1, 2, 3\}) = 100$$

and

$$\mathbf{x} = (50, 50, \ 0)$$
$$\mathbf{y} = (\ 0, 60, 40)$$
$$\mathbf{z} = (15, \ 0, 85)$$

then

$$\mathbf{x} \succ_{\{1, 2\}} \mathbf{y} \succ_{\{1, 3\}} \mathbf{z} \succ_{\{2, 3\}} \mathbf{x}$$

so that

$$\mathbf{x} \succ \mathbf{y} \succ \mathbf{z} \succ \mathbf{x}.$$

In any game dominance cannot occur through a 1-person coalition $S = \{i\}$ or through the grand coalition $S = I$. For if $S = \{i\}$ and $\mathbf{x} \succ_S \mathbf{y}$ then $y_i < x_i \le v(\{i\})$. But in this case $\mathbf{y}$ is not an imputation since the condition of individual rationality, (4.14), is violated. If $S = I$ and $\mathbf{x} \succ_I \mathbf{y}$ then $x_i > y_i$ for every $i \in I$ so that

$$\sum_{i \in I} x_i > \sum_{i \in I} y_i = v(I).$$

Thus $\mathbf{x}$ cannot be an imputation since it does not satisfy the collective rationality condition (4.15).

Our next theorem shows that the relation of dominance through a coalition, and hence the general relation of dominance, is preserved under strategic equivalence. This is clearly necessary if dominance is to serve as the basis of a solution concept.

Theorem 4.9. If $v \sim v'$ and the imputations $\mathbf{x}'$, $\mathbf{y}'$ correspond respectively to the imputations $\mathbf{x}$, $\mathbf{y}$ under this strategic equivalence, then $\mathbf{x} \succ_S \mathbf{y}$ implies $\mathbf{x}' \succ_S \mathbf{y}'$.

Proof. Since $v \sim v'$ put

$$v'(S) = kv(S) + \sum_{i \in S} c_i \quad (k > 0, \ S \subseteq I) \ .$$

Now $\mathbf{x} \succ_S \mathbf{y}$ means

$$\sum_{i \in S} x_i \le v(S) \text{ and } x_i > y_i \quad (i \in S).$$

If $\mathbf{x}' = (x_1', \ldots, x_n')$ corresponds to $\mathbf{x} = (x_1, \ldots, x_n)$ (see §4.4) then $x_i' = kx_i + c_i$ ($i \in I$). Similarly $y_i' = ky_i + c_i$ ($i \in I$). Hence

$$\sum_{i \in S} x_i' = \sum_{i \in S} (kx_i + c_i) = k \sum_{i \in S} x_i + \sum_{i \in S} c_i$$
$$\le kv(S) + \sum_{i \in S} c_i = v'(S) \ ,$$

and

$$x_i' = kx_i + c_i > ky_i + c_i = y_i' \quad (i \in S) \ .$$

Hence $\mathbf{x}' \succ_S \mathbf{y}'$ as required.

The approach we are presently adopting is to seek the solution of a cooperative game within the set of imputations. It will, therefore, be the structure of the set of imputations under the relation of dominance which will determine the essential features of the game. This being so it may occur to the reader that strategic equivalence is not necessarily the most natural equivalence relation to impose. We could instead make the

Definition. Two classical cooperative games $\langle I, v, X \rangle$, $\langle I, v', X' \rangle$ are **isomorphic** if there exists a function $f: X \to X'$, which is one to one and onto, such that for every pair of imputations $\mathbf{x}$, $\mathbf{y} \in X$

$$\mathbf{x} \succ \mathbf{y} \text{ if and only if } f(\mathbf{x}) \succ f(\mathbf{y}).$$

Plainly isomorphism is an equivalence relation, and in effect two games are isomorphic if their imputations sets have the same dominance structure. Theorem 4.9 shows that if two games are strategically equivalent then they are isomorphic. Surprisingly enough the converse has been proved true, at least for constant sum games, by McKinsey [4]. In an ingenious but mathematically quite simple paper (which uses a -1, 0 reduced form as opposed to our 0, 1 reduced form - the two notions are equivalent) he shows that for constant sum games isomorphism implies strategic equivalence.

In Theorem 4.5 we showed that the set of imputations for an inessential game consists of a single imputation, whilst there are always infinitely many imputations for an essential game

Thus an essential game cannot be isomorphic, and so *a fortiori* cannot be strategically equivalent, to an inessential game.

The relation of dominance $\mathbf{x} \succ \mathbf{y}$ expresses the preference of the imputation $\mathbf{x}$ over $\mathbf{y}$ by some coalition S, the idea being that if the imputation $\mathbf{y}$ is proposed then S will come up with a counter proposal suggesting the imputation $\mathbf{x}$. It is therefore of some interest to determine those imputations which are not dominated by any other imputation.

Definition. The set of all undominated imputations in a cooperative game is called the **core**.

Theorem 4.10. An imputation $\mathbf{x} = (x_1, \ldots, x_n)$ is in the core of a classical coopreative game with characteristic function v if and only if

$$\sum_{i \in S} x_i \geq v(S) \text{ for every } S \subseteq I . \tag{4.27}$$

Proof. Firstly suppose $\mathbf{x}$ is *not* in the core. Then $\mathbf{x}$ is dominated by some $\mathbf{y}$, so that for some coalition S, $y_i > x_i$ for every $i \in S$. Hence

$$\sum_{i \in S} x_i < \sum_{i \in S} y_i \leq v(S) ,$$

which violates (4.27).

If on the other hand (4.27) is false, then for some coalition S

$$\sum_{i \in S} x_i < v(S)$$

Note that $S \neq \{i\}$ or I, for otherwise the conditions of individual or collective rationality on the imputation $\mathbf{x}$ would be violated.

From what has been said regarding strategic equivalence it is plainly sufficient to consider games in the 0, 1 reduced form. For these games v is a monotonic increasing function with respect to set theoretic inclusion. Hence, using (4.15),

$$\sum_{i \notin S} x_i = v(I) - \sum_{i \in S} x_i \geq v(S) - \sum_{i \in S} x_i > 0 .$$

Now choose $\varepsilon > 0$ so that

$$0 < \varepsilon < \frac{1}{|S|} \left(v(S) - \sum_{i \in S} x_i \right)$$

and define $\mathbf{y} = (y_1, ..., y_n)$ by

$$y_i = \begin{cases} x_i + \varepsilon & \text{, if } i \in S, \\ \dfrac{1}{|I \setminus S|} \left(\displaystyle\sum_{i \in I \setminus S} x_i - |S| \, \varepsilon \right) & \text{, if } i \notin S \end{cases}$$

It is routine to verify that $\mathbf{y}$ is an imputation, and that $\mathbf{y} \succ_S \mathbf{x}$. Hence $\mathbf{x}$ is not in the core and the proof is complete.

Note that condition (4.27) which characterises those imputations $\mathbf{x}$ which are in the core is exactly the condition (4.16) discussed earlier.

The theorem tells us that an imputation $\mathbf{x}$ is in the core if the components x_i satisfy a finite system of inequalities. From this one can easily see that the core of a classical cooperative game is a (possibly empty) convex polyhedral set.

We have seen in Theorem 4.5 that an inessential game has a unique imputation, so that there are no dominations possible. Thus the core of a inessential game consists of its single imputation. On the other hand for essential constant sum games we have

Theorem 4.11. The core of any essential constant sum game is empty.

Proof. Since any 2-person constant sum game is inessential we can assume $n \geq 3$. Suppose $\mathbf{x}$ is an imputation in the core. Taking $S = \{i\}$ and $S = I \setminus \{i\}$ in (4.27) we have

$$x_i \geq v(\{i\}) , \tag{4.28}$$

$$\sum_{j \neq i} x_j \geq v(I \setminus \{i\}) .$$

If we add these inequalities then on using (4.15) and Theorem 4.3 we obtain

$$v(I) = \sum_{i \in I} x_i \geq v(\{i\}) + v(I \setminus \{i\}) = v(I) .$$

Hence equality holds in (4.28) for each $i \in I$, so that

$$v(I) = \sum_{i \in I} v(\{i\}) \ .$$

It now follows from Theorem 4.2 that the game is inessential, which is contrary to hypothesis.

Example 4.3 Consider the 3-person non-constant sum game in 0,1 reduced form which has characteristic function

$$v(\phi) = v(\{1\}) = v(\{2\}) = v(\{3\}) = 0$$
$$v(\{1, 2\}) = \frac{1}{3}, \ v(\{1, 3\}) = \frac{1}{2}, \ v(\{2, 3\}) = \frac{1}{4},$$
$$v(\{1, 2, 3\}) = 1 \ .$$

Since the game is in 0,1 reduced form any imputation $x = (x_1, x_2, x_3)$ satisfies

$$x_i \geq 0 \ (i \in I), \ \sum_{i \in I} x_i = 1 \ .$$

From Theorem 4.10 x is in the core if and only if

$$x_1 + x_2 \geq \frac{1}{3}, \ x_1 + x_3 \geq \frac{1}{2}, \ x_2 + x_3 \geq \frac{1}{4} \ ,$$

or equivalently

$$x_3 \leq 1 - \frac{1}{3}, \ x_2 \leq 1 - \frac{1}{2}, \ x_1 \leq 1 - \frac{1}{4} \ .$$

The shaded region in Figure 4.2 indicates the set of imputations which satisfy $x_3 \leq \frac{2}{3}$. If we draw only the triangle of imputations and take all three inequalities into account we obtain the core as the shaded region in Figure 4.3.

By varying the values of the 2-person coalitions one can obtain a variety of possible shapes for the core of 3-person games. The core, if it is non-empty, may be a line segment, a triangle, a quadrilateral, a pentagon or a hexagon as in Figure 4.3.

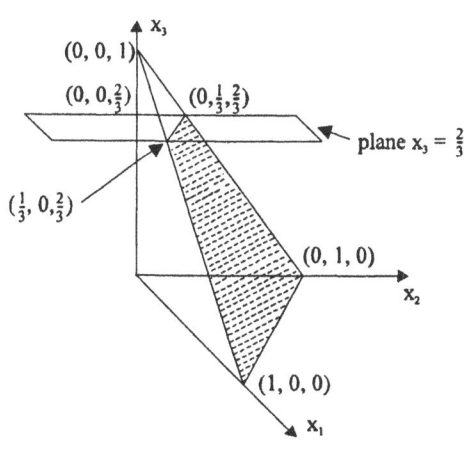

Figure 4.2 Imputations which satisfy $x_3 \leq \frac{2}{3}$.

4.6 VON NEUMANN-MORGENSTERN SOLUTIONS.

Since the core is often empty it becomes necessary to look for some other type of solution concept. Consider once again the essential 3-person constant sum game in 0,1 reduced form. For this game the solution

$$\{ \; \mathbf{x}_{12} = (\tfrac{1}{2}, \tfrac{1}{2}, 0), \; \mathbf{x}_{13} = (\tfrac{1}{2}, 0, \tfrac{1}{2}), \; \mathbf{x}_{23} = (0, \tfrac{1}{2}, \tfrac{1}{2}) \; \}$$

was advanced in §4.5. Now in what sense is this a solution? It is easy to see that none of these three imputations dominates any one of the others, but any set consisting of a single imputation would have this property. However, this set has another interesting feature: any imputation not in the set is dominated by one of the three imputations $\mathbf{x}_{ij}$.

To see this, let $\mathbf{x} = (x_1, x_2, x_3)$, $x_i \geq 0$, $x_1 + x_2 + x_3 = 1$, be an imputation not equal to any $\mathbf{x}_{ij}$. Then at most one of the x_i can be as large as $\tfrac{1}{2}$. Hence at least two components, say x_i and x_j with $i < j$, are smaller than $\tfrac{1}{2}$. This implies $\mathbf{x}_{ij} \succ \mathbf{x}$ through the coalition $\{i, j\}$.

Definition. A set V of imputations for a classical cooperative game is **stable** if

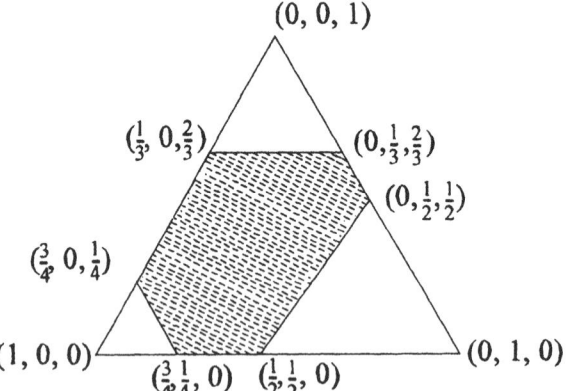

Figure 4.3 The core: those imputations which satisfy all three inequalities in Example 4.3.

(i) If $\mathbf{x}$, $\mathbf{y} \in V$ then $\mathbf{x} \succ \mathbf{y}$ *(internal stability)*.

and   (ii) If $\mathbf{y} \notin V$ there exists $\mathbf{x} \in V$ such that $\mathbf{x} \succ \mathbf{y}$ *(external stability)*.

A **von Neumann-Morgenstern** (vN-M) **solution** of a classical cooperative game is *any stable set*. A vN-M solution need not be unique; frequently there are many stable sets available.

> **Example 4.4** We have seen that for the essential 3-person constant sum game in 0,1 reduced form the symmetric set
>
> $$V = \{ \; (\tfrac{1}{2}, \tfrac{1}{2}, 0), \; (\tfrac{1}{2}, 0, \tfrac{1}{2}), \; (0, \tfrac{1}{2}, \tfrac{1}{2}) \; \}$$
>
> is stable. However, V is not the only stable set.

If we choose some fixed number c, $0 \le c \le \frac{1}{2}$, then any of the sets

$$V_1(c) = \{(c, x_2, x_3); \ x_2 + x_3 = 1 - c, \ x_2 \ge 0, \ x_3 \ge 0\},$$
$$V_2(c) = \{(x_1, c, x_3); \ x_1 + x_3 = 1 - c, \ x_1 \ge 0, \ x_3 \ge 0\},$$
$$V_3(c) = \{(x_1, x_2, c); \ x_1 + x_2 = 1 - c, \ x_1 \ge 0, \ x_2 \ge 0\}.$$

is stable.

For consider $V_3(c)$. Internal stability follows from the fact that two imputations $\mathbf{x}$, $\mathbf{y}$ $\in V_3(c)$ can differ only by having either $x_1 < y_1$ and $x_2 > y_2$, or $x_1 > y_1$ and $x_2 < y_2$. But this allows only the possibility of domination through a 1-person coalition, which can never occur.

To show $V_3(c)$ is externally stable let $\mathbf{y} = (y_1, y_2, y_3) \notin V_3(c)$. Then either $y_3 > c$ or $y_3 < c$. If $y_3 > c$, say $y_3 = c + \varepsilon$, we define $\mathbf{x}$ by

$$x_1 = y_1 + \frac{\varepsilon}{2}, \quad x_2 = y_2 + \frac{\varepsilon}{2}, \quad x_3 = c .$$

It is then easy to see that $\mathbf{x} \in V_3(c)$ and $\mathbf{x} \succ \mathbf{y}$ through $\{1, 2\}$. Now suppose $y_3 < c$. We know that either $y_1 \le \frac{1}{2}$ or $y_2 \le \frac{1}{2}$ (else $y_1 + y_2 > 1$). Say $y_1 \le \frac{1}{2}$. In this case consider $\mathbf{x} = (1-c, 0, c) \in V_3(c)$. Since $1 - c > \frac{1}{2} \ge y_1$ and $c > y_3$ we have $\mathbf{x} \succ \mathbf{y}$ through $\{1, 3\}$. If on the other hand $y_2 \le \frac{1}{2}$ then $\mathbf{x} \succ \mathbf{y}$ through $\{2, 3\}$ where $\mathbf{x} = (0, 1-c, c) \in V_3(c)$.

Similarly the sets $V_1(c)$ and $V_2(c)$ are also stable.

The solution $V_3(c)$ is said to be a **discriminatory** solution, player 3 being the discriminated player. One can imagine that players 1 and 2 have formed a coalition and agree to pay player 3 an arbitrary amount c, $0 \le c \le \frac{1}{2}$. How players 1 and 2 divide the amount $1-c$ depends upon the relative bargaining powers of the two players.

Von Neumann and Morgenstern do not say how it will be decided which player will be discriminated against or which imputation will arise. They argue that if society accepts discrimination one may find a solution of the type $V_i(c)$, where the value of c in the range $0 \le c < \frac{1}{2}$ is determined by the degree of discrimination tolerated by the society.

By varying c in the range $0 \le c < \frac{1}{2}$ we obtain continuum many stable sets each containing continuum many imputations. Indeed, every possible imputation for the essential 3-person constant sum game is included in at least one solution! This embarassing superfluity of solutions is by no means confined to 3-person games, although it does not occur in general.

The relationship between the core and any vN-M solution is given by

Theorem 4.12. If V is a vN-M solution of a classical cooperative game with core C then $C \subseteq V$.

Proof. Suppose $\mathbf{x} \in C$, then $\mathbf{x}$ is not dominated by any imputation. If $\mathbf{x} \notin V$ then $\mathbf{x}$ is dominated by some imputation in V (external stability). Hence $\mathbf{x} \in V$.

Theorem 4.13. If any vN-M solution of a classical cooperative game consists of a single imputation the game is inessential.

Proof. By Theorem 4.9 we can assume the game is in 0,1 reduced form. Let $\mathbf{x}$ be the unique imputation of a vN-M solution. If the game is essential then $|I| = n > 1$ and we can construct an imputation $\mathbf{y} = (y_1, ..., y_n)$ by setting

$$
y_j = \begin{cases} x_j + \dfrac{x_i}{(n-1)} & , \text{ if } j \neq i, \\[2mm] 0 & , \text{ if } j = i \end{cases}
$$

where x_i is some component of $\mathbf{x}$ with $x_i > 0$. Since $y_j > x_j$ $(j \neq i)$ and domination through the 1-person coalition $\{i\}$ cannot occur, it is plain that $\mathbf{x}$ does not dominate $\mathbf{y}$. Hence, since $\mathbf{y} \neq \mathbf{x}$, either $\mathbf{y}$ belongs to the vN-M solution, which is contrary to hypothesis, or there exists another imputation $\mathbf{z}$, distinct from $\mathbf{x}$, which is in the vN-M solution and dominates $\mathbf{y}$, again a contradiction. Hence the game is inessential.

For certain special types of game one can construct stable sets.

Definition. A classical cooperative game $\langle I, v, X \rangle$ in 0,1 reduced form is **simple** if for each $S \subseteq I$ either $v(S) = 0$ or $v(S) = 1$.

A simple game is one in which each coalition is either **winning**, $v(S) = 1$, or **losing**, $v(S) = 0$. Simple games have some application to political science since they include voting games as in elections etc.

Theorem 4.14. Let S be a minimal winning coalition (that is, a coalition such that $v(S) = 1$ but $v(T) = 0$ for every $T \subsetneq S$) in a simple game. Define V_S to be the set of all imputations such that $X_i = 0$ for all $i \notin S$. Then V_S is a stable set.

In essence the theorem says that every simple game has discriminatory stable sets in which a minimal winning coalition pays the remaining players nothing.

Proof. Now

$$
V_S = \{ (x_1, \ldots, x_n); \sum_{i \in S} x_i = 1, x_i \geq 0 \ i \in I, x_i = 0 \text{ if } i \notin S \}
$$

To establish internal stability suppose $\mathbf{x}, \mathbf{y} \in V_S$. Then $x_i > y_i$ implies $i \in S$, hence domination $\mathbf{x} \succ \mathbf{y}$ can only occur through a subset of S. Suppose $T \subseteq S$ and $\mathbf{x} \succ_S \mathbf{y}$, that is,

$$
\sum_{i \in T} x_i \leq v(T) , \tag{4.29}
$$

$$x_i > y_i \text{ for every } i \in T. \tag{4.30}$$

If $T \neq S$ then, since S is a *minimal* winning coalition, $v(T) = 0$ so that $x_i = 0$ for every $i \in T$, and (4.30) is impossible. If $T = S$ then (4.30) implies

$$1 = \sum_{i \in S} x_i > \sum_{i \in S} y_i = 1 ,$$

again a contradiction.

To show V_S is externally stable we can assume $S \neq I$, else V_S is the whole set of imputations and there is nothing to prove. Let $\mathbf{y} = (y_1, ..., y_n)$ be an imputation with $\mathbf{y} \notin V_S$, then $y_j > 0$ for some $j \notin S$. Put

$$\varepsilon = \sum_{i \notin S} y_i > 0,$$

and define $\mathbf{x} = (x_1, ..., x_n)$ by

$$x_i = \begin{cases} y_i + \dfrac{\varepsilon}{|S|} & , \text{ if } i \in S, \\ 0 & , \text{ if } i \notin S. \end{cases}$$

Then it is plain that $\mathbf{x} \in V_S$ and $\mathbf{x} \succ \mathbf{y}$ through S.

Definition. A classical cooperative game $\langle I, v, X \rangle$ is **symmetric** if for every $S \subseteq I$ $v(S)$ depends only on the number of elements in S.

Thus the game considered in Example 4.4 was a symmetric game which had the symmetric solution V. We now consider the non-constant sum symmetric 3-person games.

Example 4.5 An essential 3-person symmetric game in 0,1 reduced form has the characteristic function

$$\begin{aligned} v(\phi) &= v(\{1\}) = v(\{2\}) = v(\{3\}) = 0, \\ v(\{1, 2\}) &= v(\{1, 3\}) = v(\{2, 3\}) = \alpha \\ v(\{1, 2, 3\}) &= 1 , \end{aligned}$$

where $0 \leq \alpha \leq 1$. If $\alpha = 1$ we have the constant sum game, which has already been dealt with, whilst if $\alpha = 0$ we have a simple game in which every imputation belongs to the core, so that the core is the unique stable set. Thus we may suppose $0 < \alpha < 1$.

Let us try to decide heuristically what a stable set should look like for this game. If α is large (that is, close to 1) we might expect the players to behave much as in the constant sum game.

Thus they will first attempt to form a 2-person coalition. In the bargaining process of coalition formation the symmetry of the players postitions suggests that in any 2-person coalition which forms, the members will agree to divide their winnings equally.

Once a 2-person coalition has been formed its members may bargain with the other player to determine the division of the remaining $1-\alpha$ obtainable by forming the grand coalition. The third player may get any amount from nothing (grand coalition not formed) to $1-\alpha$, depending on the bargaining abilities of the three players.

This line of argument leads us to expect that, at least for large α, the set

$$W = \{ (x, x, 1-2x), (x, 1-2x, x), (1-2x, x, x) ; \frac{\alpha}{2} \le x \le \frac{1}{2} \},$$

which consists of three line segments, might be stable (see Figure 4.4). In fact W is stable if $\alpha \ge \frac{2}{3}$.

To verify internal stability we observe that since $x + x = 2x \ge \alpha = v(\{1,2\})$ any imputation $\mathbf{y} = (y_1, y_2, y_3)$ with $y_1 > x$, $y_2 > x$ is not effectively realisable by coalition $\{1,2\}$. Hence the imputation $(x, x, 1-2x)$ can be dominated only through one of the two coalitions $\{1,3\}$ and $\{2,3\}$. Take, for example, the coalition $\{1,3\}$. Suppose $\mathbf{y} \in W$ and $\mathbf{y} \succ \mathbf{x}$ through $\{1,3\}$. If $\mathbf{y} = (y, y, 1-2y)$ we have $y > x$ and $1-2y > 1-2x$, which is impossible. Similarly if $\mathbf{y} = (1-2y, y, y)$ we have $1-2y > x$

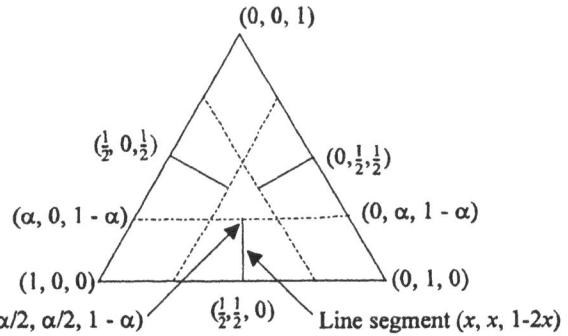

Figure 4.4 The three line segments which form W.

and $y > 1-2x$ which gives $\frac{1}{3} > y \ge \frac{\alpha}{2}$, a contradiction since $\alpha \ge \frac{2}{3}$. Finally if $\mathbf{y} = (y, 1-2y, y)$ we have $y > x$, so $2y > 2x \ge \alpha = v(\{1,3\})$, that is $\mathbf{y}$ is not effectively realisable by $\{1,3\}$. The same reasoning shows $\mathbf{y} \in W$ cannot dominate $\mathbf{x}$ through $\{2,3\}$. By symmetry each imputation of the form $(x, 1-2x, x)$, $(1-2x, x, x)$ with $\frac{\alpha}{2} \le x \le \frac{1}{2}$ is also undominated by any element of W. Hence W is internally stable.

It remains to show that W is externally stable. If $\alpha \ge \frac{2}{3}$, any imputation will have at most two components not less than $\frac{\alpha}{2}$.

Suppose an imputation $z = (z_1, z_2, z_3)$ has two such components; by symmetry we may assume that these are the first two, that is $z_1 \ge \frac{\alpha}{2}$, $z_2 \ge \frac{\alpha}{2}$. If $z_1 = z_2$ then $\mathbf{z} \in W$. Suppose $z_1 \ne z_2$. Without loss of generality we can take $z_1 > z_2$. Write $z_1 = z_2 + 3\varepsilon$, where $\varepsilon > 0$. Then $\mathbf{z}$ is dominated by the imputation $(z_2 + \varepsilon, z_2 + \varepsilon, z_3 + \varepsilon)$ through the coalition $\{2,3\}$. Moreover

$(z_2 + \varepsilon, z_2 + \varepsilon, z_3 + \varepsilon)$ is of the form $(x, x, 1-2x)$ for $x = z_2 + \varepsilon \geq \alpha/2$, and so is in W.

If, however, $\mathbf{z}$ has less than two components not less than $\alpha/2$ we may assume by symmetry that $z_1 < \alpha/2$ and $z_2 < \alpha/2$. Then $\mathbf{z}$ is dominated by $(\alpha/2, \alpha/2, 1-\alpha) \in W$ through $\{1,2\}$.

If $\alpha < 2/3$, the above argument fails. It is easy to see that such games will possess a core consisting of all imputations $\mathbf{x} = (x_1, x_2, x_3)$ with $x_i \leq 1-\alpha$, $i \in I$. One can then check that the union of the core and the three line segments of W is a stable set (Figure 4.5(a)). For $\alpha \leq 1/2$ the three line segments will all be subsets of the core; in this case the core can be shown to be the unique stable set (Figure 4.5(b)).

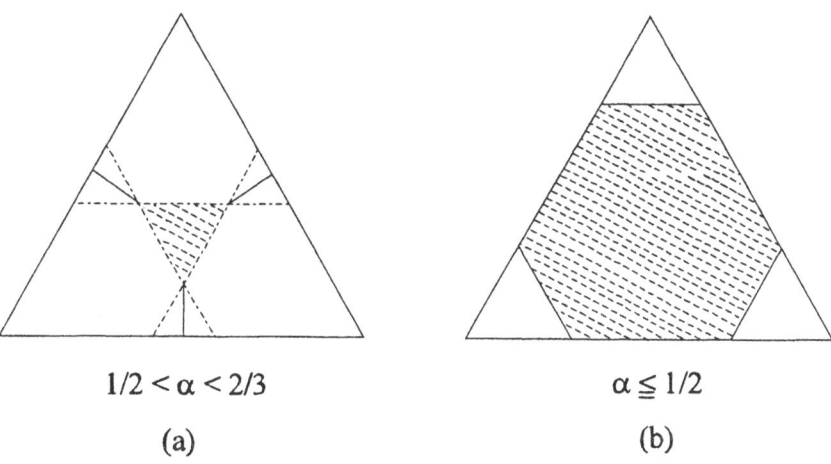

$$1/2 < \alpha < 2/3 \qquad\qquad \alpha \leq 1/2$$

$$(a) \qquad\qquad\qquad (b)$$

Figure 4.5 Stable sets for small α.

The solution of the general non-constant sum 3-person game in 0,1 reduced form is rather similar but more involved, see for example Vorobev ([5], §4.15).

For each value of $n \geq 4$ qualitatively new phenomena appear; the complexity of the stable sets encountered increases rapidly. In 1952 Shapley [6] showed just how pathological stable sets can be. The exact nature of his result is not easily described, but it has been paraphrased by Nash as follows: Shapleys theorem shows that there is a 5-person game in which your signature (assumed two-dimensional and compact) appears as a disconnected part of the stable set.

In view of this it may seem surprising that the basic problem with the vN-M solution concept is in the other direction, namely the existence of stable sets. It was proved by von Neumann and Morgenstern that every 4-person constant sum game has a vN-M solution and in particular cases all solutions are known [7].In the area of non-constant sum 4-person games and constant sum 5-person games Nering [8] has found some solutions. There is also quite an extensive literature on simple games. Bott [9] introduced a special class of simple games called (n, k)-majority games defined by

$$v(S) = \begin{cases} 0, & \text{if } |S| \leq k, \\ 1, & \text{if } |S| > k. \end{cases}$$

where $k > {}^{n}/_{2}$, and studied the symmetric solutions of such games. In [10] Gilles studied the non-symmetric, or discriminatory, solutions to (n, k)-majority games. Various other classes of games have been studied but, for example, the solubility of an arbitrary 5-person game remains an open question. It is known that a positive proportion (in the sense of Lebesgue measure) of all n-person games have unique stable sets consisting of the core.

Thus the discovery in 1967 by Lucas [11] of a cooperative game which has no vN-M solution dealt a real body blow to the theory. Lucas proof rests on earlier work of Gilles [12]. Let $\langle I, v, X \rangle$ be a game having all the properties of a classical cooperative game except for superadditivity of v. We call it a **generalised** game. For any coalition $S \subseteq I$ we define $\wp_S$, a partition of S, to be a collection of mutually disjoint subsets of S whose union is S. We next define

$$v^*(S) = \max_{\wp_S} \sum_{T \in \wp_S} v(T) .$$

This amounts to a new evaluation of the strength of S as the maximum that its members can obtain by forming mutually disjoint subcoalitions. The function v^* is superadditive and therefore $\langle I, v^*, X \rangle$ is a classical cooperative game with the same set of imputations as the generalised game $\langle I, v, X \rangle$. Gilles proved [12] that the domination $\mathbf{x} \succ_S \mathbf{y}$ in $\langle I, v, X \rangle$ implies $\mathbf{x} \succ_{S*} \mathbf{y}$ for some $S^* \subseteq S$ in $\langle I, v^*, X \rangle$ (the converse is trivial). Thus a set which is stable in one game is stable in the other (stability does not depend on superadditivity).

Lucas showed that the 10-person game with

$$v(\{1,2\}) = v(\{3,4\}) = v(\{5,6\}) = v(\{7,8\}) = v(\{9,10\}) = 1,$$
$$v(\{3,5,7\}) = v(\{1,5,7\}) = v(\{1,3,7)\} = 2,$$
$$v(\{3,5,9\}) = v(\{1,5,9\}) = v(\{1,3,9)\} = 2,$$
$$v(\{1,4,7,9\}) = v(\{3,6,7,9\}) = v(\{2,5,7,9\}) = 2,$$
$$v(\{3,5,7,9\}) = v(\{1,5,7,9\}) = v(\{1,3,7,9\}) = 3,$$
$$v(\{1,3,5,7,9\}) = 4, v(I) = 5,$$
$$v(S) = 0 \text{ for all other coalitions } S \subseteq I,$$

has no vN-M solution, which impleis the existence of a 10-person classical cooperative game with no vN-M solution.

Whilst one cannot but regard this as a serious defect in the vN-M solution concept it by no means demolishes the theory which is now directed towards defining classes of games, particularly classical cooperative games, having vN-M solutions and to classifying games according to the different types of insolubility which they possess. We shall consider an alternative solution concept in §4.8, but first we give an application of the classical cooperative game to economic theory.

4.7 THE EDGEWORTH TRADING MODEL.

Although the world in which we live is very far from being stable, upon reflection the degree of stability around us is rather surprising. Economists in particular have given much thought to the question: 'How is it that the pursuit of individual self interest leads to societies that are not characterised by chaos?'. The general equilibrium problem, which lies at the heart of competitive economic theory, can be roughly formulated as follows. If we assume that producers and consumers each act in individually selfish ways, how is it that market forces operate to create an equilibrium in which prices allocate resources and goods to those who are able and prepared to pay for them? A detailed analysis of this argument, which is remarkable in its ramifications, can become very involved [13]. However, we are now able to consider a very simple model.

The notion of the core, discussed in the preceding two sections, is related to that of the *contract curve*, a concept introduced by Edgeworth [14] in 1881. Long before the modern era of game theory, with remarkable penetration, Edgeworth described a market in which, in the first instance, there are two consumers A and B (usually referred to as Robinson and Friday) and two commodities. Each consumer has an endowment of one or both commodities, and the only choices the consumers have are (a) to consume their respective endowments or (b) to trade with one another and then consume their respective holdings after trade.

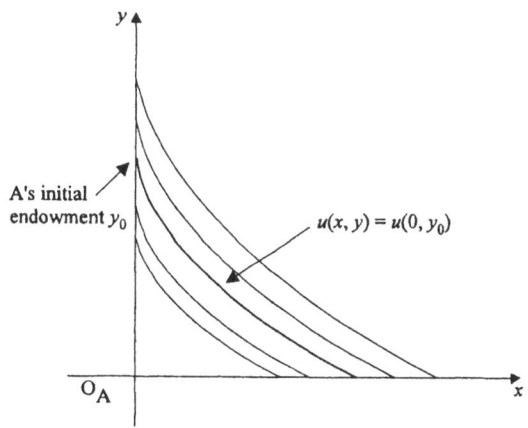

Figure 4.6 Indifference curves.

We suppose that A and B have preferences over **goods bundles,** that is pairs (x, y) where x represents an amount of the first commodity and y an amount of the second. It now becomes possible to consider all bundles (x, y) which have equal utility for A to some given bundle. This defines an **indifference curve** for A. We assume these to be smooth convex curves as in Figure 4.6.

Indifference curves do not intersect, and higher curves (to the north-east) represent strictly preferred bundles. Thus the problem for A is to trade for a bundle as far to the 'north-east' as possible. For simplicity we assume that As initial endowment is $(0, y_0)$ and Bs initial endowment is $(x_0, 0)$.

If we draw an analogous figure for B, rotate it through 180° and put it together with Figure 4.6 we obtain Figure 4.7, the 'Edgeworth box', with the origin for B in the top right corner. The height of the box represents As initial endowment and the width Bs initial endowment. The traders begin with the goods bundles $(0, y_0)$, $(x_0, 0)$ represented by the point P. Any point in the box is a feasible allocation of the total endowment, that is, a possible trade. The coordinates as measured from O_A represent As new bundle and those measured from O_B Bs new bundle.

Plainly, A will not contract for a trade below the curve PQ, nor will B contract for a trade above PR. If a trade T_1 makes at least one of the traders strictly better off (in utlity terms) and leaves the other no worse off than a trade T_2, we say T_1 **dominates** T_2. Thus in Figure 4.7 T_1 dominates T_2, moreover T_1 is itself *undominated by any trade* since it is a point of tangency to one curve from each family. In Figure 4.8 we construct the locus of all undominated trades CD which is known as the **contract curve**.

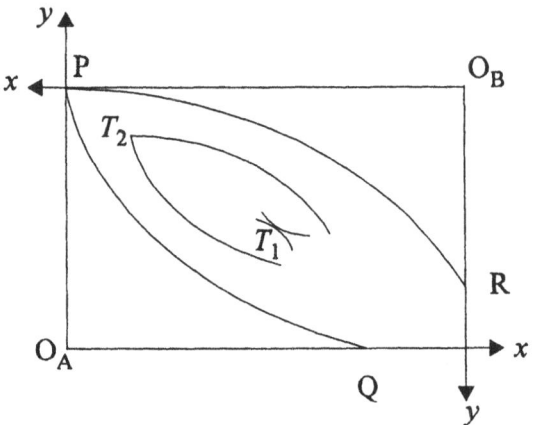

Figure 4.7 - The Edgeworth Box.

Now the contract curve has obvious similarities with the core, but notice that we have not even assumed comparable utilities for the goods being traded, whereas in the classical cooperative game we assumed all payoffs were in a comparable and transferable unit which for all the world behaved like money.

Edgeworth considered the way in which the contract curve changes when the market is made larger by replicating the traders. He considered having n traders *exactly* like A, and n *exactly* like B. 'Exactly like' means both having the same preferences and the same endowment.

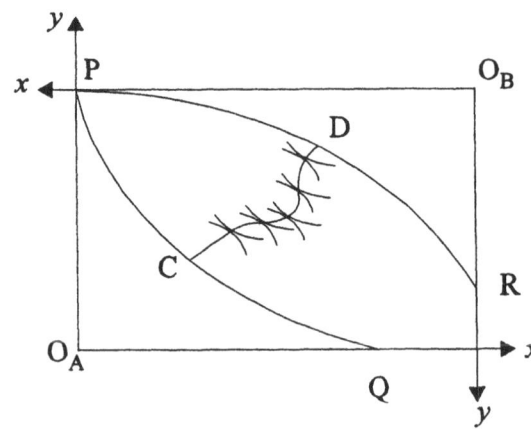

Figure 4.8 The contract curve CD.

This way of enlarging the market is ingenious because it remains possible to represent the market in a two-dimensional diagram, and for any n the contract 'curve' in the enlarged market becomes a proper subset of CD. It was thus conjectured by Edgeworth that as the number of traders increased without limit there would remain a single undominated trade. This distribution of the goods would correspond to the market price under pure competition.

To formalise Edgeworths trading model as a classical cooperative game will involve making certain simplifying assumptions. But in this way Shubik [15] was able to prove Edgeworth's conjecture (in a limited sense) and open up a whole new area of mathematical economics.

We consider a set of players $I = M \cup N$, where each trader belonging to M has initial endowment $(a, 0)$ and each trader belonging to N has $(0, b)$. Our principle simplification is to assume that all traders have identical utility functions $u(x, y)$. We assume that $u(x, y)$ is a strictly *concave* function of the point (x, y). This corresponds to the hypothesis that the curves of constant utility are convex in the (x, y) plane. If $|M| = m$ and $|N| = n$, $u(x, y)$ is defined over the rectangle $0 \leq x \leq ma, 0 \leq y \leq nb$. We suppose that u has continuous partial derivatives up to the third order, and because we shall allow the market to become large (that is, m or $n \to \infty$) we assume that for each fixed y and each fixed x

$$\lim_{x \to \infty} u(x, y) < \infty , \quad \lim_{y \to \infty} u(x, y) < \infty$$

respectively.

Because of these restrictions we can easily write down the explicit characteristic function for any game of this type. If there is no joint gain to be derived from trade, the game is inessential and the value of any coalition is the sum of the individual utilities. If the game is essential and trade takes place then

$$v(S) = |S| \max_{\substack{0 \leq x \leq s_m a \\ 0 \leq y \leq s_n b}} u(x, y)$$

where $s_m = |S \cap M|$, $s_n = |S \cap N|$. Note that $s_m + s_n = |S|$. Now for acceptable trades

$$u(x, y) \geq u(a, 0) \text{ and } u(x, y) \geq u(0, b) .$$

Thus by concavity of $u(x, y)$, the maximum value of utility that the coalition S can guarantee its members individually is estimated by

$$\frac{s_m}{|S|} u(a, 0) + \frac{s_n}{|S|} u(0, b) \leq u\left(\frac{s_m a}{|S|}, \frac{s_n b}{|S|}\right) .$$

At this point all players in the coalition will obtain equal quantities of the same goods (an economist would say that the marginal rate of substitution between the two commodities being traded will be the same for all players trading). Thus we take

$$v(S) = |S| u\left(\frac{s_m a}{|S|}, \frac{s_n b}{|S|}\right) .$$

We call the game with this characteristic function the game $[m,n]$.

The characteristic function for the 2-person game [1,1] is then

$$v(\phi) = 0, \; v(\{1\}) = u(a, 0), \; v(\{2\}) = u(0, b), \; v(\{1,2\}) = 2u(\frac{a}{2}, \frac{b}{2}) \; ,$$

which in general gives an essential 2-person non-constant sum game. Since domination of imputations is never possible through coalitions $\{i\}$ or I this game has a unique stable set, indentical with its core, consisting of *all* imputations. Denoting an imputation by (z_1, z_2), where z_i is the payoff in utility to player i, we can specify the imputation set as all pairs (z_1, z_2) which satisfy

$$z_1 + z_2 = 2u(\frac{a}{2}, \frac{b}{2}), \quad z_1 \geq u(a, 0), \quad z_2 \geq u(0, b) \; . \qquad (4.31)$$

The next step in our analysis is to show that the game $[n,n]$ has a stable set solution analogous to that in (4.31). We shall revert to our notation that x_i denotes the payoff to player $i \in I$ in utility, this should not cause confusion since it will no longer be necessary to refer to goods bundles explicitly.

Theorem 4.15. For any $n \geq 1$ the game $[n,n]$ has a stable set solution V consisting of all imputations $\mathbf{x} = (x_1, ..., x_{2n})$ for which $x_i = z_1$ if $i \in M$, $x_j = z_2$ if $j \in N$, where z_1, z_2 satisfy (4.31).

Proof. Firstly V is internally stable, for in order that $\mathbf{x} \succ_S \mathbf{y}$ we require that $x_i > y_i$ for every $i \in S$. If $S \cap N = \phi$ or $S \cap M = \phi$ this requirement on $\mathbf{x}$ is impossible to fulfil, since trading of one good for the other must take place if a strict increase in utility is to result. Thus if $\mathbf{x} \succ_S \mathbf{y}$ we have $S \cap N \neq \phi$ and $S \cap M \neq \phi$. Hence if $\mathbf{x}, \mathbf{y} \in V$ and $\mathbf{x} \succ \mathbf{y}$, then $x_i > y_i$ for some $i \in M$. But, by definition of V, this means $x_j < y_j$ for *all* $j \in N$, which contradicts our assertion that the dominance must take place through a coalition S with $S \cap N \neq \phi$.

Now suppose $\mathbf{y}$ is an imputation which is not in V. Then for some $i \in M$, $j \in N$ $y_i + y_j < 2u(^a/_2, ^b/_2)$. Choose z_1, z_2 to satisfy (4.31) and $z_1 > y_i$, $z_2 > y_j$. Then the imputation $\mathbf{x} \in V$ corresponding to (z_1, z_2) dominates $\mathbf{y}$ through $\{i,j\}$. Thus V is externally stable.

It is apparent that the stable set solution V has no tendency to 'shrink' as n is increased. Such a tendency, however, is shown by the core. We shall consider two cases: *monopoly*, the game $[1,n]$, and *pure competition*, the game $[m,n]$ where both m and n are large.

Theorem 4.16. In the game $[1,n]$ take $M = \{1\}$, $N = \{2, 3, ..., n+1\}$, then the imputation $\mathbf{x} = (x_1, ..., x_{n+1})$ such that $x_j = u(0, b)$ for all $j \in N$ belongs to the core. Moreover, given any $\varepsilon > 0$ there exists $n = n(\varepsilon)$ such that for all $n \geq n(\varepsilon)$ no imputation $\mathbf{y}$ with $y_1 \leq x_1 - \varepsilon$ can belong to the core.

Given the values of x_j for $j \in N$ we can quickly compute the value of x_1; it is

$$x_1 = (n + 1) u\left(\frac{a}{n + 1}, \frac{nb}{n + 1} \right) - nu(0, b) .$$

Notice that x_1 is as large as possible, so there are no imputations y with $y_1 > x_1$. Thus as $n \to \infty$ the theorem shows that the core approaches a single imputation at which the single trader of the first type acts as a perfectly discriminating monopolist and obtains all of the gain from trading.

Proof. As in the proof of Theorem 4.15, domination of x is only possible through a coalition S with $1 \in S$. Since x_1 is as large as possible it follows that x is in the core.

The second part of the proof is a little more involved. We first show that for every $j \in N$ there exists a $n(\varepsilon)$ such that

$$| v(I) - v(I\backslash\{j\}) - u(0, b) | < \frac{\varepsilon}{n} \quad \text{for all } n \geq n(\varepsilon) . \tag{4.32}$$

To accomplish this we expand the expression $v(I) - v(I\backslash\{j\}) - u(0, b)$ as a Taylor series about $(0, b)$. Here we use the smoothness hypothesis on u. Let subscripts 1 and 2 denote partial differentiation with respect to the first and second variable respectively. Then

$$v(I) = (n + 1) u\left(\frac{a}{n + 1}, \frac{nb}{n + 1} \right) = (n + 1) u\left(0 + \frac{a}{n + 1}, b - \frac{b}{n + 1} \right)$$

$$= (n + 1)u(0, b) + au_1 - bu_2 + \frac{1}{2}\left[\frac{a^2}{n + 1} u_{1,1} - \frac{2ab}{n + 1} u_{1,2} + \frac{b^2}{n + 1} u_{2,2} \right] + R^{(1)}$$

where $|R^{(1)}| \leq c/n^2$. for some $c > 0$ by the form of the remainder term and the continuity of the third order derivatives. Similarly

$$v(I\backslash\{j\}) = nu\left(\frac{a}{n}, \frac{(n - 1)b}{n} \right) = nu\left(0 + \frac{a}{n}, b - \frac{b}{n} \right)$$

$$= nu(0, b) + au_1 - bu_2 + \frac{1}{2}\left[\frac{a^2}{m} u_{1,1} - \frac{2ab}{n} u_{1,2} + \frac{b^2}{n} u_{2,2} \right] + R^{(2)}$$

where again $|R^{(2)}| \leq d/n^2$. for some $d > 0$. Hence

$$n \, | \, v(I) - v(I \backslash \{j\}) - u(0, b) \, | \leq$$

$$\frac{1}{2} \, | \, \frac{na^2}{n+1} u_{1,1} - a^2 u_{1,1} \, | + \, | \, \frac{nab}{n+1} u_{1,2} - abu_{1,2} \, |$$

$$+ \frac{1}{2} \, | \, \frac{nb^2}{n+1} u_{2,2} - b^2 u_{2,2} \, | + n \, | \, R^{(1)} - R^{(2)} \, | \, ,$$

where $n \, | \, R^{(1)} - R^{(2)} \, | \leq e/n$, for some $e > 0$. For n sufficiently large all terms on the right can be made arbitrarily small, which proves (4.32).

Now suppose $\mathbf{y}$ is an imputation with $y_1 \leq x_1 - \varepsilon$. Since

$$\sum_{i \in I} y_i = x_1 + n \, u(0, b),$$

It follows that

$$\sum_{i \neq I} y_i \geq n \, u(0, b) + \varepsilon \, .$$

Hence there exists $j \in N$, that is $j \neq 1$, such that

$$y_j \geq u(0, b) + \frac{\varepsilon}{n} \, .$$

Assuming $n \geq n(\varepsilon)$ and using (4.32) this gives

$$\sum_{i \in I \backslash \{j\}} y_i \leq v(I) - u(0, b) - \frac{\varepsilon}{n} < v(I \backslash \{j\}) \, ,$$

where $j \neq 1$. Suppose

$$v(I \backslash \{j\}) - \left(v(I) - u(0, b) - \frac{\varepsilon}{n} \right) = \delta > 0 \, .$$

Then the imputation obtained from $\mathbf{y}$ by adding δ/n to every component except the jth and subtracting δ from the jth component, dominates $\mathbf{y}$ through the coalition $\Lambda\{j\}$. Hence $\mathbf{y}$ is not in the core.

For $[m,n]$ games with m and n large, the case of pure competition, the situation is more complicated, and we shall only briefly discuss it. In the game $[m,n]$ the function

$$u\left(\frac{s_m a}{|S|}, \frac{s_n b}{|S|}\right) \quad (s_m + s_n = |S|)$$

represents the average return to all members of S. For m, n not fixed we can consider this as a function of the rational number s_m/s_n by writing

$$u\left(\frac{s_m a}{|S|}, \frac{s_n b}{|S|}\right) = u\left(\frac{(s_m/s_n)a}{1 + s_m/s_n}, \frac{b}{1 + s_m/s_n}\right).$$

Now $u(x, y)$ is a continuous function in rectangle $0 \le x \le a$, $0 \le y \le b$, so that

$$\sup_{\xi \ge 0} u\left(\frac{\xi a}{1 + \xi}, \frac{b}{1 + \xi}\right)$$

exists but is not necessarily attained. The cases where the sup occurs at $\xi = 0$ or as $\xi \to \infty$, giving values $u(0, b)$ or $u(a, 0)$ respectively, correspond to a situation where no coalition can improve upon the initial endowment of its members by trading, that is, to an inessential game. Thus if we suppose that the game is essential the sup is attained at some $\xi = \theta$, $0 < \theta < \infty$. Moreover, θ is unique since $u(x, y)$ is a strictly concave function. It becomes necessary to distinguish two cases according as θ is rational or irrational.

If θ is rational, say $\theta = p/q > 0$, then the maximum possible average return to members of a coalition S occurs when the ratio

$$\frac{s_m}{s_n} = \frac{|S \cap M|}{|S \cap N|} = \frac{p}{q}. \tag{4.33}$$

In an $[m,n]$ game with $m = kp$, $n = kq$ for some integer $k \ge 1$, the grand coalition I will be able to achieve this maximum return. Thus the imputation

$$\left(\frac{v(I)}{m + n}, \dots, \frac{v(I)}{m + n}\right)$$

will be undominated and so in the core.

If, however, we fix some other ratio $p'/q' \ne p/q$ and consider $[m,n]$ games with $m = kp'$, $n = kq'$ ($k \ge 1$). Then for sufficiently large k there will exist coalitions $S \ne I$ for which (4.31) is satisfied, so that such S can obtain maximal average return. Because of this Shubik suggests that we can expect the core of such games to be empty. This seems rather hard to justify in all cases; however, it is likely to be true for a reasonably large class of utility functions $u(x, y)$ and constants a, b.

It is therefore not possible to give an unqualified justification of Edgeworths conjecture, in the case of pure competition, given the classical cooperative game theory approach we have adopted. However, we could consider Edgeworth justified in a weaker sense if it were possible to construct an infinite sequence of games $[m,n]$ with non-empty cores which 'shrink' to a single point as m and n become large. Shubik [15] stated the following theorem.

Theorem. If θ is rational, say $\theta = p/q$, then for the game $[m,n]$, where $m = kp$, $n = kq$, the imputation

$$\left(\frac{v(I)}{m+n}, \ldots, \frac{v(I)}{m+n} \right)$$

is always in the core. Moreover, given any $\varepsilon > 0$ where exists $k = k(\varepsilon)$ such that for all $k \geq k(\varepsilon)$ no imputation $\mathbf{y}$ with any $y_j \leq v(I)/(m+n) - \varepsilon$ can belong to the core.

The proof is rather similar to that of Theorem 4.16, and we do not give the details. It seems likely that if θ is irrational, but not badly approximable, by using results on diophantine approximation it may still be possible to construct an infinite sequence of games $[m,n]$ having the required property.

4.8 THE SHAPLEY VALUE.

The reader who has conscientiously worked through the preceding sections may understandably feel a certain dissatisfaction with the classical theory of cooperative games. Not only is the vN-M solution non-unique in the sense that it is a set of imputations rather than a single one, but frequently many such stable sets are available. Moreover, knowing the theory will by no stretch of the imagination actually help in playing the game! Long before the discovery by Lucas of a game with no stable set solution, game theorists were searching for other solution concepts; a detailed discussion of the relative merits of several of these can be found in Luce and Raiffa [1]. In this section we examine one such idea due to Shapley [16].

Our starting point is the idea that a players evaluation of a classical cooperative game should be a real number $\phi_i(v)$, where i denotes the player index and v is the characteristic function of the game. The number $\phi_i(v)$ is the be the ith players evaluation in the sense that it represents a '*fair payoff*' to player i, taking into account that players particular strength or weakness as reflected by the characteristic function.

It is by no means obvious which function of the characteristic function we should select; certainly any arbitrary definition would be subject to question. An alternative procedure, one of extreme power and persuasiveness which is commonly used in mathematics but less known in the social sciences, is to specify the function by certain properties one feels it should have, that is, axiomatically. This was the approach adopted by Shapley; he listed three *apparently* weak conditions and was then able to show that these uniquely determine a vector

$$\Phi(v) = (\phi_1(v), \ldots, \phi_n(v)).$$

He called the function $\phi_i(v)$ the **value** of the game to the ith player.

Symmetry is certainly one of the main features of 'fairness'; players who have the same strengths and weaknesses should receive equal payoffs.

Definition. Any permutation of players $\pi : I \to I$ is called an **automorphism** of the classical cooperative game $\langle I, v, X \rangle$ if

$$v(\pi S) = v(S) \text{ for every } S \subseteq I.$$

Shapleys first condition can now be expressed as

Axiom of symmetry. For any automorphism π of the game $\langle I, v, X \rangle$,

$$\phi_i(v) = \phi_{\pi i}(v) \text{ for every } i \in I.$$

We do not need to suppose that $\Phi(v)$ is an imputation (although it eventually emerges that this is the case), in fact it is sufficient to require

Axiom of effectiveness.
$$\sum_{i \in I} \phi_i(v) = v(I) \, .$$

This axiom is subject to the same criticism as condition (4.15) on imputations. It reflects the assumption that the grand coalition I will form, and as such represents "an n-fold combination of wishful thinking".

The final axiom is based on the fact that if v and w are characteristic functions of two games with the same set of players I, then it is easy to see that $v + w$ defined by

$$(v + w)(S) = v(S) + w(S) \text{ for every } S \subseteq I$$

is also a characteristic function over I. If player $i \in I$ is participating in two simultaneous games with characteristic functions v and w, which we can think of as a single game $v + w$, he or she would have an evaluation $\phi_i(v + w)$, and we might expect that $\phi_i(v + w) = \phi_i(v) + \phi_i(w)$.

Axiom of aggregation. If v and w are two characteristic functions over I, then $\phi_i(v + w) = \phi_i(v) + \phi_i(w)$ for every $i \in I$.

Although the first two axioms are quite reasonable, the third axiom is more difficult to accept. As Luce and Raiffa ([1], 284) assert:

> "For although $v + w$ is a game composed from v and w, we cannot in general expect it to be played as if it were two separate games. It will have its own structure very different from those for v and w. Therefore one might very well argue that its *a priori* value should not necessarily be the sum of the

values of the two component games. This strikes us as a flaw in the concept
of value, but we have no alternative to suggest."

An explicit formula for the Shapley vector is given by

Theorem. The unique vector $\Phi(v) = (\phi_1(v), ..., \phi_n(v))$ which satisfies the above axioms is given
by

$$\phi_i(v) = \sum_{S \ni i} \frac{(n - |S|)! \, (|S| - 1)!}{n!} (v(S) - v(S \backslash \{i\})), \qquad (4.34)$$

where the sum is taken over all coalitions S which contain i.

A detailed proof can be found in Shapley [16] or Vorobev ([5], §4.17).

(Note. In both cases the axiom of effectiveness is stated in a slightly different form from that
given above; however, the three versions are easily seen to be equivalent.)

Example 4.6 For the 3-person symmetric game with

$$v(\phi) = v(\{1\}) = v(\{2\}) = v(\{3\}) = 0,$$
$$v(\{1,2\}) = v(\{1,3\}) = v(\{2,3\}) = \alpha,$$
$$v(\{1,2,3\}) = 1,$$

we have

$$\phi_1 = \frac{2! \, 0!}{3!} (v(\{1, 2, 3\}) - v(\{2,3\})) + \frac{1! \, 1!}{3!} (v(\{1,2\}) - v(\{2\}))$$

$$+ \frac{1! \, 1!}{3!} (v(\{1,3\}) - v(\{3\})) + \frac{0! \, 2!}{3!} (v(\{1\}) - v(\{\phi\}))$$

$$= \frac{2}{6} (1 - \alpha) + \frac{1}{6} \alpha + \frac{1}{6} \alpha + \frac{2}{6} .0 = \frac{1}{3} .$$

In this way we obtain $\Phi(v) = (\, ^1/_3, \, ^1/_3, \, ^1/_3)$, which in view of the prefect symmetry of the game
is the desired answer.

Suppose $i \in S$. The coalition $S \backslash \{i\}$ is capable of being formed in $(|S| - 1)!$ ways, and the
$n - |S|$ players not in S can be arranged in $(n - |S|)!$ ways. Thus the number of ways in
which player i can join $S \backslash \{i\}$ is $(n - |S|)!(|S| - 1)!$. There are $n!$ ways in which the grand
coalition I can form, and if we assume that each of these is equally probable we can interpret

$$\frac{(n - |S|)!(|S| - 1)!}{n!}$$

as the probability that player i will enter the coalition $S \backslash \{i\}$. Thus the theorem demonstrates that
Shapleys axioms are equivalent to the proposal that player i should receive the average of all
his or her contributions to coalitions S which contain i.

Several interesting applications of n-person game theory and Shapley value to the social sciences can be found in the literature. For example in 1964 the Nassau County Legislature of New York used weighted voting for its six members, the weights being 31:31:28:21:2:2. For the 115 votes the majority was 58. We can treat this as a 6-person simple game with $v(S) = 1$ if the total weight of coalition S is at least 58 and $v(S) = 0$ otherwise. Calculation of the Shapley vector gives a measure of the distribution of power between the legistlators and in fact $\Phi(v) = (\,{}^1/_3,\, {}^1/_3,\, {}^1/_3,\, 0,\, 0,\, 0)$. It is somewhat bizarre that the legislator with 21 votes was *never* a critical member of a winning coalition (even that persons addition to the coalition 31:2:2 creates only 56 votes)! In a paper of Mann and Shapley ([17], 151) the Shapley value is applied to the Electoral College of the United States (the results are less amazing).

> **Example 4.7** *United Nations Security Council.* At the present time (1978) U. N. Security Council has 15 members, five of whom have vetoes. For a substantive resolution to pass it is necessary to have nine affirmative votes and no vetoes.
>
> To treat this as a 15-person simple game take $I = \{1, 2, ..., 15\}$ and suppose the first five have the veto. We define a coalition S to be winning, $v(S) = 1$, if it can *defeat* a substantive resolution, otherwise $v(S) = 0$. Thus the characteristic function is
>
> If $i \in S$ for some i, $1 \le i \le 5$, $v(S) = 1$.
>
> $$\text{Otherwise } v(S) = \begin{cases} 1, & \text{if } |S| \ge 7, \\ 0, & \text{if } |S| < 7. \end{cases}$$

Fortunately it is only necessary to calculate $\phi_1 = \phi_1(v)$, since the symmetries of the game will then permit easy computation of ϕ_i for $i > 5$. If we ignore those terms in (4.34) for which $v(S) - v(S\backslash\{i\}) = 0$ we obtain

$$
\begin{aligned}
\phi_1 = \ & \frac{14!\,0!}{15!}\,(v(\{1\}) - v(\phi)) + \frac{13!\,1!}{15!}\,(v(\{16\}) - v(\{6\}))\binom{10}{1} \\
& + \frac{12!\,2!}{15!}\,(v(\{167\}) - v(\{67\}))\binom{10}{2} + \frac{11!\,3!}{15!}\,(v(\{1678\}) - v(\{678\}))\binom{10}{3} \\
& + \frac{10!\,4!}{15!}\,(v(\{16789\}) - v(\{6789\}))\binom{10}{4} \\
& + \frac{9!\,5!}{15!}\,(v(\{1678910\}) - V(\{678910\}))\binom{10}{5} \\
& + \frac{8!\,6!}{15!}\,(v(\{167891011\}) - v(\{67891011\}))\binom{10}{6}
\end{aligned}
$$

Here the expression $\binom{k}{j}$ denotes the number of ways of choosing j members from k members, and

$$\binom{k}{j} = \frac{k!}{j!(k-j)!}$$

Thus terms like $v(\{167\}) - v(\{67\})$ will occur in (4.34) $\binom{10}{2}$ times, since there are 10 members who do not have the veto and so $\binom{10}{2}$ ways of choosing coalitions such as $\{67\}$ for whom $v(\{67\}) = 0$.

Hence

$$\phi_1 = \frac{1}{15} + \frac{10}{15\times14} + \frac{10\times9}{15\times14\times13} + \frac{10\times9\times8}{15\times14\times13\times12} + \frac{10\times9\times8\times7}{15\times14\times13\times12\times11}$$
$$+ \frac{10\times9\times8\times7\times6}{15\times14\times13\times12\times11\times10} + \frac{10\times9\times8\times7\times6\times5}{15\times14\times13\times12\times11\times10\times9} .$$

By courtesy of an electronic abacus this easily gives $\phi_1 = 0.19627...$

Hence

$$\phi_i = \begin{cases} 0.19627... & \text{if } 1 \le i \le 5, \\ 0.00186... & \text{if } 6 \le i \le 15 . \end{cases}$$

From this it is apparent that 98.1% of the power is in the hands of the five permanent members and individually a permanent member is more than 105 times as powerful as an ordinary member!

PROBLEMS FOR CHAPTER 4.

1.   a) Find the characteristic function of the 3-person game with the normal form

	Player 2			Player 2	
	A	B		A	B
	A(1, 1, 0)	(4, -2, 2)		(-3, 1, 2)	(0,1, 1)
Player 1	B(1, 2, -1)	(3, 1, -1)		(2, 0, -1)	(2, 1, -1)
	C(-1, 0, 1)	(-2, 1, -1)		(0, -1, 3)	(-3, 2, 1)
	Player 3 chooses A			*Player 3 chooses B*	

b) Give a simple example to show that condition (4.13) of Theorem 4.3 can also be satisfied by a non-constant sum game.

2. Prove that the core of an n-person classical cooperative game is a convex set.

3. Let **x** be an imputation in a 3-person game. Draw a diagram showing those imputations **y** which satisfy (4.26) for $S = \{1,2\}$, $\{2,3\}$ or $\{1,3\}$.

By taking (4.25) into consideration prove that if the core of a 3-person game intersects each bounding line of the imputation set it is the unique vN-M solution. [Hint: it may be helpful to refer to Figure 4.3.].

4. For the 4-person constant sum game in 0,1 reduced form for which $v(\{i,j\}) = \frac{1}{2}$ if $i \neq j$, $1 \leq i, j \leq 4$, consider the set J of imputations, where

$$J = \{\ \gamma_i = (\gamma_{i1}, \gamma_{i2}, \gamma_{i3}, \gamma_{i4}); \ 1 \leq i \leq 4, \ \gamma_{ij} = \frac{1}{3} \text{ if } i \neq j \ \}.$$

Show that no element of J dominates another. Determine whether J is a stable set for this game, and justify your conclusion.

5. Consider the cooperative simple game $\Gamma = \langle N, v, X \rangle$, where the set of players $N = \{1, 2, 3, ...\}$ is countably infinite and for $S \subseteq N$

$$v(S) = \begin{cases} 1, & \text{if } N \backslash S \text{ is finite}, \\ 0, & \text{if } N \backslash S \text{ is infinite}. \end{cases}$$

The set of imputations X is taken to be the set of all infinite sequences $(x_1, x_2, x_3, ...)$ with $x_i \geq 0$ and for all $i \in N$ and

$$\sum_{i \in N} x_i = 1.$$

Dominance and stable sets are defined as before. Prove Γ has no stable set.

6. Prove that the core of an n-person symmetric game is empty if and only if there exists a coalition $S \subset I$ such that

$$\frac{v(S)}{|S|} > \frac{v(I)}{|I|}.$$

[Hint: To prove that the core is empty if such a coalition S exists, for any imputation **x** consider the $|S|$ smallest components of **x**].

Apply this result to Example 4.5.

CHAPTER REFERENCES

[1] Luce, R. D and Raiffa, H., *Games and Decisions*, J. Wiley and Sons (1957), New York.

[2] KcKinsey, J. C. C., *Introduction to the Theory of Games*. McGraw-Hill (1952), New York.

[3] Gilles, D. B., *Some theorems on n-Person Games*, Ph. D. thesis (1953), Department of Mathematics, Princeton University, Princeton.

[4] McKinsey, J. C. C., 'Isomorphisms of games, and strategic equivalence'. *Contributions to the Theory of Games*, I, (Ann. Math. Sutdies No. 24)117-130 (1950), Princeton.

[5] Vorobev, N. N., *Game Theory, Lectures for Economists and Systems Scientists*. Springer-Verlag (1977), New York.

[6] Shapley, L. S., *n-person games V: stable set solutions including an arbitrary closed component*. The RAND Corporation, Research Memorandum RM-1005, 1952.

[7] Mills, W. H., 'The four person game - edge of the cube'. *Annals of Math.*, **59**, 367-378 (1954).

[8] Nering, E. D., *Report of an Informal Conference on the Theory of n-person Games*. Ed. by H. W. Kuhn, Logistics Research Project (1953), Department of Mathematics, Princeton University, Princeton.

[9] Bott, R., 'Symmetric solutions to majority games'. *Contributions to the Theory of Games*, II, (Ann. Math. Studies No. 28) 319-323 (1953), Princeton.

[10] Gilles, D. B., 'Discriminatory and bargaining solutions to a class of symmetric n-person games'. *Contributions to the Theory of Games*, II, (Ann. Math. Studies No. 28) 325-342 (1953), Princeton.

[11] Lucas, W. F., *A game with no Solution*. The RAND Corporation, Research Memorandum RM-5518-PR, 1967.

[12] Gilles, D. B., 'Solutions to general non-zero-sum games'. *Contributions to the Theory of Games*, IV, (Ann. Math. Studies No. 40) 47-85 (1959), Princeton.

[13] Arrow, K. and Hahn, F., *General Competitive Analysis*. Oliver and Boyd (1971), Edinburgh.

[14] Edgeworth, F. Y., *Mathematical Psychics*. C. Kegan Paul and Co. (1881), London.

[15] Shubik, M., Edgeworth market games. *Contributions to the Theory of Games*, IV, (Ann Math. Studies No. 40) 267-278 (1959), Princeton.

[16] Shapley, L. S., A value for *n*-person games. *Contributions to the Theory of Games*, II

(Ann. Math. Studies No. 28) 307-317 (1953), Princeton.

[17] Shubik, M., ed., *Game Theory and Related Approaches to Social Behavior*. J. Wiley and Sons (1964), New York.

5

Bargaining Models

This may hurt you more than it hurts me...

5.1 INTRODUCTION.

A 2-person bargaining situation generally involves two individuals who have the opportunity to collaborate for mutual benefit in more than one way, although in degenerate situations it may be the case that only one, or neither, of the players can actually gain by cooperation. Thus the economic situations of trading between two nation states or of negotiation between employer and labour union may be regarded as bargaining problems.

If we were to assume that payoffs are in monetary terms, utility is linear in money, interpersonal comparisons of utility are meaningful, and monetary side payments are allowed, then the 2-person bargaining problem would become a 2-person cooperative game as considered in Chapter 4. However, except in the special case of a 2-person zero sum game, the classical theory of n-person cooperative games does not attempt to find a unique allocation of payoffs for a given n-person game, that is to determine what it is worth to each player to have the opportunity to engage in the game. One such solution concept, the Shapley vector, was briefly discussed in §4.8, and this should serve to illustrate the sort of thing we now require. The Shapley vector is open to a number of objections, some of which were mentioned at the time. In particular it would appear to take no account of the threats which one player or coalition may level against another in the course of preplay negotiation. One feels that a player who commands a particularly powerful threat is somehow in a stronger bargaining position than one who does not. For this reason we return to the simplest case, that of the cooperative 2-person non-zero sum game, and examine it as a bargaining problem.

In general terms we shall idealise the bargaining problem by supposing that the two individuals are highly rational, and

(i)	Payoffs are in utility terms.
(ii)	No interpersonal comparisons of utilities are assumed.

Moreover we also suppose that each player has full knowledge of the personal utility of the other. It should be pointed out that this last assumption, which is quite basic to game theory and to bargaining models in particular, is extremely dubious. For suppose that in a bargaining situation the players agree to submit to an arbiter who is committed to some prespecified method of finding a solution, then as Luce and Raiffa ([1], 134) point out

"To resolve the conflict the arbiter must first ascertain their utility functions; hence the situation deteriorates into a game of strategy where each player tries to solve the problem of how best to falsify or exaggerate his true tastes. In most situations, a player's preferences are only partially known to his adversary, and falsification of one's true feelings is an inherent and important bargaining strategy. An arbiter, to be successful, must skilfully ferret out at least a part of the truth. This reality is seriously idealised in game theory, and thereby the theory is seriously restricted. This is not to say it is useless in all situations, but only that there is always the fear that the real problem may have been abstracted away."

In practice, people are often confused about what their own objectives are, let alone what those of others may be (see, for example, the Crossman diaries [2]). As Luce and Raiffa observe, it is in the nature of things that any bargaining situation will provide some at least of the players with an incentive to conceal or misrepresent their true preferences, and this raises further technical and conceptual difficulties in applications. Some interesting work [3] has been done on this in the context of welfare economics. Here one is concerned with the question of deciding what government action, for example in respect of taxation, will in theory induce economic agents, acting in their own best interests, to reveal their true preferences so that non-distorted welfare policies can be operated. The flavour of these ideas can be obtained from the following example.

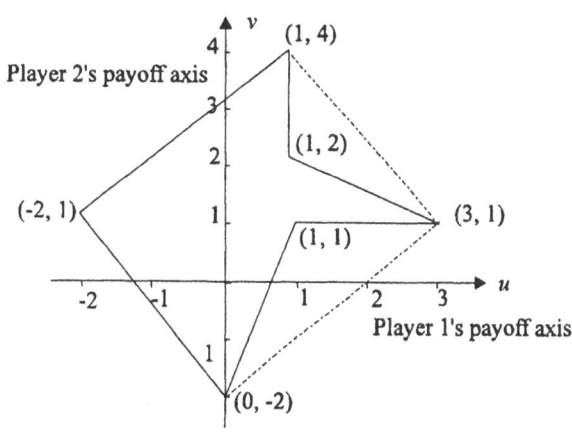

Figure 5.1 Payoff polygon.

An auctioneer announces that an object is to be sold by secret tender to the highest bidder but at the price of the *second* highest bid. The idea is that under these circumstances the players will tender their true valuations of the object.

Notwithstanding these difficulties, many different approaches to bargaining problems have been developed. We shall begin by examining the ideas developed by J. F. Nash [4, 5].

5.2 GRAPHICAL REPRESENTATION OF GAMES AND STATUS QUO POINTS.

In analysing such problems it is often convenient to have various features of the game displayed graphically. For example for the game

$$\begin{array}{c c c c} & \tau_1 & \tau_2 & \tau_3 \\ \sigma_1 & (\ 1, \ 4) & (-2, \ 1) & (\ 1, \ 2) \\ \sigma_2 & (\ 0,-2) & (\ 3, \ 1) & (\ 1, \ 1) \end{array}$$

we begin by drawing the diagram in Figure 5.1.

This is a non-convex polygon with six sides. If the players agree to coordinate their strategies appropriately, any payoff pair (u, v) which lies in the convex hull of the polygon can be achieved. This observation is an immediate consequence of the linearity of each player's utility function, that is the *expected utility hypothesis* discussed in §4.1. For example, if player 1 agrees to play the pure strategy σ_1 then by a suitable choice of mixed strategy $y = (y_1, y_2, y_3)$, $0 \leq y_i \leq 1$, $\Sigma y_i = 1$, player 2 can achieve any payoff pair (u, v) which lies in the triangle with vertices $(1, 4)$, $(-2, 1)$, $(1, 2)$. For using linearity of utility we have

$$\begin{aligned} (u(\sigma_1, y), v(\sigma_1, y)) &= (y_1 u(\sigma_1, \tau_1) + y_2 u(\sigma_1, \tau_2) + y_3 u(\sigma_1, \tau_3) , \\ & \quad y_1 v(\sigma_1, \tau_1) + y_2 v(\sigma_1, \tau_2) + y_3 v(\sigma_1, \tau_3)) \\ &= (y_1 - 2 y_2 + y_3 , \quad 4 y_1 + y_2 + 2 y_3) \\ &= y_1 (1, 4) + y_2 (-2, 1) + y_3 (1, 2) \end{aligned}$$

and as the y_i vary, subject to the conditions $0 \leq y_i \leq 1$, $\Sigma y_i = 1$, this point varies over the triangle with vertices $(1, 4)$, $(-2, 1)$, $(1, 2)$ as asserted. Similarly if player 1 agrees to play σ_2 and player 2 randomises appropriately between τ_1, τ_2 and τ_3, then any point in the triangle with vertices $(0, -2)$, $(3, 1)$, $(1, 1)$ can be achieved. However, not all points in the convex hull can be reached by *independently* randomised previously agreed mixed strategies. For consider points on the line segment joining $(3, 1)$ and $(1, 4)$. To obtain an expected payoff pair on this line it is necessary for the players to randomise between the pure strategy *pairs* (σ_1, τ_1), (σ_2, τ_2), as opposed to player 1 and player 2 independently randomising between σ_1, σ_2 and τ_1, τ_2 respectively. Nevertheless we permit this sort of agreement and define the **feasible region** (or **payoff polygon**) to be the convex hull of all payoff pairs defined by the original game. For bimatrix games the feasible region is always a closed bounded convex *polygon*, and in general we shall restrict our attention to games where the feasible region is a closed bounded convex set.

We now see that given a 2-person non-zero sum cooperative game there is a certain convex subset of $\mathbb{R}^2$, called the feasible region, consisting of all payoff pairs (u, v) which can be achieved when strategic cooperation is allowed. Our problem is to decide which point of the feasible region to single out as 'the solution' of the game. In general, the more one player gets, the less the other player will be able to get, though this need not always be the case. The question is how much will one player give the other, how little will a player be willing to accept, as the price of cooperation?

Without formalising the negotiation procedure too precisely we shall suppose the existence of a **status quo point** (u^*, v^*). Geometrically, a status quo point is on or within the boundary of the feasible region. It represents the payoff to each player if they do not agree, or in

'negotiations break down' as the
saying goes. If threats have been
presented in the course of
negotiations, in the event that
negotiations break down it is
essential that players be *compelled*
to carry out their threats. Otherwise,
since the players are supposed to be
rational beings, a threat would have
little meaning.

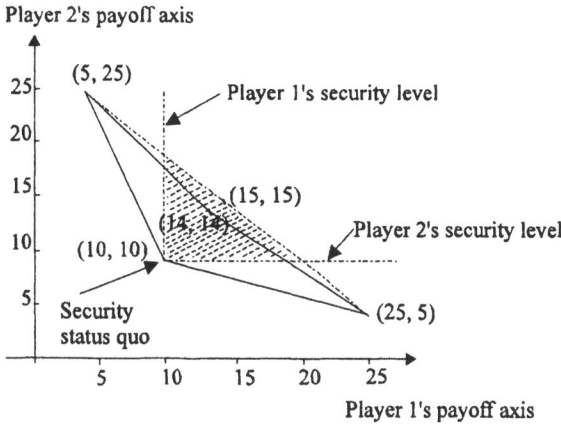

Figure 5.2 Graph for Example 5.1.

What sort of points can we consider
to be status quo points? One
possibility commonly used is to
take as the status quo that point
which corresponds to each player's
security level. The **security level** of a player is the maximum payoff which that player can
obtain regardless of what the other player does. It is easily found by ignoring the other player's
payoffs and treating the resulting matrix as a zero sum game.

Example 5.1 Consider the game

$$
\begin{array}{cc}
 & \tau_1 \qquad \tau_2 \\
\begin{array}{c} \sigma_1 \\ \sigma_2 \end{array} &
\left(
\begin{array}{cc}
(10,\ 10) & (25,\ \ 5) \\
(\ 5,\ 25) & (14,\ 14)
\end{array}
\right)
\end{array}
$$

To find the security level for player 1 we examine the game

$$
\left(
\begin{array}{cc}
10 & 25 \\
5 & 14
\end{array}
\right)
\quad (\text{Payoffs to player } 1)
$$

This game has a saddle point at (σ_1, τ_1) and hence value 10, which is therefore player 1's
security level.

Similarly for player 2 we examine the game

$$
\left(
\begin{array}{cc}
10 & 5 \\
25 & 14
\end{array}
\right)
\quad (\text{Payoffs to player } 2)
$$

Warning! This shows payoffs to the *column* player, and hence in seeking a saddle point etc. one
must consider the minimum of the row maxima and the maximum of the column minima, rather
than the usual way round. Again the game has a saddle point at (σ_1, τ_1) which gives player 2's
security level also as 10.

Thus in this game (10, 10) is the status quo point defined by the players' respective security levels. The situation is illustrated graphically in Figure 5.2.

Since the security levels represent payoffs which are guaranteed to the players without cooperation, it can be argued that the point with these coordinates constitutes a status quo point from which negotiations between the players can begin. Thus for the game above, the point (10, 10) could constitute a status quo point, and negotiations could then proceed to find a solution somewhere on, or within, the boundary of the shaded triangle in Figure 5.2. In this case a fairly obvious procedure is for the players to agree to coordinate their strategies so that the pure strategy pairs (σ_1, τ_2), (σ_2, τ_1) are each chosen for half of the playing time, then the outcome of the game would be the point (15, 15) shown on the most north-easterly boundary of the payoff polygon in Figure 5.2.

There are many games where the adoption of security levels as a status quo point is not satisfactory, because security levels give no indication of the bargaining strengths of the players.

Example 5.2.

$$
\begin{array}{cc}
 & \tau_1 \qquad\quad \tau_2 \\
\begin{array}{c} \sigma_1 \\ \sigma_2 \end{array} &
\left(\begin{array}{cc}
(\ \ 5,\ 20) & (\ -7,-19) \\
(-16,-4) & (\ 20,\ \ \ 5)
\end{array} \right)
\end{array}
$$

In this game a cooperative agreement to randomise equally between (σ_1, τ_1) and (σ_2, τ_2) would give each player a gain of $12\frac{1}{2}$. However, unlike the previous example, the present game offers player 2 a bargaining advantage. In preplay negotiations player 2 can threaten to play τ_1 all the time. In this circumstance the best that player 1 could do would be to play σ_1, which leads to the outcome (5, 20) precisely that desired most by player 2. If, on the other hand, player 1 counters τ_1 by using σ_2 to inflict a loss of 4 on player 2, then of course 1 loses 16. Thus τ_1 is said to be a **threat strategy** which player 2 can use to obtain a bargaining advantage. To find the best threat strategy is usually not as simple as in this game, especially where the best threat strategy is mixed rather than pure, and we shall return to this question in §5.5.

If both players use their best threat strategies, although the threat of σ_2 by player 1 is hardly effective, then (-16, -4) will be the threat status quo point. The threat and security status quo points for this game are shown in Figure 5.3. If the status quo point is taken as (-16, -4) then we should seek a solution within the shaded area.

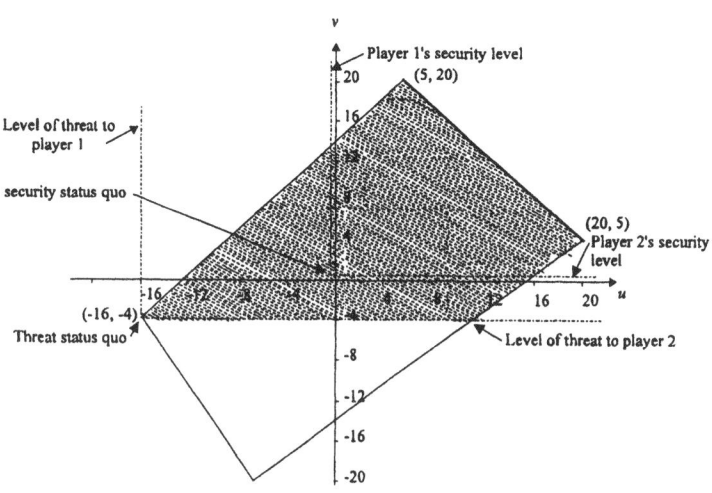

Figure 5.3 Graph for Example 5.2.

Thus the choice of the status quo point (u^*, v^*) may vary with different assumptions about the attitudes, behaviour and bargaining strengths of the players. But whatever status quo point is adopted, it will not in itself generate a solution although it will obviously play a part in finding it.

5.3 THE NASH BARGAINING MODEL.

We have seen that a 2-person bargaining problem is characterised by a closed bounded convex subset S, say, of the plane called the feasible region together with a status quo point $(u^*, v^*) \in S$. We can define a **bargaining solution,** or **arbitration scheme,** to be a function ϕ which maps such a triple (S, u^*, v^*) into a unique outcome $(\bar{u}, \bar{v}) \in S$, that is

$$\phi(S, u^*, v^*) = (\bar{u}, \bar{v}).$$

Our problem is: how is such a function ϕ to be defined? Nash suggested the following axioms as being reasonable conditions to require of any such function.

Axiom 1. (**Individual rationality**). $\bar{u} \geq u^*, \bar{v} \geq v^*$.

This merely requires that the arbitrated outcome must be at least as good as the status quo.

Axiom 2. (**Feasibility**). $(\bar{u}, \bar{v}) \in S$.

Axiom 3. (**Pareto optimality**). If $(u, v) \in S$, $u \geq \bar{u}$ and $v \geq \bar{v}$, then $(u, v) = (\bar{u}, \bar{v})$.

Thus $(\bar{u}, \bar{v})$ is not bettered by any other feasible point.

Axiom 4. (**Independence of irrelevant alternatives**). If $(\bar{u},\ \bar{v}) \in T \subset S$ and $(\bar{u},\ \bar{v}) = \phi(S, u^*, v^*)$ then $(\bar{u},\ \bar{v}) = \phi(T, u^*, v^*)$.

This axiom states that if the arbitrated outcome $(\bar{u}, \bar{v})$ of the larger bargaining problem (S, u^*, v^*) is actually a feasible outcome of a smaller problem (T, u^*, v^*), then it shall also be the arbitrated outcome of the smaller problem. To put it another way, if certain new feasible outcomes are added to a bargaining problem but the status quo point remains unchanged, either the arbitrated solution is also unchanged or it becomes one of the new outcomes. Despite its innocent appearance this axiom has been subjected to some criticism, although usually in a context other than game theory ([1], 132).

Axiom 5. (**Invariance with respect to utility transformations**). Let T be obtained from S by the transformation

$$u' = a_1 u + b_1 \quad (a_1 > 0),$$
$$v' = a_2 v + b_2 \quad (a_2 > 0).$$

Then if $\phi(S, u^*, v^*) = (\bar{u}, \bar{v})$ we require that

$$\phi(T, a_1 u^* + b_1,\ a_2 v^* + b_2) = (a_1 \bar{u} + b_1,\ a_2 \bar{v} + b_2).$$

Since linear utility is only defined up to a linear transformation of this type and no interpersonal comparison of utility is assumed, this axiom is unobjectionable. If, however, interpersonal comparison of utility is possible, there are obvious objections to demanding that the solution be the same for a player if the payoff is made in pence as opposed to pounds, say, when the other player's payoff *continues* to be made in pounds.

Axiom 6. (**Symmetry**). If (S, u^*, v^*) is such that

(i) $(u, v) \in S$ implies $(v, u) \in S$,
(ii) $u^* = v^*$,
and   (iii) $\phi(S, u^*, v^*) = (\bar{u}, \bar{v})$,

then $\bar{u} = \bar{v}$.

This means that if an abstract version of a bargaining problem places the players in completely symmetric roles, the arbitrated outcome shall give them equal utility payoffs, where utility is measured in the units which made the game symmetric.

It is a remarkable fact that given these six axioms the function ϕ, and therefore the arbitration procedure which solves the bargaining problem $\phi(S, u^*, v^*)$, is uniquely determined. In fact we have

Theorem 5.1 (Nash). Suppose there exist points $(u, v) \in S$ with $u > u^*$ and $v > v^*$, and that the maximum of

$$g(u, v) = (u - u^*)(v - v^*)$$

over this set is attained at $(\bar{u}, \bar{v})$. Then the point $(\bar{u}, \bar{v})$ is uniquely determined, and the function $\phi(S, u^*, v^*) = (\bar{u}, \bar{v})$ is the unique function which satisfies axioms 1 to 6.

If the hypothesis of the theorem is *not* satisfied, the problem is very much easier. For suppose there are no points $(u, v) \in S$ with $u > u^*$ and $v > v^*$. By the convexity of S, if there is a point $(u, v) \in S$ with $u > u^*$ and $v = v^*$, there can be no point $(u, v) \in S$ with $v > v^*$. In this situation we simply define $(\bar{u}, \bar{v})$ to be that point in S which maximises u, subject to the constraint $v = v^*$. Similarly, if there is a point $(u, v) \in S$ with $u = u^*$ and $v > v^*$, there can be no point $(u, v) \in S$ with $u > u^*$; we then define $(\bar{u}, \bar{v})$ to be that point in S which maximises v, subject to the constraint $u = u^*$. It is easily checked that these solutions satisfy the axioms, and that axioms 1 to 3 ensure uniqueness.

Recall that S is a closed bounded subset of the plane, and hence the intersection of S with the region defined by $u \geq u^*$, $v \geq v^*$ is also closed bounded. Since g is continuous it attains its maximum on this set. Moreover, by the hypothesis of the theorem, there exist points $(u, v) \in S$ with $u > u^*$ and $v > v^*$, that is with $g(u, v) > 0$, consequently the maximum will not occur on the lines $u = u^*$ or $v = v^*$ where of course $g(u, v) = 0$. Having disposed of the existence of the maximum we next deal with the uniqueness.

Lemma 5.1. If there exist points $(u, v) \in S$ with $u > u^*$ and $v > v^*$ then

$$\max_{\substack{u > u^*,\ v > v^* \\ (u, v) \in S}} g(u, v)$$

is attained at a unique point $(\bar{u}, \bar{v})$.

Proof. This is a simple consequence of the convexity of S. Suppose there are two distinct points (u', v'), (u'', v'') which maximise $g(u, v)$, and let M be the maximum value. Since $M > 0$, $u' = u''$ implies $v' = v''$. Hence without loss of generality we can suppose that $u' < u''$, in which case $v' > v''$. Put

$$(u, v) = \frac{1}{2}(u', v') + \frac{1}{2}(u'', v'') .$$

Then $u > u^*$, $v > v^*$ and, since S is convex, $(u, v) \in S$.

Now

$$g(u, v) = (\frac{1}{2}(u' + u'') - u^*)(\frac{1}{2}(v' + v'') - v^*)$$

$$= \frac{(u' - u^*) + (u'' - u^*)}{2} \cdot \frac{(v' - v^*) + (v'' - v^*)}{2}$$

$$= \frac{(u' - u^*)(v' - v^*)}{2} + \frac{(u'' - u^*)(v'' - v^*)}{2} + \frac{(u' - u'')(v'' - v')}{4}$$

$$= M + \frac{(u' - u'')(v'' - v')}{4}$$

Since $u' < u''$ and $v' > v''$, the last expression is positive, hence $g(u, v) > M$, which contradicts the maximality of M. Hence the point $(\bar{u}, \bar{v})$ which maximises $g(u, v)$ is unique.

Lemma 5.2. Let (S, u^*, v^*) and $(\bar{u}, \bar{v})$ be as above. Let

$$h(u, v) = (\bar{v} - v^*)u + (\bar{u} - u^*)v .$$

Then for every point $(u, v) \in S$ we have $h(u, v) \leq h(\bar{u}, \bar{v})$.

This lemma shows that the line through $(\bar{u}, \bar{v})$, whose slope is the negative of the line joining $(\bar{u}, \bar{v})$ to (u^*, v^*), is a support line for S, that is S lies entirely on or below the line.

Proof. Suppose $(u, v) \in S$ with $h(u, v) > h(\bar{u}, \bar{v})$. Then, since h is linear,

$$h(u - \bar{u}, v - \bar{v}) = h(u, v) - h(\bar{u}, \bar{v}) = \Delta > 0$$

Let $0 < \theta < 1$ and define

$$(u', v') = \theta(u, v) + (1 - \theta)(\bar{u}, \bar{v}) = (\bar{u}, \bar{v}) + \theta(u - \bar{u}, v - \bar{v}) .$$

If we now expand $g(u', v')$ about the point $(\bar{u}, \bar{v})$, using Taylor's theorem, we obtain the identity

$$g(u', v') = g(\bar{u}, \bar{v}) + (u' - \bar{u})(\bar{v} - v^*) + (v' - \bar{v})(\bar{u} - u^*) + (u' - \bar{u})(v' - \bar{v}$$

which gives

$$g(u', v') = g(\bar{u}, \bar{v}) + \theta(u - \bar{u})(\bar{v} - v^*) + \theta(v - \bar{v})(\bar{u} - u^*) + \theta^2(u - \bar{u})(v - \bar{v})$$
$$= g(\bar{u}, \bar{v}) + \theta h(u - \bar{u}, v - \bar{v}) + \theta^2(u - \bar{u})(v - \bar{v})$$
$$= g(\bar{u}, \bar{v}) + \theta\Delta + \theta^2(u - \bar{u})(v - \bar{v}) .$$

Next choose θ, $0 < \theta < 1$, sufficiently small to ensure that $\theta|(u - \bar{u})(v - \bar{v})| < \Delta$, then $\theta\Delta + \theta^2(u - \bar{u})(v - \bar{v}) > 0$. Hence $g(u', v') > g(\bar{u}, \bar{v})$. But this contradicts the maximality of $g(u, v)$. Hence $h(u, v) \leq h(\bar{u}, \bar{v})$ for all $(u, v) \in S$.

We can now prove the theorem.

Proof of Theorem 5.1. Let $(\bar{u}, \bar{v})$ be the unique point which maximises $g(u, v)$ and whose existence is assured by Lemma 5.1. This point clearly satisfies axioms 1 and 2 by construction. It also satisfies axiom 3, for if $u \geq \bar{u}$ and $v \geq \bar{v}$ but $(u, v) \neq (\bar{u}, \bar{v})$, then $g(u, v) > g(\bar{u}, \bar{v})$, which contradicts the maximality of $g(\bar{u}, \bar{v})$. Axiom 4 is plainly satisfied, for if $(\bar{u}, \bar{v})$ maximises $g(u, v)$ over $S \cap \{(u, v); u \geq u^*, v \geq v^*\}$, it also maximises over the smaller $T \cap \{(u, v); u \geq u^*, v \geq v^*\}$. With regard to axiom 5 we observe that if $u' = a_1 u + b_1$, $v' = a_2 v + b_2$, then

$$g'(u', v') = (u' - (a_1 u^* + b_1))(v' - (a_2 v^* + b_2)) = a_1 a_2 g(u, v)$$

Hence if $(\bar{u}, \bar{v})$ maximises $g(u, v)$ it follows that $(a_1 u + b_1, a_2 v + b_2)$ maximises $g'(u', v')$. Finally, $(\bar{u}, \bar{v})$ satisfies axiom 6, for suppose that (S, u^*, v^*) is symmetric in the sense of axiom 6. Then $u^* = v^*$ and $(\bar{v}, \bar{u}) \in S \cap \{(u, v); u \geq u^*, v \geq v^*\}$, hence $g(\bar{v}, \bar{u}) = (\bar{v} - u^*)(\bar{u} - v^*) = (\bar{v} - v^*)(\bar{u} - u^*) = g(\bar{u}, \bar{v})$. Since $(\bar{u}, \bar{v})$ is the unique point which maximises $g(u, v)$ it follows that $(\bar{v}, \bar{u}) = (\bar{u}, \bar{v})$, that is, $\bar{u} = \bar{v}$.

We now know that the function $\phi(S, u^*, v^*) = (\bar{u}, \bar{v})$ satisfies the six axioms, and we must show it is the unique function which satisfies them. Consider the closed half plane

$$H = \{(u, v); h(u, v) \leq h(\bar{u}, \bar{v})\},$$

where $h(u, v)$ is defined in Lemma 5.2. By Lemma 5.2 $S \subset H$. Let H' be obtained from H by the linear transformation

$$u' = \frac{u - u^*}{\bar{u} - u^*}, \qquad v' = \frac{v - v^*}{\bar{v} - v^*}. \qquad (5.1)$$

It is readily verified that H' is closed half plane

$$H' = \{(u', v'); u' + v' \leq 2\}.$$

Now since $\bar{u} > u^*$ and $\bar{v} > v^*$, the transformation (5.1) is of the type considered in axiom 5. Under this transformation $H \to H'$ and

$$u^* \to 0, \qquad v^* \to 0,$$
$$\bar{u} \to 1, \qquad \bar{v} \to 1.$$

Consider now the bargaining problem $(H', 0, 0)$. Since H' is symmetric by axiom 6, the solution must lie on the line $u' = v'$. Combining this with axiom 3 it follows that $(1, 1)$ is the unique solution to $(H', 0, 0)$. Hence by axiom 5, applied to the inverse of the transformation (5.1), $(\bar{u}, \bar{v})$ is the unique solution to (H, u^*, v^*). But since $S \subset H$ and $(\bar{u}, \bar{v}) \in S$, it now follows that $(\bar{u}, \bar{v})$ is the *unique* solution to (S, u^*, v^*).

Let us apply this method to the game discussed in Example 5.2, where $(u^*, v^*) = (-16, -4)$. To determine the Nash solution we have to find the point $(\bar{u}, \bar{v})$ which maximises $(u + 16)(v + 4)$ over the shaded region in Figure 5.3. However, we do not need to investigate *every* point in the region: it is apparent from the nature of the problem, or from axiom 3, that $(\bar{u}, \bar{v})$ cannot be an interior point and that it must lie on the line segment $(5, 20)$ and $(20, 5)$. Hence we need only maximise $(u + 16)(v + 4)$ on the line $v = 25 - u$ $(5 \le u \le 20)$. In fact the function $(u + 16)(29 - u)$ assumes a maximum at $u = 6\frac{1}{2}$. Hence the Nash solution is $(6\frac{1}{2}, 18\frac{1}{2})$. That player 2 gets so much more than player 1 is a reflection of the fact that player 2 has the greater threat. If we were to assume that the security levels represented the status quo point, that is $(u^*, v^*) = (-\frac{1}{4}, \frac{1}{2})$, then the solution would be $(12\frac{1}{8}, 12\frac{7}{8})$.

> **Example 5.3.** Two men are offered $500 if they can agree how to divide it, otherwise they each get nothing. The first man is very rich with a utility for $x given by $u = cx$, where $c > 0$ is small. The second man is poor, having only $100 capital, and his utility for a sum of money is proportional to its logarithm, thus his utility for $x is $v = \log(100 + x) - \log 100$. How should the money be divided?

In this example the status quo is $(u^*, v^*) = (0, 0)$ should they fail to agree. Suppose they agree to pay player 1 $y and player 2 $(500 - y)$. The utility to each player is then

$$u = cy, \quad v = \log(600 - y) - \log 100 = \log\frac{600 - y}{100},$$

respectively. Thus the set $S \cap \{(u, v); u \ge u^*, v \ge v^*\}$ will be the convex hull of $(0, 0)$ and the arc with the equation

$$v = \log\left(\frac{600 - \frac{u}{c}}{100}\right)$$

see Figure 5.4.

We seek the point in this region which maximises $g(u, v) = uv$, that is the value of u which maximises

$$g = u \log\left(\frac{600 - \frac{u}{c}}{100}\right).$$

If we differentiate and set equal to zero, we obtain the equation

$$\frac{\frac{u}{c}}{600 - \frac{u}{c}} = \log\left(\frac{600 - \frac{u}{c}}{100}\right),$$

which has the approximate solution $y = u/c = 390.60. Thus player 1 receives this amount, and player 2 should receive \$109.40; in the words of an old song "the rich get richer and the poor get poorer". This may seem strange, but of course our analysis contains no ethical considerations. The fact is that player 2's utility for money decreases rapidly, whilst player 1's does not. Thus player 2, being poor and very eager to get at least something, can be 'bargained down' by player 1, and the solution reflects this.

We can think of the working of the bargaining model as a three-stage process: first the players agree to an outcome satisfying the six Nash axioms for a given pair of threats, second they each choose a threat strategy which gives the threat status quo point, finally the payoffs are found which satisfy the six axioms. This last stage is accomplished by using Theorem 5.1 as illustrated above.

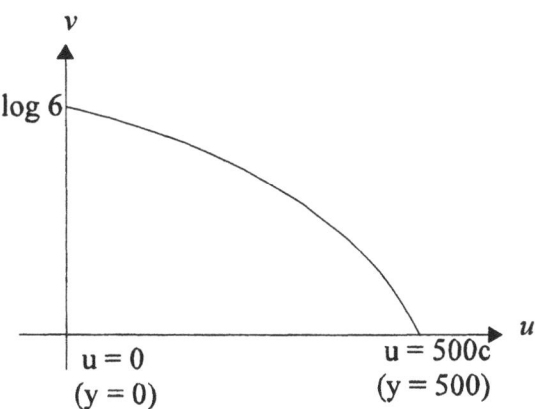

Figure 5.4 Graph for Example 5.3.

5.4 THE THREAT GAME.

The question arises how the players are to choose their threat strategies. Each pair of threaat strategies (x, y), $x \in X_1$, $y \in X_2$ leads to a unique determination of the final payoffs. Thus the original cooperative game, with mixed strategy sets X_1, X_2 for each player respectively, has been reduced to a *non-cooperative* game in which each player chooses a threat strategy, and the payoff is then determined by the Nash scheme. We call this underlying non-cooperative game the **threat game**. A pure strategy for player 1, say, in the threat game is to choose a mixed strategy $x \in X_1$. Hence the threat game is in fact an *infinite* game. On the face of it this reduction of the whole problem to an infinite non-cooperative game is not very helpful. The appropriate solution concept is that of an equilibrium point, and generally we should expect different equilibria to give different payoffs. In these circumstances it would make no sense to speak of a *best* threat, and it would certainly make a difference to the final payoffs if one player were to choose a threat first and then inform the other.

The striking feature of Nash's analysis is that none of these gloomy forebodings comes to pass. The threat game always has a pure strategy equilibrium point, corresponding to a mixed strategy pair (x, y) in the original cooperativ game, *moreover every equilibrium point gives rise to the same final payoff pair $(\bar{u}, \bar{v})$*. Thus it does make sense to speak of an optimal threat, and whilst an optimal threat need not be unique, all optimal threats yield the same solution point. Hence the right to issue the first threat is not after all an advantage. Our next task is to see how it comes about that the threat game has these peculiarly pleasant properties.

By axiom 3 any solution $(\bar{u}, \bar{v})$ to the bargaining game (S, u^*, v^*) lies in the subset S^0 of S consisting of all points $(u, v) \in S$ such that there is no $(u', v') \in S$ with $u' > u$ and $v' \geq v$, or $u' \geq u$ and $v' > v$. Since S is convex, S^0 is just those points $(u, v) \in S$ for which there is no $(u', v') \in S$ with $u' > u$ and $v' > v$. We call S^0 the **Pareto optimal boundary** of S.

In the language of §4.7 S^0 is just the set of undominated pairs (u, v). The Pareto optimal boundaries of all the games so far considered in this chapter have been indicated on their graphs by a heavy line. There is a special relationship, given by Lemma 5.1, between the threat and solution points. It is that if we consider the slope of the straight line from the threat point to the solution point, then the line through the solution point which has the negative of this slope is a support line to S. This line will be a tangent to S if the Pareto optimal boundary is smooth at the solution point. This is illustrated in Figure 5.5, where if any other point on the line TP, such as U, happened to be the threat point, the solution would be unchanged because the same slope condition would be met.

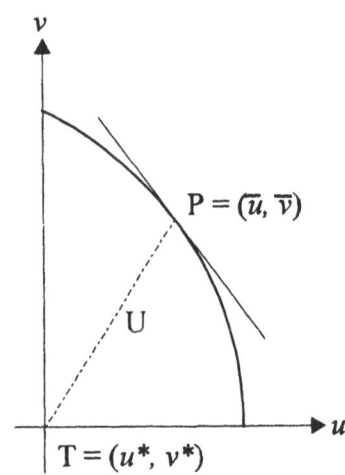

Figure 5.5 Smooth Pareto optimal boundary.

For a bimatrix game the set S is a closed bounded polygon as in Figure 5.6. Here each point U on a straight line such as T_1P_1, if taken as a threat point, has as its associated solution the end point P_1 of the line which lies on the Pareto optimal boundary. The slope of T_1P_1 is the negative of the slope of BC. For a point like C the associated points in S which, as threat points, would have C as their solution, are all the points on the line T_2C, all the points on the line T_3C and all the points in S which lie between these two lines, such as V.

Plainly the more the threat point is high and to the left, the more player 2 is benefited. Conversely, player 1 is better off as the threat point moves down and to the right. To put it another way: since the final payoffs of the game are determined by the position of P on the Pareto optimal boundary of S, which is a 'negatively sloping' curve, each player's payoff is a monotonic decreasing function of the other's.

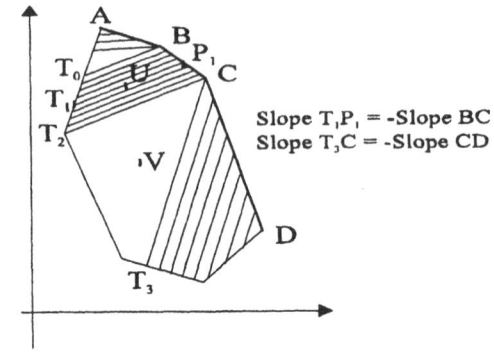

Figure 5.6 Polygonal Pareto optimal boundary ABCD.

Thus the threat game has a special character which puts it closer to a zero sum game than to a general non-cooperative game.

Definition[12]. A two person non-cooperative game with payoffs P_1, P_2 to player 1 and 2 respectively is a game of **pure conflict** or an **antagonistic** game if all outcomes are Pareto optimal, that is if (P_1', P_2'), (P_1'', P_2'') are two possible outcomes and $P_1'' \leq P_1'$, then $P_2'' \geq P_2'$.

[12]Such games are also sometimes called **strictly competitive**.

The threat game is therefore a game of pure conflict, and so the following theorem applies.

Theorem 5.2. The solution point for a game of pure conflict is unique.

Proof. Let (x', y'), (x'', y'') be two equilibrium points for such a game. We have to prove that

$$(P_1(x', y'), P_2(x', y')) = (P_1(x'', y''), P_2(x'', y'')).$$

Without loss of generality we can suppose that $P_1(x', y') \leq P_1(x'', y'')$.

Since (x'', y'') is an equilibrium point

$$P_2(x'', y') \leq P_2(x'', y''),$$

and as the game is a game of pure conflict it follows that

$$P_1(x'', y') \geq P_1(x'', y'').$$

Also since (x', y') is an equilibrium point,

$$P_1(x'', y') \leq P_1(x', y').$$

The last two inequalities yield $P_1(x', y') \geq P_1(x'', y'')$; hence $P_1(x', y') = P_1(x'', y'')$ as required. Doing the same exercise for player 2 establishes the theorem. (Incidentally, we have also proved that the equilibria form a rectangular set - see problems 2 and 3 of Chapter 2.)

It remains to show that there exists a pure strategy equilibrium point (corresponding to a pair of mixed strategies in the original cooperative game) for the threat game. We sketch the proof for the case where the original cooperative game is a bimatrix game, although the method can be generalised.

Theorem 5.3. Let Γ be a cooperative bimatrix game in which the players choose threat strategies to determine a threat status quo point, the outcome then being determined by the Nash bargaining model. Such a game has an equilibrium pair of threat strategies (x_t, y_t).

Proof[13]. The solution point $P = (\bar{u}, \bar{v})$ varies continuously as a function of the threat point (u^*, v^*), and the threat point varies continuously with the threat strategy pair (x, y); hence the solution point is a continuous function of (x, y).

Now if one player's threat is held fixed, say player 1's at x, then the position of P is a function of the other player's threat y. The coordinates (u^*, v^*) are linear functions of y. Hence the function ϕ which takes y to (u^*, v^*) defined by this situation is a linear transformation of the

[13]This should be omitted on a first reading.

space X_2 of player 2's mixed strategies into the feasible region S.

Consider the line segments which relate threat points to solution points, as illustrated in Figure 5.6. That part of $\phi(X_2)$ which intersects the most favourable (for player 2) line segment will contain the points $(u(x, y), v(x, y))$ for which y is a best reply to player 1's fixed threat x. This set of best replies, which we denote by $B_2(x)$, must be a convex closed bounded subset of X_2 because of the linearity and continuity of ϕ.

Now it can be shown that the continuity of P as a function of (x, y) ensures that the set valued function defined by $x \to B_2(x)$ is an upper semicontinuous function of x (see Appendix 1). Similarly $y \to B_1(y)$ is upper semicontinuous, where $B_1(y)$ is the set of player 1's best replies to y. It follows that the function defined by

$$(x, y) \to B_1(y) \times B_2(x)$$

is upper semicontinuous. Moreover $B_1(y) \times B_2(x)$ is a convex subset of $X_1 \times X_2$. We can now use the *Kakutani fixed point theorem* (see Appendix 1) to tell us that there exists some pair $(x_t, y_t) \in B_1(y_t) \times B_2(x_t)$, which amounts to saying that each threat is a best reply to the other, that is (x_t, y_t) is an equilibrium pair of threat strategies.

Unfortunately as this proof depends on the Kakutani fixed point theorem it is of a non-constructive nature. Finding best threat strategies is frequently rather difficult.

5.5 SHAPLEY AND NASH SOLUTIONS FOR BIMATRIX GAMES.

Let (a_{ij}, b_{ij}) be a cooperative $m \times n$ bimatrix game, where a_{ij} and b_{ij} denote the payoffs to players 1 and 2 respectively. We can apply the results of §5.3 in two distinct ways to produce solutions to such a game.

Following Luce and Raiffa we call the first method the **Shapley procedure** because it is a slight generalisation (non-comparable utility) of a special case (two players only) of the Shapley value of an n-person game discussed in §4.8. The Shapley procedure is quite simple: it consists of taking the status quo point (u^*, v^*) as the security status quo and then applying Theorem 5.1 to solve the bargaining game (S, u^*, v^*). This method takes no account of the threat possibilities.

The second method, which produces the Nash solution, is to find best threats for each player and to take the status quo point to be the threat point (u^*, v^*). Theorem 5.1 is applied in the usual way to solve (S, u^*, v^*). Now computing optimal threat strategies can, in general, be quite complicated, since the final solution point depends not only on the threat point (u^*, v^*) but also on the Pareto optimal boundary of S. If S is an *arbitrary* closed bounded convex subset of the plane the Pareto optimal boundary S^0 could exhibit quite pathological behaviour. However, for a bimatrix game, S is a closed bounded convex polygon and so S^0 is just a union of line segments. Even so we shall see that difficulties can occur if S^0 consists of more than one line segment.

For the game illustrated in Figure 5.6 the Pareto optimal boundary is ABCD, and we know that the final arbitrated solution will lie on one of the line segments AB, BC, and CD. Provided that the solution does not turn out to be at a vertex we can always find it and the associated threat strategies by checking each line segment in turn. Consider for example the line segment BC. Suppose the slope of BC is -ρ ($\rho > 0$). If $T = (u^*, v^*)$ is taken as the threat point and *if the line from T with slope* ρ *intersects* BC, say at $P = (\bar{u}, \bar{v})$, then P is a candidate for the final arbitrated solution. Let the coordinates of B be (u_B, v_B). Solving for the intersection of the line BC and the line T of slope ρ, we find

$$\bar{u} = \frac{v_B - v^* + \rho(u_B + u^*)}{2\rho}, \quad \bar{v} = \frac{v_B + v^* + \rho(u_B - u^*)}{2} \quad (5.2)$$

Thus in selecting threat strategies player 1 seeks to maximise $\rho u^* - v^*$, and player 2 to minimise this quantity. If the original game has bimatrix (a_{ij}, b_{ij}) then setting $A = (a_{ij})$, $B = (b_{ij})$ the payoff functions u, v are given by

$$u(x, y) = xAy^T, \quad v(x, y) = xBy^T, \quad (x \in X_1, \ y \in X_2)$$

where y^T denotes the transpose of the row vector y. Hence

$$\rho u - v = x(\rho A - B)y^T.$$

Therefore in seeking the best threat strategies it is as if each player were playing the zero sum game with matrix $\rho A - B$. Let (x_t, y_t) be a pair of optimal strategies for this game. Interpreted geometrically the situation is this. If player 1 chooses x_t, player 2 cannot choose y so that $(u(x_t, y), v(x_t, y))$ lies above the line through $(u^*, v^*) = (u(x_t, y_t), v(x_t, y_t))$ with slope ρ, and similarly if player 2 chooses y_t, player 1 cannot force $(u(x, y_t), v(x, y_t))$ below this line. We can verify this directly. For the condition that $(u(x_t, y), v(x_t, y))$ lies on or below the line is just

$$(u(x_t, y) - u^*, v(x_t, y) - v^*)(1, -\frac{1}{\rho})^T \geq 0,$$

since $(1, -^1/_\rho)$ is a downward normal to the line. This condition is just

$$u(x_t, y) - u(x_t, y_t) - \frac{1}{\rho}(v(x_t, y) - v(x_t, y_t)) \geq 0,$$

that is,

$$x_t Ay^T - x_t Ay_t^T - \frac{1}{\rho}(x_t By^T - x_t By_t^T) \geq 0,$$

that is

$$x_t(\rho A - B)y^T \geq x_t(\rho A - B)y_t^T.$$

Similarly the condition that $(u(x, y_t), v(x, y_t))$ lies on or above the line is

$$x(\rho A - B)y_t^T \le x_t(\rho A - B)y_t^T .$$

Taken together the last two inequalities assert that (x_t, y_t) is an optimal pair of strategies for the game with matrix $\rho A - B$.

To summarise: if the slope of BC is $-\rho$ ($\rho > 0$) we solve the game $\rho A - B$, say an optimal pair is (x_t, y_t). Provided that the line through $(u(x_t, y_t), v(x_t, y_t))$ with slope ρ passes through BC we can take

$$(u^*, v^*) = (u(x_t, y_t), v(x_t, y_t))$$

as the threat point. If the line does not pass through BC we can repeat the procedure for each remaining line segment of the Pareto optimal boundary.

Turning again to Figure 5.6 we observe that this process may fail if all pairs of optimal threat strategies are associated with payoff pairs (u^*, v^*) which lie in an unruled region such as T_2CT_3. If this were the case the final arbitrated solution would lie at a vertex such as C but apart from inspired guesswork there seems to be no simple method to determine the threat strategies.

Example 5.4.

$$
\begin{array}{cc}
 & \tau_1 \qquad\quad \tau_2 \\
\begin{array}{c}\sigma_1 \\ \sigma_2\end{array} &
\left(\begin{array}{cc}
(-3, \ 0) & (-1, -2) \\
(\ 2, \ 1) & (\ 1, \ 3)
\end{array}\right)
\end{array}
$$

The graph for this game is given in Figure 5.7. The Pareto optimal boundary is just the line segment joining (1, 3) and (2, 1).

Shapley solution. To find the security level for player 1 we solve the game

$$
\begin{array}{c}
\quad\ \ \tau_1\ \ \tau_2 \\
\begin{array}{c}\sigma_1 \\ \sigma_2\end{array}
\left(\begin{array}{cc} -3 & -1 \\ 2 & 1 \end{array}\right)
\end{array}
\text{ (Payoffs to player 1)}
$$

which has a saddle point at (σ_2, τ_2) giving player 1's security level as 1. Similarly for player 2 the game

$$
\begin{array}{c}
\quad\ \ \tau_1\ \ \tau_2 \\
\begin{array}{c}\sigma_1 \\ \sigma_2\end{array}
\left(\begin{array}{cc} 0 & -2 \\ 1 & 3 \end{array}\right)
\end{array}
\text{ (Payoffs to player 2)}
$$

has a saddle point at (σ_1, τ_1) so that player 2's security level is 0.

For the Shapley solution we therefore take $(u^*, v^*) = (1, 0)$ to be the status quo point and seek the point $(\bar{u}, \bar{v})$ which maximises $(u - u^*)(v - v^*) = (u - 1)v$ on the line $v = -2u + 5$ ($1 \le u \le 2$). This gives $(\bar{u}, \bar{v}) = (^7/_4, ^3/_2)$ as the Shapley solution.

Nash solution. The Pareto optimal boundary is a single line segment of slope -2. Taking $\rho = 2$ we have to solve the zero sum game with matrix $2A - B$, that is

$$\begin{pmatrix} -6 & 0 \\ 3 & -1 \end{pmatrix}$$

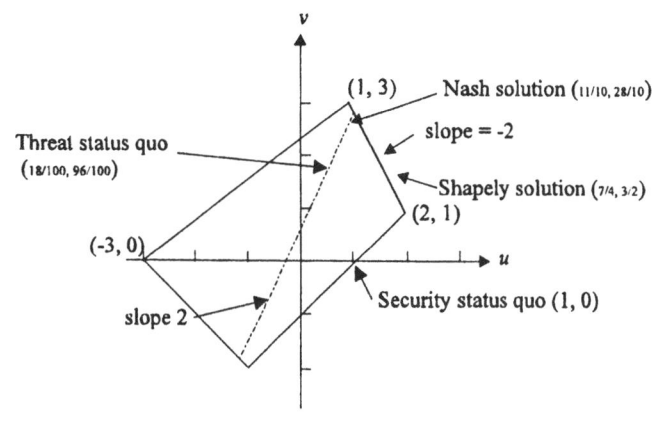

Figure 5.7 Diagram for Example 5.4.

This game has optimal strategies $(^4/_{10}, ^6/_{10})$ for player 1 and $(^1/_{10}, ^9/_{10})$ for player 2. These are the threat strategies. The threat point (u^*, v^*) is given by

$$u^* = (\frac{4}{10}, \frac{6}{10}) \begin{pmatrix} -3 & -1 \\ 2 & 1 \end{pmatrix} \begin{pmatrix} \frac{1}{10} \\ \frac{9}{10} \end{pmatrix} = \frac{18}{100}$$

$$v^* = (\frac{4}{10}, \frac{6}{10}) \begin{pmatrix} 0 & -2 \\ 1 & 3 \end{pmatrix} \begin{pmatrix} \frac{1}{10} \\ \frac{9}{10} \end{pmatrix} = \frac{96}{100}$$

To find the Nash solution we can either maximise $(u - ^{18}/_{100})(v - ^{96}/_{100})$ on the line $v = -2u + 5$ ($1 \le u \le 2$) or use (5.2) directly, with $(u_B, v_B) = (1, 3)$, $\rho = 2$ and $(u^*, v^*) = (^{18}/_{100}, ^{96}/_{100})$. Either way we find $(\bar{u}, \bar{v}) = (^{11}/_{10}, ^{28}/_{10})$ as the Nash solution.

Example 5.5.

$$\begin{array}{c c c c} & \tau_1 & \tau_2 & \tau_3 \\ \sigma_1 & (1, 4) & (-2, 0) & (4, 1) \\ \sigma_2 & (0, -1) & (3, 3) & (-1, 4) \end{array}$$

We first draw the graph in Figure 5.8. The Pareto optimal boundary is ABC. Write

$$A = \begin{pmatrix} 1 & -2 & 4 \\ 0 & 3 & -1 \end{pmatrix}, \quad B = \begin{pmatrix} 4 & 0 & 1 \\ -1 & 3 & 4 \end{pmatrix}.$$

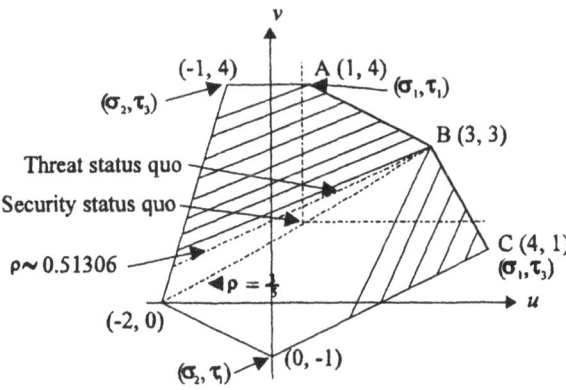

Figure 5.8 Diagram for Example 5.5.

Shapley solution. For the game with matrix A we find, by the methods of §1.6,

	$5/6$	$1/6$	0	
$1/2$	1	-2	4	$1/2$
$1/2$	0	3	-1	$1/2$
	$1/2$	$1/2$	$3/2$	$v = 1/2$

which gives player 1's security level as $1/2$. Similarly the game with matrix B, where payoffs are to the *column* player, has value $3/2$ which is player 2's security level.

We next maximise $(u - 1/2)(v - 3/2)$ on ABC. Along AB $v = -1/2 u + 9/2$ $(1 \le u \le 3)$ and $(u - 1/2)(-1/2 u + 3)$ has positive derivatives so this maximum is attained at $u = 3$. Along BC, $v = -2u + 9$ $(3 \le u \le 4)$ and $(u - 1/2)(-2u + 15/2)$ has negative derivatives, so this maximum is again attained at $u = 3$. Hence the required point is $(\bar{u}, \bar{v}) = (3, 3)$ and this is the Shapley solution.

Nash solution. For the line segment AB with slope $-1/2$ we take $\rho = 1/2$ and consider the game $\rho A - B$ which has matrix

$$\begin{pmatrix} -\dfrac{7}{2} & -1 & 1 \\[2mm] 1 & -\dfrac{3}{2} & -\dfrac{9}{2} \end{pmatrix}$$

This is easily solved to give $(^{11}/_{20}, {}^{9}/_{20})$ and $(^{11}/_{20}, 0, {}^{9}/_{20})$ as optimal strategies for player 1 and 2 respectively. The corresponding threat point would be

$$u = (\frac{11}{20}, \frac{9}{20}) \begin{pmatrix} 1 & -2 & 4 \\ 0 & 3 & -1 \end{pmatrix} \begin{pmatrix} \dfrac{11}{20} \\[1mm] 0 \\[1mm] \dfrac{9}{20} \end{pmatrix} = \frac{109}{100} ,$$

$$v = (\frac{11}{20}, \frac{9}{20}) \begin{pmatrix} 4 & 0 & 1 \\ -1 & 3 & 4 \end{pmatrix} \begin{pmatrix} \dfrac{11}{20} \\[1mm] 0 \\[1mm] \dfrac{9}{20} \end{pmatrix} = \frac{2}{5} .$$

However, the line through $(^{109}/_{100}, {}^{2}/_{5})$ with slope $^{1}/_{2}$ does *not* intersect AB, so we discard this possibility.

For the line segment BC with slope -2 we take $\rho = 2$ and consider the game $\rho A - B$ which has matrix

$$\begin{pmatrix} -2 & -4 & -7 \\ 1 & 3 & -6 \end{pmatrix}$$

This game has a saddle point at (σ_2, τ_3) and the line through (-1, 4) with slope 2 does not, of course, intersect BC.

It therefore remains to consider the unruled regions in Figure 5.8. Here we observe that if player 2 selects τ_2 then the possible payoff pairs (u, v) are confined to the line segment joining (-2, 0) and (3, 3). By randomising between σ_1 and σ_2 player 1 can control the location of (u, v) on this line but cannot force (u, v) below the line. This suggests, but does not prove, that the final Nash solution is at B. The other possibility is A and, although it seems unlikely, it is conceivable that player 2 has a threat strategy which will enforce A.

Let us look at the game assocated with player 2's threat τ_2. The line segment joining (-2, 0) and (3, 3) has slope $\rho = {}^{3}/_{5}$, and the associated game $\rho A - B$ has the solution

	$15/26$	0	$11/26$	
$7/13$	$-17/5$	$-6/5$	$-7/5$	$-89/65$
$6/13$	1	$-6/5$	$-23/5$	$89/65$
	$-89/65$	$-78/65$	$-89/65$	$v = -89/65$

This gives

$$u = \left(\frac{7}{13}, \frac{6}{13}\right) \begin{pmatrix} 1 & -2 & 4 \\ 0 & 3 & -1 \end{pmatrix} \begin{pmatrix} \dfrac{15}{26} \\ 0 \\ \dfrac{11}{26} \end{pmatrix} = \frac{347}{338} \approx 1.03 \ ,$$

$$v = \left(\frac{7}{13}, \frac{6}{13}\right) \begin{pmatrix} 4 & 0 & 1 \\ -1 & 3 & 4 \end{pmatrix} \begin{pmatrix} \dfrac{15}{26} \\ 0 \\ \dfrac{11}{26} \end{pmatrix} = \frac{671}{338} \approx 1.98 \ .$$

Now the point $(^{347}/_{338}, {}^{671}/_{338})$ lies *above* the line segment joining $(-2, 0)$ and $(3, 3)$, which shows that player 2 can achieve a higher threat point by randomising between τ_1 and τ_3 than by using τ_2. We should therefore investigate the possibility of player 2 using a threat strategy of the form $(\mu, 0, 1-\mu)$, $0 \le \mu \le 1$.

Against σ_1 the strategy $(\mu, 0, 1-\mu)$ gives the payoff pair

$$\mu(1, 4) + (1 - \mu)(4, 1) = (4 - 3\mu, 3\mu + 1),$$

and against σ_2

$$\mu(0, -1) + (1 - \mu)(-1, 4) = (-1 + \mu, -5\mu + 4).$$

The equation of the line adjoining these two points is

$$v + 5\mu - 4 = \frac{-8\mu + 3}{4\mu - 5}(u + 1 - \mu) . \tag{5.3}$$

From Figure 5.8 it is apparent that if player 2 uses $(\mu, 0, 1-\mu)$ then by randomising between σ_1 and σ_2 player 1 can control the position of the point (u, v) on this line but cannot force the point below the line. The fact that $(^{347}/_{338}, {}^{671}/_{338})$ is only *slightly* above the line joining $(-2, 0)$ and $(3, 3)$ strengthens our view that the final Nash solution should be B. The line (5.3) passes through B = (3, 3) if

$$3 + 5\mu - 4 = \frac{-8\mu + 3}{4\mu - 5}(3 + 1 - \mu)$$

which gives

$$12\mu^2 + 6\mu - 7 = 0 .$$

The positive root of this equation is

$$\mu = \frac{-6 + \sqrt{372}}{24} \approx 0.55364 .$$

The slope of the corresponding line is

$$\rho = \frac{-8\mu + 3}{4\mu - 5} \approx 0.51306 .$$

The solution for the associated game $\rho A - B$ is to five decimal places

	0.55364	0	0.44636	
0.54844	-3.48694	-1.02612	1.05224	-1.46083
0.45156	1	-1.46082	-4.51306	-1.46081
	-1.46082	-1.22241	-1.46082	$v \sim -1.46082$

The threat point corresponding to these strategies is

$$u^* = (0.54844, 0.45156)\begin{pmatrix} 1 & -2 & 4 \\ 0 & 3 & -1 \end{pmatrix}\begin{pmatrix} 0.55364 \\ 0 \\ 0.44636 \end{pmatrix} \approx 1.08129 ,$$

$$v^* = (0.54844, 0.45156)\begin{pmatrix} 4 & 0 & 1 \\ -1 & 3 & 4 \end{pmatrix}\begin{pmatrix} 0.55364 \\ 0 \\ 0.44636 \end{pmatrix} \approx 2.01558 .$$

Now the point (1.08129, 2.01558) *does* lie on the line through (3, 3) having slope $\rho \approx 0.51306$. Thus player 1 cannot force the threat point below this line, whilst player 2 cannot force it above. Hence the Nash solution is indeed the vertex B.

5.6 OTHER BARGAINING MODELS.

For the case where no interpersonal comparison of utility is assumed the Nash solution was independently arrived at by Raiffa [6] who used a different rationalisation. Both Raiffa [6] and Braithwaite [7] offer operationally different schemes which satisfy all the Nash axioms saving the independence of irrelevant alternatives. Both solutions are independent of separate changes in utility scales. Nevertheless in each case the justification of the procedure involves a somewhat arbitrary change of utility scale in order to establish a *tacit* comparability of utility. This could be construed as a weakness, although each scheme has its attractive features. A brief discussion of both can be found in [1].

If side payments are permitted a tacit comparison of utility is introduced. This also has the effect of enlarging the feasible region S. If utility is assumed to be linear with money, the convexity of S is preserved. By the time this stage is reached one may as well suppose that payoffs are in monetary units and utility is linear with money.

> **Example 5.6** (The battle of the sexes). This is a game which was given a name before the advent of Women's Liberation, and it should hastily be explained that it is seldom analysed to the advantage of the man. A husband and wife have not been married so long that they do not still find joy in each other's company. On the other hand they have been married long enough to be interested in other things as well. Each Saturday evening the husband wishes to go to the boxing and the wife to the ballet, say. If agreement cannot be reached each goes to their preferred choice alone but neither deems this outcome as satisfactory. The payoffs are given below.

		Wife	
		Boxing	Ballet
Husband	Boxing	(2, 1)	(-1, -1)
	Ballet	(-1, -1)	(1, 2)

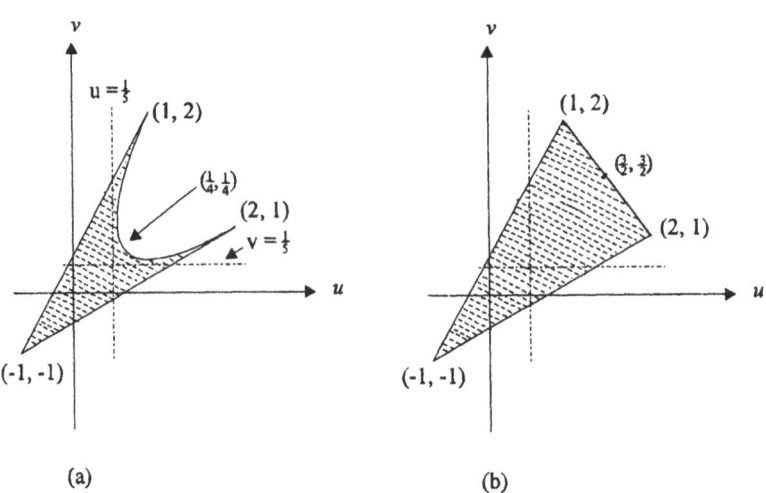

(a)   (b)

Figure 5.9 Battle of the sexes.

In Figure 5.9 we give feasible regions for this game under three different hypotheses. In Figure 5.9(a) the game is regarded as non-cooperative. The security level of each player is $1/5$.

There are two pure strategy equilibria, corresponding to payoffs (2, 1) and (1, 2) respectively, neither of which is likely to be viewed as a satsifactory solution by both parties. There is also a mixed strategy equilibrium which gives both players a payoff of $1/5$. This is achieved if the husband uses the mixed strategy $(3/5, 2/5)$ and the wife $(2/5, 3/5)$. But this too is doubtful since both players can *guarantee* themselves a payoff of $1/5$ by using their security level strategies.

Treated as a cooperative game with no comparison of utilities the feasible region is given in Figure 5.9(b), and the Pareto optimal boundary is just the line segment joining (1, 2) and (2, 1). Both the Nash and the Shapley solution correspond to the point $(3/2, 3/2)$. This could be realised in practise by husband and wife agreeing to toss a coin to decide on boxing or ballet. The more cynical reader might regard this as a triumph of commonsense over mathematics!

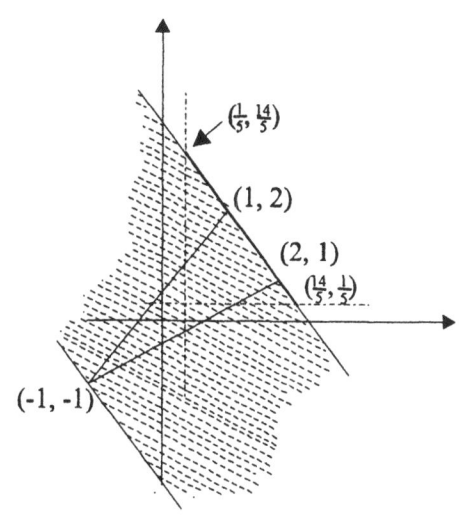

Figure 5.10 BoS regarded as a cooperative game.

If the game is regarded as cooperative with side payments we may as well regard the payoffs in the game bimatrix to represent monetary satifaction. Figure 5.10 is then obtained by observing that if the payoff pair (u, v) is achievable without side payments then $(u - w, v + w)$ is achievable with side payments. The Pareto optimal boundary can then be taken to be the line segment joining $(1/5, 14/5)$ and $(14/5, 1/5)$. Again the Nash and Shapley solutions are at $(3/2, 3/2)$.

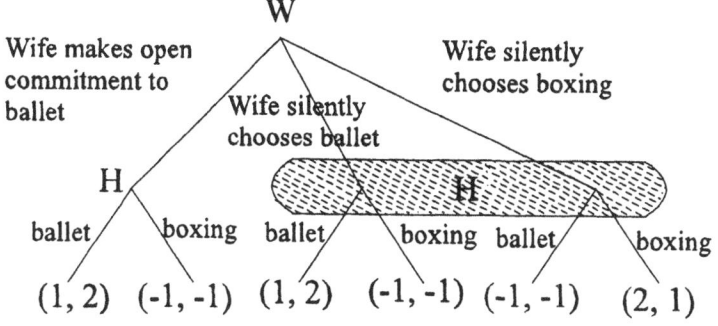

Figure 5.11 Alternative negotiation scheme.

To emphasise that the Nash bargaining model can be deemed to produce a satisfactory outcome *only if both players agree to use it* let us consider a preplay situation in which the wife has the opportunity to speak first and make a commitment, that is a binding statement of intent, of going to the ballet. In the Nash model, threats are implemented only if negotiations break down, whereas here we are saying that if the wife threatens to go the the ballet she must actually go. The simplest negotiation scheme within which this idea makes sense is illustrated in Figure 5.11.

In this negotiation game we can identify three pure strategies for the wife and four for the husband.

	Wife's pure strategies
W1	Make commitment to ballet.
W2	Silently choose ballet.
W3	Silently choose boxing.

	Husband's pure strategies
H1	Choose boxing regardless of wife's action.
H2	Choose ballet regardless of wife's action.
H3	Choose boxing if wife silent; otherwise ballet.
H4	Choose ballet if wife silent; othewise boxing.

The payoff bimatrix is given below.

	W1	W2	W3
H1	(-1, -1)	(-1, -1)	(2, 1)
H2	(1, 2)	(1, 2)	(-1, -1)
H3	(1, 2)	(-1, -1)	(2, 1)
H4	(-1, -1)	(1, 2)	(-1, -1)

As a non-cooperative game this has a superfluity of equilibria but two are of special interest. Considering payoffs only to the husband we notice that H3 dominates H1 and H2 dominates H4. If we eliminate the first and fourth row and now consider only payoffs to the wife, new dominances appear: W1 dominates W2 and W3. Thus the equilibria at (H2, W1) and (H3, W1) are natural solutions, especially in view of the fact that both give the same payoff (1, 2).

What this amounts to is that the wife gets in first with a unilateral commitment to the ballet and the husband acquiesces. Where side payments are permitted she can not only get her husband to the ballet but also extract up to 80 cents from him for the privilege - point ($^1/_5$, $^{14}/_5$) in Figure 5.10.

The negotiations procedure examined above is exceedingly simple but nevertheless embodies the essential features of any procedure within which binding commitments are possible and one or the other player has the opportunity to speak first. If the wife speaks first the outcome is (1, 2), and if the husband speaks first it is (2, 1). The problem of who is to speak first does not occur in the Nash model. Assuming a prior commitment to this model, the threats are implemented only if one player or the other will not accept the arbitrated outcome. Moreover, as we have seen in §5.4, the theory shows that it is irrelevant who issues the first threat.

Harsanyi [8] has developed an n-person bargaining model with non-comparable utility which generalises the Nash model. If payoffs are taken in monetary units and utility is linear with money, Harsanyi's solution for the n-person case looks remarkably like formula (4.34) giving the Shapley solution. The difference occurs because the Shapely solution uses security levels whereas the Harsanyi solution uses threat levels. This is of course precisely the difference between the 2-person Shapley and Nash solutions discussed in this chapter. It is worth observing that Shapley's model is formulated in terms of the characteristic function form of the game whereas the Nash or Harsanyi models are based on the normal form.

PROBLEMS FOR CHAPTER 5.

1. Consider the cooperative game

$$
\begin{array}{c c}
 & \tau_1 \qquad \tau_2 \\
\begin{array}{c} \sigma_1 \\ \sigma_2 \end{array} &
\left(\begin{array}{cc} (-1,\ -1) & (\ 1,\ \ 1) \\ (\ 2,\ -2) & (-2,\ \ 2) \end{array} \right)
\end{array}
$$

Show graphically the payoff polygon, the security level for each player and the Pareto optimal boundary. Find the Shapley and Nash solutions and discuss player 2's threat strategies. Which player has the better bargaining position?

2. For the cooperative game

$$
\begin{array}{c c}
 & \tau_1 \qquad \tau_2 \\
\begin{array}{c} \sigma_1 \\ \sigma_2 \end{array} &
\left(\begin{array}{cc} (\ 2,\ 5) & (\ 3,\ 2) \\ (\ 1,\ 0) & (\ 6,\ 1) \end{array} \right)
\end{array}
$$

a) Show graphically the payoff polygon, the security level for each player and the Pareto optimal boundary. Find the Shapley and Nash solutions.

b) Discuss the game as a non-cooperative game. Your discussion should include consideration of security levels adn equilibrium points, and why you do or do not favour one of the equilibrium points as the likely outcome. What do you think should be the non-cooperative solution of the game?

3. Consider the cooperative game

$$\begin{array}{cc} & \tau_1 \qquad \tau_2 \\ \begin{array}{c}\sigma_1 \\ \sigma_2\end{array} & \left(\begin{array}{cc} (\ 1,\ \ 0) & (-a,\ -b) \\ (-c,\ -d) & (\ 0,\ \ 1) \end{array}\right), \end{array}$$

where $0 < a,\ b,\ c,\ d < 1$. Find the Shapley and Nash solutions and discuss the effect of varying each of the parameters $a,\ b,\ c,\ d$.

4. What conclusions can be drawn from the observation that the optimal threat strategies in Example 5.5 involve irrational numbers as probability components?

CHAPTER REFERENCES

[1] Luce, R. D. and Raiffa, H., *Games and Decisions*. J. Wiley and Sons (1957), New York.

[2] Crossman, R. H. S., *Diaries of a Cabinet Minister*, (1964-70) 3 vols. Hamish Hamilton and Jonathan Cape (1975-77), London.

[3] Green, J. and Lafont, J. J., 'Characterisation of satisfactory mechanisms for the revelation of preferences for public goods.' *Econometrica*, **45**, 427-438 (1977).

[4] Nash, J. F., 'The bargaining problem'. *Econometrica*, **18**, 155-162 (1950).

[5] Nash, J. F., 'Two Person Cooperative Games'. *Econometrica*, **21**, 128-140 (1953).

[6] Raiffa, H., 'Arbitration schemes for generalised two-person games'. *Contributions to the Theory of Games*, II, (Ann. Math. Studies No. 28) 361-387 (1953), Princeton.

[7] Braithwaite, R. B., *Theory of Games as a Tool for the Moral Philosopher*. Cambridge University Press (1955), Cambridge.

[8] Harsanyi, J. C., 'A bargaining model for the cooperative *n*-person game'. *Contributions to the Theory of Games*, IV, (Ann. Math. Studies No. 40) 325-355 (1959), Princeton.

Appendix 1

Fixed Point Theorems

Let f be a continuous function of the closed interval $[-1, 1]$ into itself. Figure A 1 suggests that the graph of f must touch or cross the indicated diagonal, or, more precisely, that there must exist a point $x_0 \in [-1, 1]$ with the property that $f(x_0) = x_0$. The proof is easy. We consider the continuous function F defined on $[-1, 1]$ by $F(x) = f(x) - x$ and we observe that $F(-1) \geq 0$ and $F(1) \leq 0$. It then follows from the Weierstrass intermediate value theorem that there exists $x_0 \in [-1, 1]$ such that $F(x_0) = 0$, or $f(x_0) = x_0$. We say $x_0 \in [-1, 1]$ is a **fixed point** of f.

Let us now try to replace $[-1, 1]$ by any subset S of $\mathbb{R}^n$ and ask is it true that every continuous function $f : S \to S$ has a fixed point? For example, let S be a circle in $\mathbb{R}^2$. Consider a function f which consists of a rotation of S about its centre, then f is a continuous function $f : S \to S$, but plainly if f is anything other than a full rotation through $360°$ it will not have a fixed point. The problem here is that a circle in $\mathbb{R}^2$ is not convex. Now take $S = \mathbb{R}$ and define $f(x) = x + 1$ for every $x \in \mathbb{R}$. Plainly f has no fixed point, and the problem in this case is that S is not compact (compact is equivalent to closed bounded in $\mathbb{R}^n$). It turns out that the convexity and compactness of S are the crucial conditions, in fact we have

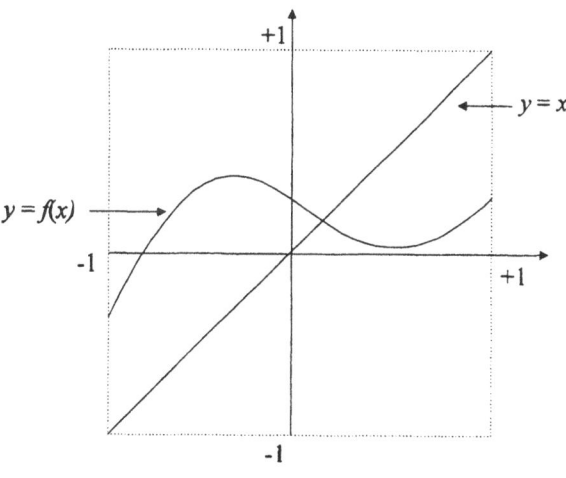

Figure A 1 Illustrating the existance of a fixed point for a continuous function on $[0, 1]$.

Brouwer's fixed point theorem. Let S be a convex compact subset of $\mathbb{R}^n$ and f a continuous function $f : S \to S$, then f leaves some point of S fixed.

There are several proofs of this classic result, but since they all depend on the methods of algebraic topology, we refer the reader to Bers ([1], 80).

Brouwer's theorem itself is a special case of:

Schauder's fixed point theorem. Let S be a convex compact subset of a Banach space and f

a continuous function $f:S \to S$, then f leaves some point of S fixed.

For a proof, together with a discussion of other related results, see Bers ([1], 93-97). Fixed point theorems are of considerable interest in analysis and have a wide range of applications in other areas of mathematics.

We now turn to Kakutani's generalisation of the Brouwer fixed point theorem; for a detailed discussion see [2], and for an interesting general discussion see [3]. This has emerged as an indispensable tool in game theory and mathematical economics. Before stating this result we clarify a few terms.

If $S \subseteq \mathbb{R}^n$, a function ϕ which takes every $x \in S$ to a subset $\phi(x) \subseteq S$ is called a **point-to-set** map of S. A point-to-set map ϕ is called **upper semi-continuous** if $x_n \to x_0$, $x_n \in S$, and $y_n \to y_0$, $y_n \in \phi(x_0)$, together imply $y_0 \in \phi(x_0)$. This definition of upper semi-continuous for a point-to-set map reduces to that for an ordinary function $\phi:S \to \mathbb{R}$ given in 2.8, provided that we interpret the containing relation $y \in \phi(x)$ as $y \leq \phi(x)$ in this special case.

A point-to-set map ϕ is said to be **lower semi-continuous** if for every $y \in \phi(x_0)$ whenever $x_n \to x_0$ there exists $y_n \in \phi(x_n)$ such that $y_n \to y_0$.

It is also possible to express these properties of point-to-set maps in terms of neighbourhood concepts. For example ϕ is upper semi-continuous at x_0 if for any open set U with $\phi(x_0) \subseteq U$ there exists a $\delta = \delta(U) > 0$ such that $|x - x_0| < \delta$ implies $\phi(x) \subseteq U$.

Kakutani's fixed point theorem. Let S be a convex compact subset of $\mathbb{R}^n$ and ϕ an upper semi-continuous point-to-set map which takes each $x \in S$ to a convex set $\phi(x) \subseteq S$, then there exists $x_0 \in S$ such that $x_0 \in \phi(x_0)$.

For a proof see [2].

REFERENCES

[1]   Bers, L., *Topology*, lecture notes, New York University Institute of Mathematical Sciences (1957), New York.

[2]   Kakutani, S., 'A generalisation of Brouwer's fixed point theorem'. *Duke Math. Journal*, **8**, 457-58 (1941).

[3]   Bing, R. H., 'The elusive fixed point property'. *Amer. Math. Monthly*, **76**, 119-132 (1969).

Appendix II

Some Poker Terminology

Occasional use has been made of certain gambling terms, particularly with reference to poker. Whilst it seems safe to assume a passing familiarity with games such as chess, I am assured that it is unreasonable to expect universal aquaintance with the rules and terminology of poker. Admittedly this shook my preconceptions to the core, but I will endeavour to sketch the general idea! For complete details of poker and other gambling card games the reader is referred to Scarne [1]. In fact if you intend to play poker for money (and it is a pointless game to play otherwise) I *strongly urge* a careful study of Scarne. It is undoubtedly the best book I know on the practical aspects of gambling card games. Another, more mathematical, treatment of games of chance is given in Epstein [2].

There are many variants of poker, but I shall describe *Straight* or *Draw* poker which is played with between two and six players and a standard 52-card pack.

All players first pay an equal amount into the pot, and this is known as an **ante**. Each player is then dealt five cards face down, and the remaining cards are put to one side for future use in drawing. The players then examine their cards.

Poker hands are classified into ten types, the best hand being the least probable. From highest to lowest the possible hands and their respective probabilities of being dealt in the first five cards are:

Royal flush (0.0000015); the five highest cards of a given suit, namely AKQJ10.

Straight flush (0.0000138); any five cards of the same suit in numerical sequence, such as 109876 of spades. This straight flush is called a **ten high**. A ten high beats a nine high etc.

Four of a kind (0.00024); any four cards of the same denomination, for example AAAA2. The odd card is irrelevant and does not affect the rank of the hand.

Full house (0.00144); three of one kind and two of another, for example 33322.

Flush (0.00196); any five cards of the same suit but not in numerical sequence.

Straight (0.00392); five cards in numerical sequence but not of the same suit.

Three of a kind (0.0211); three cards of the same numerical value plus two different and irrelevant cards that are not paired, for example KKK54.

Two pairs (0.0475); two different pairs plus one odd card, for example 1010554. This example is called **tens up**.

One pair(0.422); two cards of the same denomination plus three indifferent (unmatched) cards, for example 1010973.

No pair (0.501); a hand which contains five indifferent cards not of the same suit, not in sequence and falling into none of the above types.

The essence of poker lies in the betting rules, and these too vary considerably. Having examined their cards the players now begin to bet. Working clockwise from the player on the dealer's left, each player has an opportunity to make the opening bet. The player to do so, known as the **opener**, must hold at least a pair of jacks and will be required to prove this if he or she later wins the pot. A player who does not open when given the opportunity is said to **pass** or **check**. Once the betting has been opened a player can either:

> **Fold**: by placing his or her cards face down in the centre of the table and taking no further part in the play of this round. Of course players who fold lose their contribution to the pot.
>
> or   **Play**; by putting an amount into the pot consisting of the opening bet.
>
> or   **Raise**; by putting an amount into the pot consisting of the opening bet plus an amount for the raise.

The remaining players may now either play by putting an amount into the pot equal to the total amount of the raiser or, if they have already put the opening bet into the pot, they merely add the amount of the raise. A player may **reraise** by matching the raiser's total bet plus an amount for the reraise, or a player may fold any time it is his or her turn to bet. To remain active every player must match the individually largest contribution to the pot. This procedure of folding, playing, raising and reraising is continued until the players stop raising.

If all the players but one fold, this player wins the pot. Should this player be the opener then he or she must show at least a pair of jacks, otherwise the winner is not required to expose the hand.

If more than one player remains active, and there are no uncovered bets in the pot, the remaining players may if they desire be dealt in turn one, two or three cards in an attempt to improve their hand, or they may **stand pat** (draw no cards). This procedure is known as **the draw**. A player must discard the appropriate cards *before* the dealer deals their replacements. Cards once discarded cannot be taken back.

The player who opened, if still active, has the first opportunity to bet after the draw, and play goes on clockwise to each active player who can pass, open, fold, raise and reraise as before. This process continues until:

> 1. All players pass, and a **showdown** is therefore reached immediately after the draw. In this case the opener exposes his or her hand first and the remaining players expose their hands continuing clockwise with the first player to the opener's left.
>
> or   2. All players have folded except one, in whch case this player wins the pot and, unless the opening player, is not required to expose any part of his or her hand.
>
> or   3. All players have matched the highest bet and ceased to raise or reraise (it is not permitted to reraise ones own raise). Now the last person to raise is being **called by** the remaining players, and a showdown takes place. In this case the showdown then rotates clockwise.

In the event of a showdown the highest hand wins the pot. In the case of ties the tied players split the pot equally.

REFERENCES

[1]   Scarne, J., *Scarne on cards*. Revised, augmented edition, Constable and Comp Ltd. (1965), London.

[2]   Epstein, R. A., *The Theory of Gambling and Statistical Logic*. Academic Press (1977), New York.

Solutions to problems

CHAPTER 1.

1. *A's pure strategies*
I Choose Head.
II Choose Tail.

B's pure strategies
I Choose Head.
II Choose Tail.

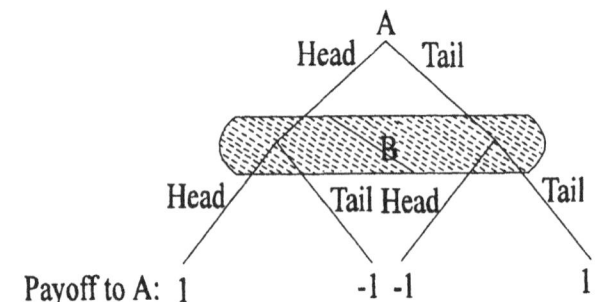

Figure S 1 Matching Pennies by choosing.

Solution (can be found graphically):

A's optimal strategy: $(^1/_2, ^1/_2)$,
B's optimal strategy: $(^1/_2, ^1/_2)$,
Value = 0.

Matrix form

		B	
		I	II
A	I	1	-1
	II	-1	1

2. Neither player has any decisions to make as the outcome is determined by a sequence of chance events: we therefore have what is known as a stochastic model. There are no pure strategies and no normal form.

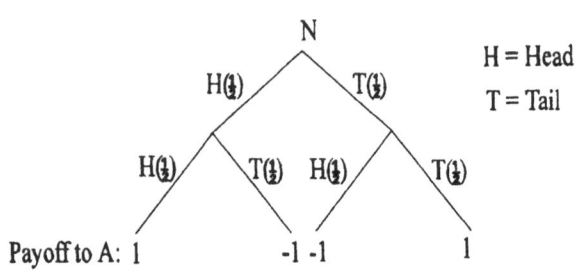

Figure S 2 Matching Pennies by tossing.

3.

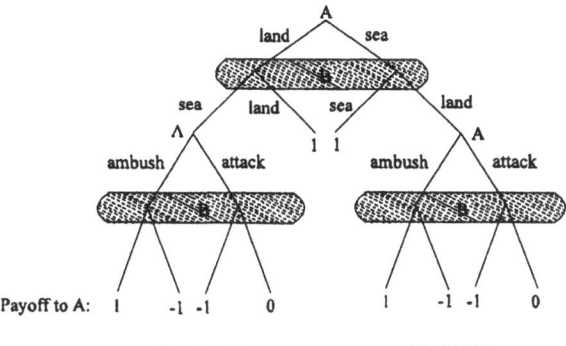

Figure S 3 Attack on Concord.

A = American   B = British

	A's pure strategies			*B's pure strategies*
I	Land; then ambush.		I	Sea; then leave.
II	Land; then attack.		II	Sea; then wait.
III	Sea; then ambush.		III	Land; then leave.
IV	Sea; then attack.		IV	Land; then wait.

Check that a solution is:

A's optimal strategy: $(\frac{1}{6}, \frac{1}{3}, \frac{1}{6}, \frac{1}{3})$,
B's optimal strategy: $(\frac{1}{6}, \frac{1}{3}, \frac{1}{6}, \frac{1}{3})$,
Value = $\frac{1}{3}$.

Matrix form

		B			
		I	II	III	IV
A	I	1	-1	1	1
	II	-1	0	1	1
	III	1	1	1	-1
	IV	1	1	-1	0

4. Take A = Al, B = Bill, R = real fly, F = fake fly, S = swat, N = not swat.

A's pure strategies
I   Real fly.
II   Fake, then real.
III   Fake, then fake.

B's pure strategies
I   Swat.
II   Not swat, then swat.
III   Not swat, then not swat.

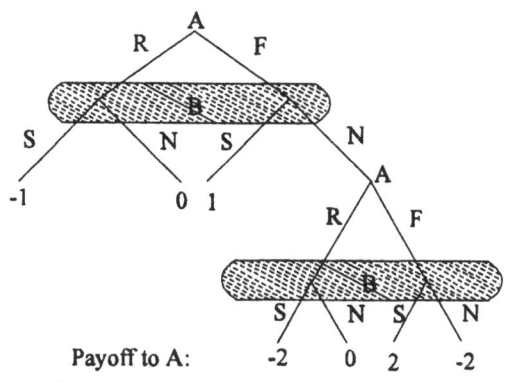

Payoff to A: -2   0   2   -2

Figure S 4 Peace Corps v Flies.

Check that a solution is:

A's optimal strategy: $(^5/_8, \, ^1/_4, \, ^1/_8)$,
B's optimal strategy: $(^1/_4, \, ^1/_4, \, ^1/_2)$,
Value $= -^1/_4$.

Matrix form

		B		
		I	II	III
	I	-1	0	0
A	II	1	-2	0
	III	1	2	-2

5. Take A = Ann, B = Bea, K = King, J = Joker.

A's pure stratgies
I   Ace, then resign.
II   Ace, then Ace.
III   Ace, then Queen.
IV   Queen.

B's pure strategies
I   King.
II   Joker, then King.
III   Joker, then Joker.

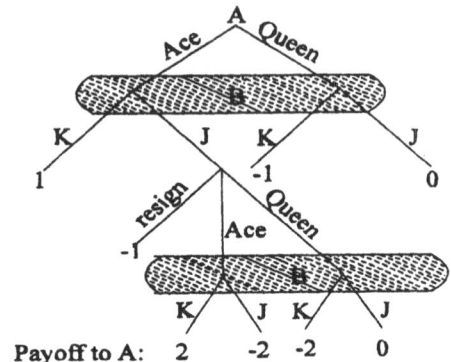

Payoff to A: 2   -2 -2   0
Figure S 5 Aces and Queens.

This game has a similar matrix to the previous one, except that A has an additional pure strategy, AI. It can readily be verified that a solution is

A's optimal strategy: $(0, \, ^1/_8, \, ^1/_4, \, ^5/_8)$,
B's optimal strategy: $(^1/_4, \, ^1/_4, \, ^1/_2)$,
Value $= \, ^1/_4$.

Hence the previous game is in some sense a subgame of this one.

Matrix form

		B		
		I	II	III
	I	1	-1	-1
A	II	1	2	-2
	III	1	-2	0
	IV	-1	0	0

6. (a) Since $4 > -4, \, 8 > -2, \, 2 > -2, \, 6 > 0$ we have $A(2) > A(4)$. Similarly $A(2) > A(1)$. We delete rows (1) and (4). Since $4 < 8, \, -2 < 0$ we have $B(1) > B(2)$. Similarly $B(1) > B(4)$. We delete columns (2) and (4). The game is now reduced to

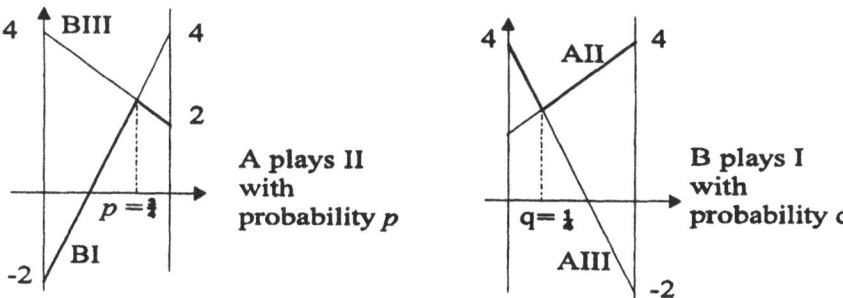

		B	
		I	III
A	II	4	2
	III	-2	4

We can solve this graphically:

Figure S 6 Graphical solution for Q1-6.

In this way we find for the original game

A's optimal strategy: $(0, {}^3/_4, {}^1/_4, 0)$,
B's optimal stragegy: $({}^1/_4, 0, {}^3/_4, 0)$,
Value $= {}^5/_2$.

Checking against the original matrix shows this to be correct. Moreover, since each dominance used in the reduction was strict, it is actually the case that we have found all optimal strategies for both players.

(b) $A(1) > A(2)$, so delete row (2).
$B(2) > B(1)$ so delete column (1).
Now $A(1) \geq A(4)$, so delete row (4).
Now $B(4) > B(2)$, so delete column (2). The game is reduced to the form on the right, which is easily solved as in (a) to give a final solution of the original game

		B	
		III	IV
A	I	6	11
	III	12	-6

A's optimal strategy: $({}^{18}/_{23}, 0, {}^5/_{23}, 0)$,
B's optimal stragegy: $(0, 0, {}^{17}/_{23}, {}^6/_{23})$,
Value $= {}^{168}/_{23}$.

(c) A(2) > A(1), A(2) > A(4), so
delete rows (1) and (4). Now B(4) >
B(1), B(4) > B(2), so delete columns
(1) and (2). The game is reduced to
the form on the right, which again is
easily solved, as before, to give a
final solution

		B	
		III	IV
A	II	8	9
	III	11	7

A's optimal strategy: $(0, \frac{4}{5}, \frac{1}{5}, 0)$,
B's optimal strategy: $(0, 0, \frac{2}{5}, \frac{3}{5})$,
Value $= \frac{43}{5}$.

(d) Whether you arrive at the conclusion by a long chain of dominances, or you have quite
rightly been checking for saddle points, you will doubtless have observed the saddle point in
the (3, 4) position, giving the game the value 2.

7.

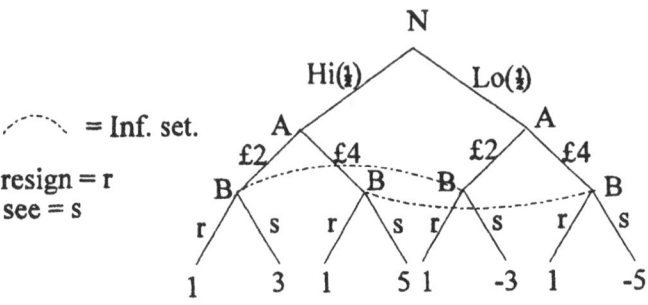

Figure S 7 Hi-Lo.

	A's pure strategies			B's pure strategies	
	If Hi, bet.	If Lo, bet.		If A bets £4	If A bets £2
I	2	2	I	Resign	See
II	2	4	II	See	See
III	4	2	III	Resign	Resign
IV	4	4	IV	See	Resign

Initial Table

Nature	Hi ($^1/_2$)				Lo ($^1/_2$)			
	BI	BII	BIII	BIV	BI	BII	BIII	BIV
I	3	3	1	1	-3	-3	1	1
II	3	3	1	1	1	-5	1	-5
III	1	5	1	5	-3	-3	1	1
IV	1	5	1	5	1	-5	1	-5

Plainly BII ≥ BIII. This is as far as one can go with simple elimination of one pure strategy by domination by another. However, we might notice that ½AIII + ½AIV ≥ AI, that is AI is dominated by a mixed strategy depending on AIII and AIV. The resulting 3 × 3 game is not soluble by the techniques of this chapter. However, you can easily check that

A's optimal strategy: $(0, 0, ^1/_3, ^2/_3)$,
B's optimal strategy: $(^1/_3, ^2/_3, 0, 0)$,
Value = $^1/_3$,

Matrix form

B

A		I	II	III	IV
	I	0	0	1	1
	II	2	-1	1	-2
	III	-1	1	1	3
	IV	1	0	1	0

is a solution for the original game. Thus A should always bet four dollars on Hi, and bet four dollars on Lo $^2/_3$ of the time, whilst B should always 'see' if A bets two dollars and 'see' $^2/_3$ of the time if A bets four dollars.

8.

A's pure strategies
I   Go to - ||
II   Go to |, |
III   Go to |, -

B's pure strategies
I   Take 0 if A takes 1
II   Take 1 if A takes 1
III   Take 2 if A takes 1

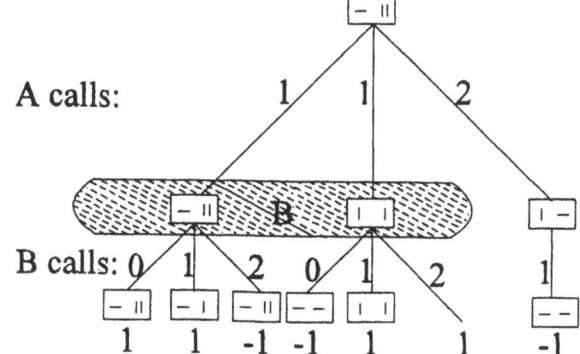

Figure S 8 Payoff to A in 1-2 Nim.

Note that if A takes 2 then B knows the state, takes the last match and wins. There is no real need to include this in the strategy statements.

Plainly AII (or AI) $\geq$ AIII so the
matrix becomes

$$\begin{pmatrix} 1 & 1 & -1 \\ -1 & 1 & 1 \end{pmatrix}$$

But now BI (or BIII) $\geq$ BII. The
resulting 2×2 game is easily solved
to give:

Matrix form

		B		
		I	II	III
	I	1	1	-1
A	II	-1	1	1
	III	-1	-1	-1

A's optimal strategy: $(\frac{1}{2}, \frac{1}{2}, 0)$,
B's optimal stragegy: $(\frac{1}{2}, 0, \frac{1}{2})$,
Value = 0,

as the solution to the original game.

9. (a) A(2) > A(4), A(2) > A(5) so delete rows (4) and (5). The game is reduced to

$$\begin{pmatrix} 4 & -3 \\ 2 & 1 \\ -1 & 3 \end{pmatrix}$$

If B plays I with probability q we
obtain the graph:

This gives us B's optimal strategy
$(\frac{2}{5}, \frac{3}{5})$ and the value of the game
$\frac{7}{5}$. Inspection of the graph also tells
us that the 2×2 subgame involving
AII and AIII should produce an
optimal strategy for A. If A plays AII
with probability p and AIII with
probability $1 - p$, we find from the
reduced payoff matrix

Figure S 9 Graph for B in Q1-9(a).

$$2p - (1 - p) = 3p - 1 \geq \frac{7}{5},$$

$$p + 3(1 - p) = -2p + 3 \geq \frac{7}{5},$$

which give $p = \frac{4}{5}$. Thus a full solution is:

A's optimal strategy: $(0, \frac{4}{5}, \frac{1}{5}, 0, 0)$,
B's optimal stragegy: $(\frac{2}{5}, \frac{3}{5})$,
Value = $\frac{7}{5}$.

(b) No pure strategies can be eliminated by straightforward domination by another pure
strategy, so we proceed immediately to the graph. Let A play I with probability p. We obtain

A carefully drawn graph reveals that A's gain floor can be held to its maximum if B uses some probability combination of BII and BV. This gives us A's optimal strategy as $(^7/_{13}, ^6/_{13})$ and the value $-^3/_{13}$. If B uses II with probability q and V with probability $1 - q$ we find

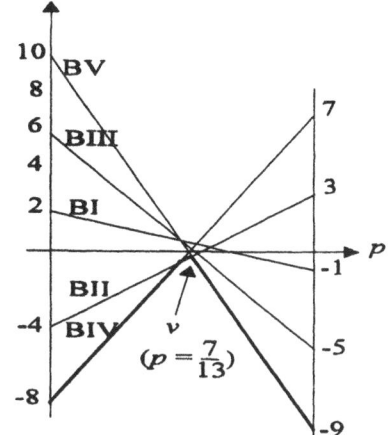

Figure S 10 Graph for A in Q1-9(b).

$$3q - 9(1 - q) = 12q - 9 \le -\frac{3}{13}$$

$$-4q + 10(1 - q) = -14q + 10 \le -\frac{3}{13}$$

which gives $q = ^{19}/_{26}$. Thus the full solution is:

> A's optimal strategy: $(^7/_{13}, ^6/_{13})$,
> B's optimal strategy: $(0, ^{19}/_{26}, 0, 0, ^7/_{26})$,
> Value $= -^3/_{13}$.

(c) $\text{AIV} \ge \text{AII}$, $\text{BIV} \ge \text{BII}$. The payoff matrix becomes

	I	III	IV
I	-3	0	2
III	1	-2	0
IV	0	3	-1

which becomes

	I	III	IV
IV	0	3	-1
I	-3	0	2
III	1	-2	0

upon rearranging the rows. It is now the matrix of a symmetric game with $a > 0$, $b < 0$, $c > 0$ so by §1.6 the solution is:

$$P_{IV} = \frac{2}{6}, \quad P_I = \frac{1}{6}, \quad P_{III} = \frac{3}{6} \text{ for A},$$

$$P_I = \frac{2}{6}, \quad P_{III} = \frac{1}{6}, \quad P_{IV} = \frac{3}{6} \text{ for B}.$$

Thus the full solution to the original game is:

> A's optimal strategy: $(^1/_6, 0, ^3/_6, ^2/_6)$,
> B's optimal strategy: $(^2/_6, 0, ^1/_6, ^3/_6)$,
> Value $= 0$.

10.

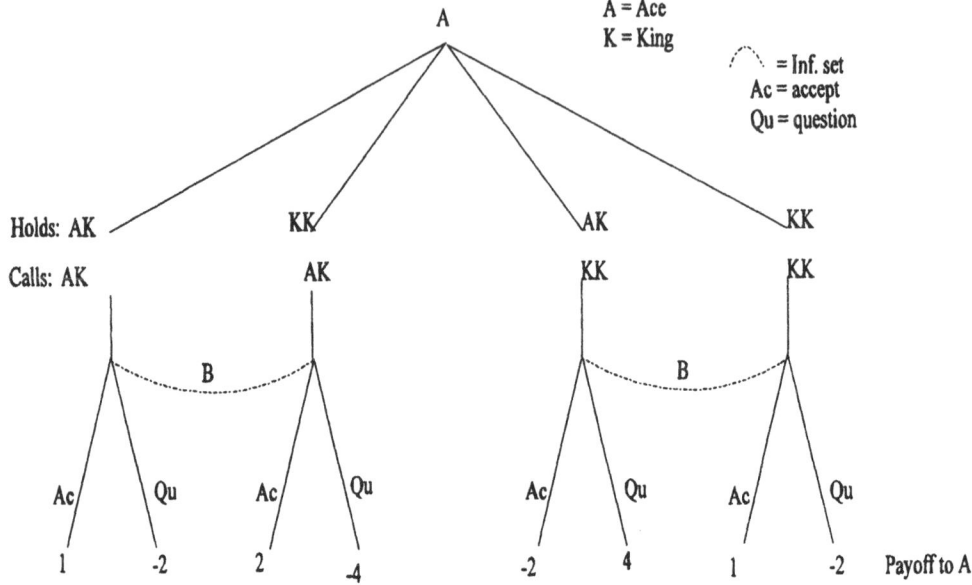

Figure S 11 Extensive form for Kings and Aces.

	A's pure strategies	
	Hold	Call
I	AK	AK
II	KK	AK
III	AK	KK
IV	KK	KK

	B's pure strategies	
	If A calls AK	If A calls KK
I	Accept	Accept
II	Accept	Question
III	Question	Accept
IV	Question	Question

Now B(4) ≥ B(2), B(3) ≥ B(1), so delete columns (2) and (1). We now find A(1) > A(2), A(3) ≥ A(1) so delete rows (2) and (1).

Matrix form

		B			
		I	II	III	IV
A	I	1	1	-2	-2
	II	2	2	-4	-4
	III	-2	4	-2	4
	IV	1	-2	1	-2

We are left with the game on the right, which is easily solved to give as a final solution:

A's optimal strategy: $(0, 0, \frac{1}{3}, \frac{2}{3})$,
B's optimal stragegy: $(0, 0, \frac{2}{3}, \frac{1}{3})$,
Value = 0 (the game is fair).

	B	
	III	IV
III	-2	4
IV	1	-2

(A on left of III/IV rows)

11. We first construct a 2×3 game graphically which has the desired solution set. This can be done in a variety of ways; one is given below.

This corresponds to the game

	B		
	I	II	III
I	3	-1	$\frac{1}{2}$
II	-1	3	$\frac{1}{2}$

(A on left)

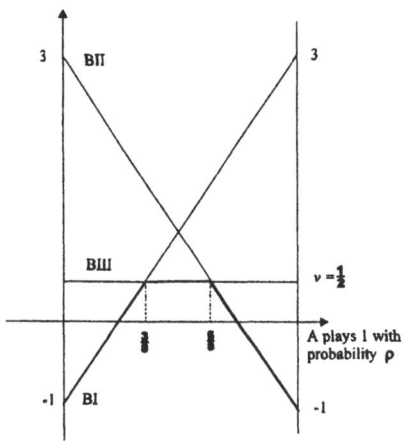

Figure S 12 Graphical solution for Q1-11.

To obtain a solution to the question we just invent any payoffs for AIII as long as AI or AII *strictly* dominate AIII. For example,

$$\begin{pmatrix} 3 & -1 & \frac{1}{2} \\ -1 & 3 & \frac{1}{2} \\ 2 & -2 & 0 \end{pmatrix}$$

is a possible solution.

12. Let D = Daughter, M = Mother, F = Father. We are given the ranking of the players in an individual board game as D > M > F. Let DF, DM, MF denote a Daughter-Father game, etc. Suppose the probability of D winning a DF, D winning a DM, M winning an MF is p, q, r respectively. We are told these probabilities can be assumed fixed throughout the tournament. From the ranking of the players we infer $p > r$ (the question is a little vague here, but if the daughter is a stronger player than the mother it seems reasonable that the daughter is more likely to beat her father than is her mother).

In the game tree below a single letter at some state indicates the tournament has been won by the person indicated.

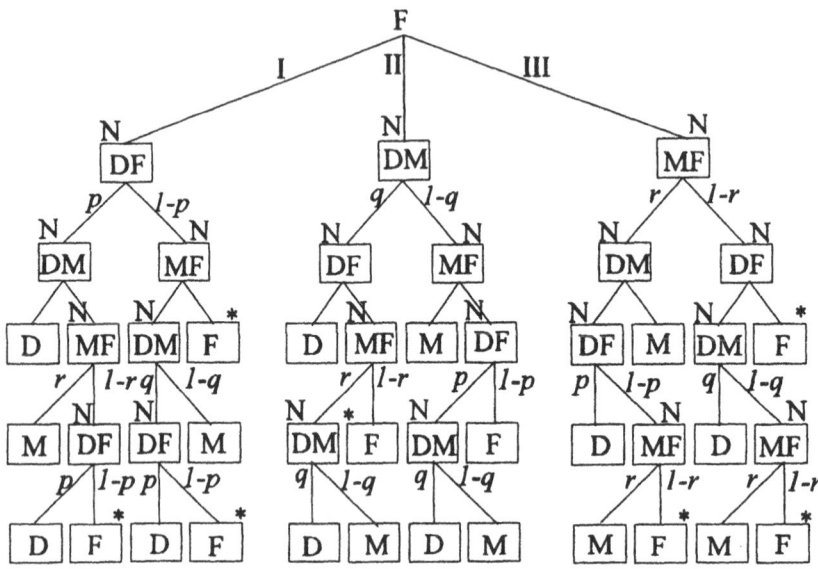

* = Terminal winning state for father

Figure S 13 Daughter-Mother-Father game.

For each pure strategy for F we calculate the total probability of arriving at a terminal winning state by chasing back up the game tree from each such state.

Thus

$$\text{I} : (1 - p)(1 - r)(1 - q)p + (1 - p)qr(1 - p) + (1 - r)(1 - p)$$

$$\text{II} : (1 - r)(1 - p)q + (1 - p)(1 - r)(1 - q) = (1 - r)(1 - p)(q + 1 - q)$$
$$= (1 - r)(1 - p)$$

$$\text{III} : (1 - r)(1 - p)qr + (1 - r)(1 - q)p(1 - r) + (1 - p)(1 - r)$$

Since the probability of winning by using II is $(1 - r)(1 - p)$ and this term appears in I and III (whilst the other terms are positive), plainly the probability of winning by using either of these is higher than when II is used. It therefore remains to prove that III is more desirable than I. We note that $(1 - r)(1 - p)$ appears in both, and for the other terms we have

$$(1 - r)(1 - q)p(1 - r) > (1 - p)(1 - r)(1 - q)p,$$
$$(1 - r)(1 - p)qr \qquad > (1 - p)qr(1 - p),$$

since $p > r$, and the proof is complete.

13. (i)

$$P(x, y) = \frac{1}{2}y^2 - 2x^2 - 2xy + \frac{7}{2}x + \frac{5}{4}y$$

$$\frac{\partial P}{\partial x} = -4x - 2y + \frac{7}{2} = 0,$$

$$\frac{\partial P}{\partial y} = -2x + y + \frac{5}{4} = 0,$$

Stationary point at $(\frac{3}{4}, \frac{1}{4})$

$$\frac{\partial^2 P}{\partial x^2} = -4,$$

$$\frac{\partial P}{\partial x \partial y} = -2, \quad P_{xy^2} - P_{xx}P_{yy} = 4 + 4 > 0.$$

$$\frac{\partial^2 P}{\partial y^2} = 1,$$

Hence the stationary point is a geometric saddle point. Inspection shows that the surface is correctly oriented for this to be a game theoretic saddle point. So a complete solution is

$$x = \frac{3}{4}, \quad y = \frac{1}{4}, \quad v = \frac{47}{32} \ (= 1.46875)$$

(ii) It is a 'not outrageously stupid' thing for A to do, because the expected value of $\xi \in [0,1]$ is:

$$\bar{\xi} = \int_0^1 \xi.3\xi^2 \, d\xi = \left[\frac{3\xi^4}{4} \right]_0^1 = \frac{3}{4}$$

that is, A's optimal pure strategy.

If A plays $p(\xi) = 3\xi^2$ and B plays any pure strategy y, the payoff to A is

$$P(p, y) = \int_0^1 (\frac{1}{2}y^2 - 2\xi^2 - 2\xi y + \frac{7}{2}\xi + \frac{5}{4}y) 3\xi^2 \, d\xi$$

$$= 3\left[\frac{1}{2}y^2 \frac{\xi^3}{3} - \frac{2\xi^5}{5} - \frac{2y\xi^4}{4} + \frac{7}{2}\frac{\xi^4}{4} + \frac{5}{4}y\frac{\xi^3}{3} \right]_0^1$$

$$= 3\left(\frac{1}{6}y^2 - \frac{2}{5} - \frac{1}{2}y + \frac{7}{8} + \frac{5}{12}y \right)$$

$$= 3\left(\frac{1}{6}y^2 - \frac{1}{12}y + \frac{19}{40} \right),$$

B wishes to minimise this; $P(p, y)$ is a minimum when

$$\frac{1}{3}y - \frac{1}{12} = 0, \text{ that is, } y = \frac{1}{4}.$$

Thus even if B knew A's mixed strategy, B could play no better pure strategy than the previously optimal $y = \frac{1}{4}$. (Of course there is undoubtedly a mixed strategy for B that would do better in this case.) Then

$$P(p, \tfrac{1}{4}) = \frac{1}{2} \cdot \frac{1}{16} - \frac{1}{4} \cdot \frac{1}{4} + \frac{57}{40} = \frac{223}{160} \quad (= 1.39375).$$

Plainly A is $0.075 = \frac{3}{40}$ worse off, which is to be expected since his mixed strategy is not optimal.

14. Technically we are using Stieltjes integrals here, but they behave very intuitively; a quick introduction can be found in McKinsey [16] Chapter 9; other references are Widder [17], [18].

We have

$$\int_0^1\int_0^1 16(x-y)^2 dF(x)dG_0(y) = \int_0^1 16(x-\tfrac{1}{2})^2 dF(x) \le \int_0^1 16 \cdot \tfrac{1}{4} dF = 4,$$

since G_0 concentrates y at $y = \frac{1}{2}$ with probability 1; $(x - \frac{1}{2})^2 \le \frac{1}{4}$ for $0 \le x \le 1$; and $\int dF = 1$.

$$\int_0^1\int_0^1 16(x-y)^2 dF_0(x)dG_0(y) = \int_0^1 16(x-\tfrac{1}{2})^2 dF_0 \quad (\text{as before})$$

$$= 16(-\tfrac{1}{2})^2 \cdot \tfrac{1}{2} + 16(\tfrac{1}{2})^2 \tfrac{1}{2} = 4$$

since F_0 concentrates x at $x = 0$ with probability $\frac{1}{2}$ and $x = 1$ with probability $\frac{1}{2}$.

$$\int_0^1\int_0^1 16(x-y)^2 dF_0(x)dG(y) = \int_0^1 (16(-y)^2 \tfrac{1}{2} + 16(1-y)^2 \tfrac{1}{2})dG$$

$$= 8\int_0^1 (2y^2 - 2y + 1)dG.$$

$$\ge 8\int_0^1 \tfrac{1}{2} dG = 4,$$

since $2y^2 - 2y + 1 \ge \frac{1}{2}$ for $0 \le y \le 1$.

15. For any given strategy y if A, B can choose the same, in which case $dx/dt = 0$ for $0 \le t \le T$, and then

$$\int_0^T x\,dt = x_0 T.$$

Thus

$$\max_x \min_z \int_0^T x\,dt \le x_0 T.$$

But since $dx/dt = (y - z)^2 \ge 0$, $x(t)$ is non-decreasing and so

$$\int_0^T x \, dt \geq x_0 T,$$

for any pair of strategies (y, z). Thus

$$\max_x \min_z \int_0^T x \, dt = x_0 T.$$

On the other hand if B chooses a z, then A can choose a y defined by

$$y(t, x) = \begin{cases} 1, & \text{if } z(t, x) \leq \frac{1}{2}, \\ 0, & \text{if } z(t, x) > \frac{1}{2}. \end{cases}$$

This ensures $dx/dt \geq \frac{1}{4}$ for $0 \leq t \leq T$. Hence $x(t) \geq x_0 + \frac{1}{4}t$, so that

$$\min_z \max_y \int_0^T x \, dt \geq x_0 T + \frac{1}{8}T^2.$$

For the strategy $z(t, x) \equiv \frac{1}{2}$ of B and for any strategy y of A, $dx/dt \leq \frac{1}{4}$ for $0 \leq t \leq T$. Thus $x(t) \leq x_0 + \frac{1}{4}t$, and so

$$\min_z \max_y \int_0^T x \, dt = x_0 T + \frac{1}{8}T^2.$$

CHAPTER 2.

1. (a) (σ_2, τ_1) and (σ_1, τ_2) are both equilibrium pairs where player 1 prefers (σ_1, τ_2) and player 2 prefers (σ_2, τ_1). However, if player 2 has any reason to think that player 1 will choose σ_1 then he or she dare not choose τ_1 for fear of getting -300, but then 1, knowing this, has every reason to choose σ_1 which gives the best return. But this argument has 'positive feedback' since now 2, having some rationalisation for 1's adoption of σ_1, has all the more reason to avoid τ_1. Thus the eqilibrium (σ_1, τ_2) 'psychologically dominates' (σ_2, τ_1).

(b) In this case (σ_2, τ_1), (σ_1, τ_2) are again equilibrium points, but now (σ_2, τ_1) gives a higher yield (12, 8) to both players. Yet player 1 may hesitate to use σ_2 on the grounds that player 2 would argue that τ_2 'psychologically dominates' τ_1, and so long as player 2 can give any rationale for 1's choosing σ_1, 2 does not dare choose τ_1. Again the argument has 'positive feedback' and it reinforces (σ_1, τ_2) even though (σ_2, τ_1) gives both players a higher yield.

2. From inequalities (2.21) and (2.22) we have

$$P(\mathbf{x}'', \mathbf{y}') \leq P(\mathbf{x}', \mathbf{y}') \leq P(\mathbf{x}', \mathbf{y}'') \leq P(\mathbf{x}'', \mathbf{y}'') \leq P(\mathbf{x}'', \mathbf{y}')$$

Hence equality holds throughout, and it follows that $(\mathbf{x}', \mathbf{y}'')$, $(\mathbf{x}'', \mathbf{y}')$ are equilibrium points.

In the non-zero sum game

$$
\begin{array}{cc}
 & \tau_1 \quad\;\; \tau_2 \\
\begin{array}{c}\sigma_1 \\ \sigma_2\end{array} &
\left(\begin{array}{cc}
(3, 2) & (0, 0) \\
(0, 0) & (2, 3)
\end{array}\right)
\end{array}
$$

the pairs (σ_1, τ_1), (σ_2, τ_2) are equilibria but (σ_2, τ_1), (σ_1, τ_2) are not.

3. Consider a pure strategy n-tuple $\sigma = (k, k, ...,k)$ with all components equal. For any $i \in I$, if $\sigma_i \neq k$ we have

$$P_i(\sigma|\sigma_i) = 0 \leq P_i(\sigma) = a_{ik} \,.$$

Hence by Theorem 2.3 σ is an equilibrium point. Next consider a pure strategy n-tuple $\sigma = (\sigma_1, ..., \sigma_n)$ which has at least three distinct components, so that $P_i(\sigma) = 0$ for every $i \in I$. In this case if any player $i \in I$ unilaterally changes strategy to $\sigma'_i \neq \sigma_i$, there are still at least two distinct components in $\sigma \parallel \sigma'_i$, so that $P_i(\sigma \parallel \sigma'_i) = 0$. Hence for any $i \in I$

$$0 = P_i(\sigma|\sigma'_i) \leq P_i(\sigma) = 0 \,,$$

and again by Theorem 2.3 σ is an equilibrium point.

It remains to consider pure strategy n-tuples $\sigma = (\sigma_1, ..., \sigma_n)$ with exactly two distinct components, the doubtful case being, $\sigma_j = k$ for $j \neq i$, $\sigma_i \neq k$, say. Then if $\sigma'_i = k$

$$P_i(\sigma \parallel \sigma'_i) = a_{ik} \nleq P_i(\sigma) = 0,$$

so by Theorem 2.3 σ is not an equilibrium point.

To show the equilibria do not form a rectangular set consider for example the pure strategy n-tuples $(1, 1, ..., 1)$, $(2, 2, ..., 2)$ both of which are equilibrium points. If the set were rectangular then $(2, 1, 1, ..., 1)$, say, would be an equilibrium point. But we have just shown that this is not the case.

4. Suppose $x_i(\sigma^j_i) > 0$. Then

$$P_i(x|\sigma^j_i)\, x_i(\sigma^{j\,\prime}_i) < P_i(x)\, x_i(\sigma^j_i) \,.$$

For all pure strategies $\sigma_i \in S_i$ with $\sigma_i \neq \sigma^j_i$ we have, since x is an equilibrium point,

$$P_i(x|\sigma_i) \leq P_i(x) \,,$$

whence

$$P_i(x|\sigma_i)\, x_i(\sigma_i) \leq P_i(x)\, x_i(\sigma_i) \,.$$

Summing the last inequality over all $\sigma_i \neq \sigma^j_i$ and adding the first inequality we obtain

$$\sum_{\sigma_i \in S_i} P_i(x \| \sigma_i) \, x_i(\sigma_i) < P_i(x) \ ,$$

since

$$\sum_{\sigma_i \in S_i} x_i(\sigma_i) = 1 \ .$$

But using (2.6) and (2.7) the left side of this inequality is just $P_i(x)$ so that

$$P_i(x) < P_i(x) \ ,$$

a contradiction.

5. We can use the techniques at the end of §2.2 where a mixed strategy 2-tuple is denoted by (x, y) meaning player 1 uses the mixed strategy $(x, 1-x)$ and player 2 $(y, 1-y)$. One finds that the equilibrium points are

(a) $(^2/_9, \ ^3/_{14})$,   (b) $(0, 0), (1, 1), (^1/_3, \ ^2/_3)$.

6. We notice that for player 1: if $\sigma_1 = 1$ then $P_1 = 0$ unless $\sigma_2 = ... = \sigma_n = 2$, in which case $P_1 = 1$; if $\sigma_1 = 2$ then $P_1 = 0$ unless $\sigma_2 = ... = \sigma_n = 1$, in which case $P_1 = 2$. Similarly for the other players.

Consider now the mixed strategy n-tuple $x = (x_1, ..., x_n)$, where $x_i = (p_i, 1 - p_i)$ for $1 \le i \le n$ and p_i is the probability of choosing $\sigma_i = 1$. From (26) or the above observation we obtain

$$P_i(x) = p_i \prod_{j \ne i} (1 - p_j) + 2(1 - p_i) \prod_{j \ne i} p_j \ .$$

Also

$$P_i(x \| \sigma_i) = \prod_{j \ne i} (1 - p_j) \quad \text{if } \sigma_i = 1 \ ,$$

$$P_i(x \| \sigma_i) = 2 \prod_{j \ne i} p_j \quad \text{if } \sigma_i = 2 \ .$$

According to Theorem 2.3, x is an equilibrium point if and only if

$$\prod_{j \ne i} (1 - p_j) \le p_i \prod_{j \ne i} (1 - p_j) + 2 (1 - p_i) \prod_{j \ne i} p_j \qquad (1)$$

and

$$\prod_{j \ne i} p_j \le p_i \prod_{j \ne i} (1 - p_j) + 2 (1 - p_i) \prod_{j \ne i} p_j \qquad (2)$$

for every $i \in I$. Rearranging (1) we have

$$\prod_j (1 - p_j) \le 2 (1 - p_i) \prod_{j \ne i} p_j$$

that is,

$$\prod_{j \neq i} (1 - p_j) \leq 2 \prod_{j \neq i} p_j \qquad (3)$$

Similarly rearranging (2) gives

$$2 \prod_{j \neq i} p_j \leq \prod_{j \neq i} (1 - p_j) . \qquad (4)$$

From (3) and (4) it follows that x is an equilibrium point if and only if

$$\prod_{j \neq i} (1 - p_j) = 2 \prod_{j \neq i} p_j \text{ for every } i \in I . \qquad (5)$$

For $n = 2$ or 3 the system of equations (5) has no solution with any $p_i = 0$ or 1, but for $n = 4$ there are several such solutions, for example $p_1 = p_4 = 1$, $p_2 = p_3 = 0$. If $n \geq 5$ we can find solutions with $p_1 = p_4 = 1$, $p_2 = p_3 = 0$ and the remaining $n - 4$ p_i arbitrary. To complete the analysis suppose $0 < p_i < 1$ for every $i \in I$.

Consider the equation (5) for $i = k$, $i = l$ where $k \neq l$. This gives

$$\prod_{j \neq k} (1 - p_j) = 2 \prod_{j \neq k} p_j \text{ and } \prod_{j \neq l} (1 - p_j) = 2 \prod_{j \neq l} p_j .$$

If we put $A = \prod(1 - p_j)$, $B = \prod p_j$, since $0 < p_i < 1$, all i, we can write these as

$$\frac{A}{1 - p_k} = \frac{2B}{p_k} , \qquad \frac{A}{1 - p_l} = \frac{2B}{p_l}$$

Since $A \neq 0$ and $B \neq 0$ we easily see that $p_k = p_l$. But k and l were arbitrary, so that *every player must use the same mixed strategy in x*. Condition (5) therefore becomes simply

$$(1 - p)^{n-1} = 2 p^{n-1} .$$

Solving for p we obtain

$$p = \frac{1}{1 + 2^{1/(n-1)}} ,$$

as required.

Remark. The process of finding equilibrium points in a bimatrix game (and hence also in a matrix game) consists in carrying out a finite number of *rational* operations on the values of the payoff matrix (see [16], [17], and [18] for algorithms for finding equilibria of bimatrix games). For $n \geq 3$ the above value of p is irrational, which shows that the situation for $n = 2$ is untypical.

7. Using the notation of (2.24) if player 1 employs the mixed strategy $x = (^1/_m, ^1/_m, ..., ^1/_m)$ then

$$P(x, \tau_j) = x \cdot \beta_j = \frac{1}{m} g_j > 0 \text{ for } 1 \leq j \leq m .$$

Hence for any strategy $\mathbf{y} = (y_1, ..., y_m)$ for player 2

$$P(\mathbf{x}, \mathbf{y}) = \sum_{j=1}^{m} y_i \, P(\mathbf{x}, \tau_j) = \sum_{j=1}^{m} y_j \, \frac{1}{m} \, g_j \geq \frac{1}{m} \, (\min g_i) \sum y_j \, .$$

But $\Sigma y_j = 1$, so

$$\min_{\mathbf{y} \in \Sigma^{(2)}} P(\mathbf{x}, \mathbf{y}) \geq \frac{1}{m} \, (\min g_i) > 0 \, .$$

Hence

$$v = \max_{\mathbf{x} \in \Sigma^{(1)}} \, \min_{\mathbf{y} \in \Sigma^{(2)}} P(\mathbf{x}, \mathbf{y}) \geq \frac{1}{m} \, (\min g_i) > 0 \, ,$$

that is, the value of any diagonal game is positive.

If $\mathbf{x}^0 = (x^0_1, ..., x^0_m)$ is optimal for player 1, let us suppose $x^0_j = 0$ for some j. If player 2 chooses pure strategy τ_j we obtain

$$v = P(\mathbf{x}^0, \tau_j) = \mathbf{x}^0 \cdot \beta_j = x_1 . 0 + ... + x_{j-1} . 0 + 0 . g_j + x_{j+1} . 0 + ... + x_m . 0 = 0$$

that is, $v \leq 0$, which contradicts $v > 0$. Hence $x^0_j > 0$ $(1 \leq j \leq m)$.

It now follows from Theorem 2.9(ii) that if $\mathbf{y} = (y_1^0, ..., y_m^0)$ is any optimal strategy for player 2

$$P(\sigma_j, \mathbf{y}^0) = \mathbf{y}_0 \cdot \alpha_j = v \quad \text{for } 1 \leq j \leq m \, .$$

But $\mathbf{y}_0.\alpha_j = y_j^0 g_j$ so that $y_j^0 = v/g_j$. Since

$$\sum_{j=1}^{m} y_j^0 = 1 \, ,$$

we obtain

$$v = \frac{1}{\displaystyle\sum_{k=1}^{m} 1/g_k}$$

Now, $y_j^0 = v/g_j$ gives

$$y_j^0 = \frac{1}{g_j} \frac{1}{\displaystyle\sum_{k=1}^{m} 1/g_k} , \quad (1 \leq j \leq m),$$

as required.

Thus $y_j^0 > 0$ $(1 \leq j \leq m)$, so using Theorem 2.9(i) we have for $j = i$

$$P(\mathbf{x}^0, \tau_i) = \mathbf{x}_0 \cdot \beta_i = x_i^0 g_i = v \ (1 \leq i \leq m) \, .$$

As before, this gives

$$x_i^0 = \frac{1}{g_i} \frac{1}{\sum\limits_{k=1}^{m} 1/g_k} \qquad (1 \le i \le m).$$

It follows that $\mathbf{x}^0$ and $\mathbf{y}^0$ are uniquely determined.

8. Let $\mathbf{x}^0$, $\mathbf{y}^0$ be optimal strategies for players 1 and 2, respectively, in the game with matrix $A = (a_{ij})$. Denote the payoff function for this game by $P(\mathbf{x}, \mathbf{y})$. Then $P(\mathbf{x}^0, \mathbf{y}^0) = v$ and

$$P(\mathbf{x}, \mathbf{y}^0) \le v \le P(\mathbf{x}^0, \mathbf{y}) \quad \text{for all } \mathbf{x} \in \Sigma^{(1)}, \mathbf{y} \in \Sigma^{(2)}. \tag{1}$$

If $Q(\mathbf{x}, \mathbf{y})$ is the payoff function for the game with matrix B we have for $\mathbf{x} \in \Sigma^{(1)}, \mathbf{y} \in \Sigma^{(2)}$

$$\begin{aligned}
Q(\mathbf{x}, \mathbf{y}) &= \sum_{i=1}^{m} \sum_{j=1}^{n} x_i (a_{ij} + k) y_j \\
&= \sum_{i=1}^{m} \sum_{j=1}^{n} x_i a_{ij} y_j + \sum_{i=1}^{m} \sum_{j=1}^{n} x_i k y_j \\
&= P(\mathbf{x}, \mathbf{y}) + k \left(\sum_{i=1}^{m} x_i \right) \left(\sum_{j=1}^{n} y_j \right) \\
&= P(\mathbf{x}, \mathbf{y}) + k .
\end{aligned}$$

Hence from (1)

$$Q(\mathbf{x}, \mathbf{y}^0) \le v + k \le Q(\mathbf{x}^0, \mathbf{y}) \quad \text{for all } \mathbf{x} \in \Sigma^{(1)}, \mathbf{y} \in \Sigma^{(2)},$$

and the conclusion follows.

9. Since $1 + 2 + 3 + \ldots + m = \frac{1}{2}m(m+1)$ it is easy to see that the mixed strategy $(^1/_m, ^1/_m, \ldots, ^1/_m)$ is optimal for both players, and $v = \frac{1}{2}(m+1)$ (This uses Corollary 2.1). In general, if every row and column of an $m \times m$ matrix sums to the same number $s > 0$, then a similar argument shows that the value of the corresponding game is s/m.

10. If we draw a graph for this game, as in §1.6, we obtain for $d = 1$,

The limiting situation corresponds to the dotted lines. Plainly any $d > 0$ will give a graph with the same general characteristics. From the graph it is apparent that the unique solution is

$\mathbf{x} = (\frac{1}{2}, \frac{1}{2})$
$\mathbf{y} = (1, 0, 0, 0, \ldots)$
$v = d$.

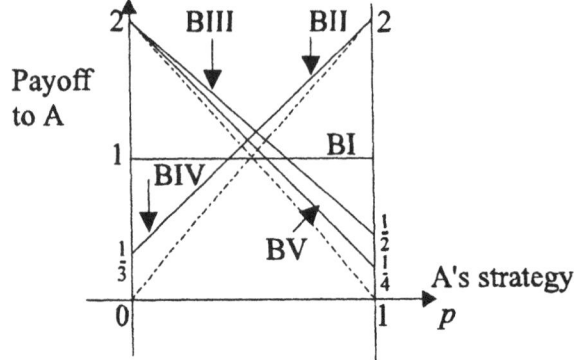

Figure S 14 Graph for Q2-10.

11. Put A = Ann, B = Bill. Each antes 1.

A can go Hi (bet 2) or Lo (bet 1).
B can Fold (bet 0) or See (match A's bet).

The extensive form is:

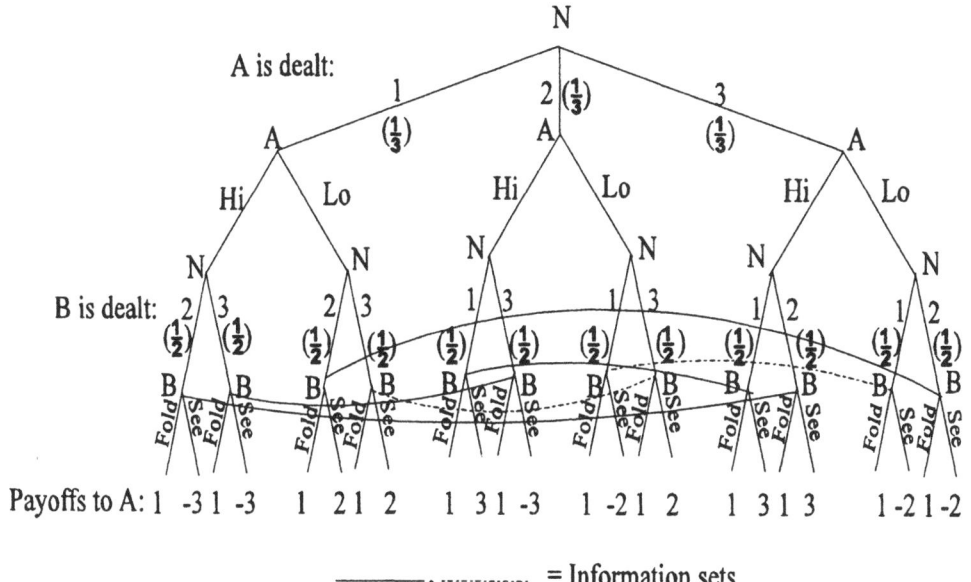

Payoffs to A: 1 -3 1 -3   1   2 1 2   1 3 1 -3   1 -2 1 2   1 3 1 3   1 -2 1 -2

———— , ·········· = Information sets

Figure S 15 Ann and Bill's simplified Poker.

A's pure strategies are all of the form:

"If hold 1 go Hi/Lo. If hold 2 go Hi/Lo. If hold 3 go Hi/Lo."

Since there are two possibilities for each situation there are $2^3 = 8$ such pure strategies.

B's pure strategies are all of the form (F = Fold, S = See)

	If A goes Hi	If A goes Lo
If hold 1	F/S	F/S
If hold 2	F/S	F/S
If hold 2	F/S	F/S

Since there are two possibilities for each situation there are $2^6 = 64$ such pure strategies. Inspection of the tree shows:

(i)   If A holds 1, A will not go Hi; similarly if she holds 3, she will not go Lo.
(ii)   If B holds 1 and A goes Hi, B will Fold; similarly if he holds 1 and A goes Lo, B will See.

(iii)   If B holds 3 and A goes Hi, B will See; similarly if he holds 3 and A goes
         Lo, B will Fold.

A's pure strategies

I   If hold 1 go Lo. If hold 2 go Lo. If hold 3 go Hi.
II   If hold 1 go Lo. If hold 2 go Hi. If hold 3 go Hi.

B's pure strategies

	Hold	A-Hi	A-Lo			Hold	A-Hi	A-Lo
I	1	F	F		II	1	F	S
	2	F	F			2	F	S
	3	S	F			3	S	F
III	1	F	S		IV	1	F	S
	2	S	F			2	S	S
	3	S	F			3	S	F

Formally, this amounts to the elimination of pure strategies from the 8×64 matrix by appealing
to dominance.

From the tree we obtain

$$E(I, I) = \frac{1}{3} (\frac{1}{2}.1 + \frac{1}{2}.2) + \frac{1}{3} (\frac{1}{2}(-2) + \frac{1}{2}.1) + \frac{1}{3} (\frac{1}{2}(1) + \frac{1}{2}.1) = \frac{1}{2}$$

$$E(I, II) = \frac{1}{3} (\frac{1}{2}.2 + \frac{1}{2}.1) + \frac{1}{3} (\frac{1}{2}(-2) + \frac{1}{2}.1) + \frac{1}{3} (\frac{1}{2}.1 + \frac{1}{2}.1) = \frac{2}{3}$$

$$E(I, III) = \frac{1}{3} (\frac{1}{2}.1 + \frac{1}{2}.1) + \frac{1}{3} (\frac{1}{2}(-2) + \frac{1}{2}.1) + \frac{1}{3} (\frac{1}{2}.1 + \frac{1}{2}.3) = \frac{5}{6}$$

$$E(I, IV) = \frac{1}{3} (\frac{1}{2}.2 + \frac{1}{2}.1) + \frac{1}{3} (\frac{1}{2}(-2) + \frac{1}{2}.1) + \frac{1}{3} (\frac{1}{2}.1 + \frac{1}{2}.3) = 1$$

$$E(II, I) = \frac{1}{3} (\frac{1}{2}.1 + \frac{1}{2}.1) + \frac{1}{3} (\frac{1}{2}.1 + \frac{1}{2}(-3)) + \frac{1}{3} (\frac{1}{2}.1 + \frac{1}{2}.1) = \frac{1}{3}$$

$$E(II, II) = \frac{1}{3} (\frac{1}{2}.2 + \frac{1}{2}.1) + \frac{1}{3} (\frac{1}{2}.1 + \frac{1}{2}(-3)) + \frac{1}{3} (\frac{1}{2}.1 + \frac{1}{2}.1) = \frac{1}{2}$$

$$E(II, III) = \frac{1}{3} (\frac{1}{2}.1 + \frac{1}{2}.1) + \frac{1}{3} (\frac{1}{2}.1 + \frac{1}{2}(-3)) + \frac{1}{3} (\frac{1}{2}.1 + \frac{1}{2}.3) = \frac{2}{3}$$

$$E(II, IV) = \frac{1}{3} (\frac{1}{2}.2 + \frac{1}{2}.1) + \frac{1}{3} (\frac{1}{2}.1 + \frac{1}{2}(-3)) + \frac{1}{3} (\frac{1}{2}.1 + \frac{1}{2}.3) = \frac{5}{6}$$

Matrix

	I	II	III	IV	ROW MIN	
I	$\frac{1}{2}$	$\frac{2}{3}$	$\frac{5}{6}$	1	$\frac{1}{2}$	← MAX OF MIN
II	$\frac{1}{3}$	$\frac{1}{2}$	$\frac{2}{3}$	$\frac{5}{6}$	$\frac{1}{3}$	
COL MAX	$\frac{1}{2}$	$\frac{2}{3}$	$\frac{5}{6}$	1		

↑ MIN OF MAX

Saddle point at (AI, BI). Value = $\frac{1}{2}$.

12. ϕ is continuous at $x = x_0$ if

$$\forall\ \varepsilon > 0 \ \exists\ \delta > 0 \text{ such that } |\phi(x) - \phi(x_0)| < \varepsilon \text{ whenever } |x - x_0| < \delta .$$

Suppose this is false. Then

$$\exists\ \varepsilon > 0 \text{ such that } \forall \text{ integer } v \geq 1 \ \exists\ x_v \in \left(x_0 - \frac{1}{v},\ x_0 + \frac{1}{v}\right)$$
$$\text{such that } |\phi(x_v) - \phi(x_0)| \geq \varepsilon .$$

Now the sequence $\phi(x_v)$ $(v = 1, 2, 3, ...)$ may not be convergent, but since $0 \leq \phi(x_v) \leq 1$ by the Bolzano-Weierstrass theorem there exists a convergent subsequence $\phi(x_{v_j})$ $(j = 1, 2, ...)$. Since

$$|\phi(x_{v_j}) - \phi(x_0)| \geq \varepsilon > 0 , \tag{1}$$

We have

$$y_1 = \lim_{j \to \infty} \phi(x_{v_j}) \neq \phi(x_0) = y_0 , \text{ say} .$$

By definition of ϕ

$$P(x_{v_j},\ \phi(x_{v_j})) < P(x_{v_j},\ y)$$

for all $y \neq \phi(x_{v_j})$. By (1) this holds for $y = y_0$, that is,

$$P(x_{v_j},\ \phi(x_{v_j})) < P(x_{v_j},\ y_0)$$

Letting $j \to \infty$ and using the continuity of P gives

$$P(x_0,\ y_1) \leq P(x_0,\ y_0)$$

where $y_1 \neq y_0$. But y_0 was the *unique* value of y which minimised $P(x_0, y)$. This gives a contradiction. Hence ϕ is continuous at every point $x \in [0,1]$.

Similarly ψ is continuous in $[0,1]$. It follows from a standard theorem in analysis that $\psi \circ \phi$

is a continuous function, being the composition of two continuous functions.

By the Brouwer fimxed point theorem, since $\psi \circ \phi : [0,1] \to [0,1]$ is a continuous mapping of a convex closed bounded subset of $\mathbb{R}$ into itself, there exists $x^0 \in [0,1]$ such that

$$x^0 = \psi(\phi(x^0)) .$$

Put $y^0 = \phi(x^0) \in [0,1]$. Then

$$x^0 = \psi(y^0), \quad y^0 = \phi(x^0) ,$$

that is,

$$\max_{x \in [0,1]} P(x, y^0) \text{ is attained at } x = x^0 ,$$

$$\min_{y \in [0,1]} P(x^0, y) \text{ is attained at } y = y^0 ,$$

from which we obtain

$$P(x, y^0) \le P(x^0, y^0) \ \forall \ x \in [0,1] ,$$
$$P(x^0, y^0) \le P(x^0, y) \ \forall \ x \in [0,1] ,$$

that is, (x^0, y^0) is a saddle point.

13. (i) If $0 \le y' < y'' \le 1$ we have

$$E(F, y') - E(F, y'') = \int_0^1 (P(x, y') - P(x, y'')) \, F'(x) \, dx$$

$$= \int_0^{y'} (K(x, y') - K(x, y'')) \, F'(x) \, dx + \int_{y'}^{y''} (-K(y', x) - K(x, y'')) \, F'(x) \, dx$$

$$+ \int_{y''}^1 (-K(y', x) + K(y'', x)) \, F'(x) \, dx .$$

Now $K(x, y)$ is continuous on the unit square, so it is certainly bounded there, say

$$|K(x, y)| \le B \ \forall \ (x, y) \in [0,1] \times [0,1] .$$

Also $K(x, y)$ is uniformly continuous on the unit square, in particular this means

$$\forall \ \varepsilon > 0 \ \exists \ \delta > 0 \text{ such that}$$
$$\max \{ \ |K(x, y') - K(x, y'')|, \ |K(y', x) - K(y'', x)| \ \} < \varepsilon$$
$$\text{whenever } |y' - y''| < \delta \text{ and } x \in [0,1] .$$

Using these facts and $|\int f| \le \int |f|$ we obtain

$$|E(F, y') - E(F, y'')| \le \varepsilon \int_0^{y'} |F'(x)| \, dx + 2B \int_{y'}^{y''} |F'(x)| \, dx + \varepsilon \int_{y''}^{1} |F'(x)| \, dx$$

provided $|y' - y''| < \delta$. Since F' is continuous on $[0,1]$ it is bounded there, say $|F'| \le C$ on $[0,1]$. Thus since $y' < y'' \le 1$

$$|E(F, y') - E(F, y'')| \le C\varepsilon + 2BC\delta + C\varepsilon$$

if $|y' - y''| < \delta$. From which it follows that $E(F, y)$ is a continuous function of y.

Now suppose $E(F, y_0) = \alpha > 0$ for some $y_0 \in (a, b)$. Then by continuity $E(F, y) \ge \alpha/2 > 0$ for all y in some interval $I = (a, b)$ about y_0. Consider

$$E(F, F) = \int\int P(x, y) \, dF(x) \, dF(y) = \int_0^1 \int_0^1 P(x, y) \, dF(x) \, dF(y)$$

$$= \int_0^1 E(F, y) \, F'(y) \, dy \ .$$

Since F is optimal for player 1, $E(F, y) \ge 0$ for all $y \in [0,1]$; also $F' \ge 0$ in $[0,1]$ by hypothesis. Hence, since $F' > 0$ in I,

$$E(F, F) \ge \int_I E(F, y) \, F'(y) \, dy \ge \int_I \frac{\alpha}{2} F'(y) \, dy > 0 \ .$$

which is a contradiction since F is supposed optimal for both players and the value of the game is zero (if it exists at all), that is, $E(F, F) = 0$. Hence

$$E(F, y) = 0 \text{ for all } y \in (a, b) \ .$$

(ii) We need the formula for differentiation under the sign of integration. If

$$G(y) = \int_{\psi(y)}^{\phi(y)} f(x, y) \, dx$$

then

$$G'(y) = \int_{\psi(y)}^{\phi(y)} \frac{\partial f}{\partial y} \, dx + f(\phi(y), y) \, \phi'(y) - f(\psi(y), y) \, \psi'(y) \ .$$

This is valid if ψ and ϕ are functions with finite derivatives, $f_y(x, y)$ is continuous and f_{yy} is bounded (see for example Apostol [19] 9 - 38, p. 220). Now

$$E(F, y) = \int_0^1 P(x, y) \, dF(x) = \int_0^y K(x, y) \, F'(x) \, dx - \int_y^1 K(y, x) \, F'(x) \, dx \ .$$

Applying the differentiation formula to each integral yields

$$\frac{\partial}{\partial y} E(F, y) = \int_0^y \frac{\partial K(x, y)}{\partial y} F'(x) \, dx + K(y, y) \, F'(y) . 1 - K(0, y) \, F'(0) . 1$$

$$- \int_y^1 \frac{\partial K(y, x)}{\partial y} F'(x) \, dx - K(y, 1) \, F'(1) . 0 + K(y, y) \, F'(y) . 1$$

$$= 2 \, K(y, y) \, F'(y) + \int_0^y \frac{\partial K(x, y)}{\partial y} F'(x) \, dx - \int_y^1 \frac{\partial K(y, x)}{\partial y} F'(x) \, dx .$$

Since $F'(x) = 0$ outside (a, b) and by (i) $E_y(F, y) = 0$ for $y \in (a, b)$, the result follows.

14. If player 1 fires at time x and 2 at time y, where $x < y$, then player 1 has probability x of hitting his opponent, giving payoff 1, and probability of $1 - x$ of missing. In the event that player 1 misses, there is a probability of y that he will be hit, giving payoff -1. hence $P(x, y) = x - (1-x) y$ if $x < y$. Similarly, $P(x, y) = -y + (1-y) x$ if $x > y$, and $P(x, y) = 0$ if $x = y$.

Thus in question 13 we take

$$K(x, y) = x - y + xy .$$

Hence

$$- 2 \, K(y, y) = - 2 y^2 ,$$

and the integral equation becomes

$$-2 \, y^2 \, F'(y) = \int_a^y (-1 + x) \, F'(x) \, dx - \int_y^b (-1 - x) \, F'(x) \, dx .$$

Differentiating both sides with respect to y, using the formula quoted above, gives

$$- 4 \, y \, F' - 2 \, y^2 \, F'' = (y - 1) \, F' + (y + 1) \, F' ,$$

that is,

$$y \, F'' + 3 \, F' = 0 .$$

This differential equation has the solution

$$F'(y) = k \, y^{-3} \qquad (k > 0) .$$

It remains to determine a, b and k.

Suppose $b < 1$. Since $E(F, y) = 0$ for all $y \in (a, b)$ and is continuous for y in $[0,1]$, it follows that $E(F, b) = 0$, that is,

$$\int_a^b (x - b + bx) \, dF(x) = 0$$

If $a < x < b < 1$ then $x(1-b) < 1-b$, that is, $-1 + x < -b + bx$. Hence

$$\int_{a}^{b} (x - 1 + x)\ \mathrm{d}F(x) < 0 ,$$

and so $E(F, 1) < 0$, a contradiction against the optimality of F. Thus $b = 1$, and $E(F, 1) = 0$.

Now

$$E(F, 1) = \int_{a}^{1} (2x - 1)\ \frac{k}{x^3}\ \mathrm{d}x = 0 \quad (k > 0)$$

gives us
$$3a^2 - 4a + 1 = 0 .$$

Solving for a gives $a = 1$ or $a = 1/3$. Since $a = 1$ is impossible we deduce $a = 1/3$. Since F is a strategy,

$$\int_{a}^{1} \mathrm{d}F = \int_{\frac{1}{3}}^{1} kx^{-3}\ \mathrm{d}x = 1 ,$$

which leads to $k = 1/4$.

Hence an optimal strategy for either player is the continuous distribution F defined by

$$F'(x) = \begin{cases} 0 , & \text{if } x < \frac{1}{3} , \\ \dfrac{1}{4x^3} , & \text{if } x > \frac{1}{3} . \end{cases}$$

One can readily check that this *is* a solution. It can also be shown that in this case the solution is unique [5].

15. If a player knows that his opponent has fired and missed he will naturally hold his fire until time $t = 1$, when he is certain to hit. Thus if player 1 chooses x, and 2 chooses y with $x < y$, we see that 1 will win with probability x and lose with probability $1-x$. In this way we obtain

$$P(x, y) = \begin{cases} 2x - 1 , & \text{if } x < y , \\ 0 , & \text{if } x = y , \\ 1 - 2y , & \text{if } x > y , \end{cases}$$

and it can readily be verified that $x = y = \frac{1}{2}$ is a saddle point for this game.

CHAPTER 3.

1. Let $\mathbb{C} = \{ \mathbf{z} \in \mathbb{R}^n;\ \mathbf{z} = \mathbf{x}A,\ \mathbf{x} \in \mathbb{R}^m \}$. It is easy to verify that $\mathbb{C}$ is closed and convex. Put

$$M = \{ \; \mathbf{x} \in \mathbf{R}^m ; \mathbf{x} A = \mathbf{b} \; \},$$
$$N = \{ \; \mathbf{y} \in \mathbf{R}^t ; A \mathbf{y}^T = \mathbf{0} \text{ and } \mathbf{y} \mathbf{b}^T = 1 \; \}.$$

If $M \neq \phi$ and $N \neq \phi$ there exist $\mathbf{x} \in \mathbf{R}^m$, $\mathbf{y} \in \mathbf{R}^n$ such that

$$1 = \mathbf{y} \mathbf{b}^T = \mathbf{b} \mathbf{y}^T = (\mathbf{x} A) \mathbf{y}^T = \mathbf{x} (A \mathbf{y}^T) = 0,$$

a contradiction.

If $M = \phi$ then $\mathbf{b} \notin \mathbf{C}$, hence by Lemma 3.1 there exists $\mathbf{y} \in \mathbf{R}^n$ such that

$$\mathbf{y} \mathbf{b}^T > \mathbf{y} \mathbf{z}^T \text{ for every } \mathbf{z} \in \mathbf{C},$$

that is,

$$\mathbf{y} \mathbf{b}^T > \mathbf{y} \mathbf{z}^T = \mathbf{z} \mathbf{y}^T = \mathbf{x} A \mathbf{y}^T \text{ for every } \mathbf{x} \in \mathbf{R}^m.$$

Put $\mathbf{c}^T = A \mathbf{y}^T$. Then

$$\mathbf{y} \mathbf{b}^T > \mathbf{x} \mathbf{c}^T = \sum_{i=1}^{m} x_i c_i \text{ for every } \mathbf{x} \in \mathbf{R}^m.$$

If $\mathbf{c} \neq \mathbf{0}$, say $c_j \neq 0$, then taking $\mathbf{x} = (0, ..., \mathbf{y} \mathbf{b}^T / c_j, 0, ..., 0)$, where $\mathbf{y} \mathbf{b}^T / c_j$ is the jth component, we obtain $\mathbf{y} \mathbf{b}^T > \mathbf{y} \mathbf{b}^T$, a contradiction. Hence $\mathbf{c} = \mathbf{0}$. Let $\mathbf{y} \mathbf{b}^T = \mu > 0$ and put $\mathbf{y}_1 = \mu^{-1} \mathbf{y}$. Then

$$\mathbf{y}_1 \mathbf{b}^T = \mu^{-1} \mathbf{y} \mathbf{b}^T = 1$$

and

$$A \mathbf{y}_1^T = A (\mu^{-1} \mathbf{y}^T) = \mu^{-1} A \mathbf{y}^T = \mu^{-1} \mathbf{c}^T = 0.$$

Hence $N \neq \phi$. it follows that precisely one of M, N is empty.

2. If we can exhibit a solution of the inequalities

$$2 y_1 + y_2 \leq 0,$$
$$3 y_1 + 2 y_2 \leq 0,$$
$$-5 y_1 - 6 y_2 \leq 0,$$
$$\text{and } -4 y_1 - 3 y_2 > 0,$$

then by Theorem 3.2 the equations of the question have no non-negative solution. The above inequalities are equivalent to

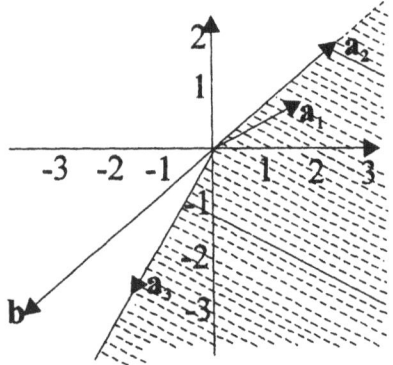

Figure S 16 Geometric interpretation for the equations of Q3-2

$$y_1 \leq -\frac{1}{2} y_2, \quad y_1 \leq -\frac{2}{3} y_2, \quad y_1 \geq -\frac{6}{5} y_2, \quad y_1 < -\frac{3}{4} y_2$$

which are satisfied by $y_1 = -1$, $y_2 = 1$.

One can see geometrically that the equations have no non-negative solution by writing them in the form

$$x_1 \mathbf{a}_1 + x_2 \mathbf{a}_2 + x_3 \mathbf{a}_3 = \mathbf{b} ,$$

where

$$\mathbf{a}_1 = \begin{pmatrix} 2 \\ 1 \end{pmatrix}, \quad \mathbf{a}_2 = \begin{pmatrix} 3 \\ 2 \end{pmatrix}, \quad \mathbf{a}_3 = \begin{pmatrix} -5 \\ -6 \end{pmatrix}, \quad \text{and } \mathbf{b} = \begin{pmatrix} -4 \\ -3 \end{pmatrix}.$$

The set of vectors $\mathbb{C} = \{ \Sigma x_i \mathbf{a}_i; \mathbf{x} \geq \mathbf{0} \}$ is represented by the shaded region $\mathbb{C}$ in the figure.

3. The dual of the given program is

$$\text{Maximise } 3 y_1 + 6 y_2 + 2 y_3 + 2 y_4$$

subject to

$$
\begin{aligned}
y_1 + 3 y_2 \quad\quad + y_4 &\leq 8 , \\
2 y_1 + y_2 \quad\quad + y_4 &\leq 6 , \\
y_2 + y_3 \quad\quad &\leq 3 , \\
y_1 + y_2 + y_3 \quad\quad &\leq 6 , \\
\end{aligned}
$$

and $y_j \geq 0 \ (1 \leq j \leq 4)$

If $\mathbf{x} = (1, 1, 2, 0)$ is a solution for the original program then since $x_1 > 0, x_2 > 0, x_3 > 0$ (3.25) shows that for any solution $\mathbf{y} = (y_1, y_2, y_3, y_4)$ of the dual, the first three inequalities above are actually equalities. Moreover since $x_1 + x_3 = 3 > 2$, in the fourth constraint, (3.26) shows that $y_4 = 0$. Thus to find a solution for the dual it suffices to solve

$$
\begin{aligned}
y_1 + 3 y_2 &= 8 , \\
2 y_1 + y_2 &= 6 , \\
y_2 + y_3 &= 3 .
\end{aligned}
$$

This yields $\mathbf{y} = (2, 2, 1, 0)$ as a solution for the dual, provided $\mathbf{x} = (1, 1, 2, 0)$ was a solution for the primal. To check we compute

$$\mathbf{x} \mathbf{c}^T = (1, 1, 2, 0)(8, 6, 3, 6)^T = 20 = (2, 2, 1, 0)(3, 6, 2, 2)^T = \mathbf{y} \mathbf{b}^T .$$

Hence $\mathbf{x}$ and $\mathbf{y}$ are solutions by Corollary 3.1.

4. The dual of the given program is

$$\text{Minimise } 8 x_1 + 6 x_2 + 6 x_3 + 9 x_4$$

subject to

$$x_1 + 2x_2 \qquad + \; x_4 \geq 2 \, ,$$
$$3x_1 + \; x_2 + \; x_3 + \; x_4 \geq 4 \, ,$$
$$x_3 + \; x_4 \geq 1 \, ,$$
$$x_1 \qquad + \; x_3 \qquad \geq 1 \, ,$$

and $x_j \geq 0 \; (1 \leq j \leq 4)$

Assume that $y = (2, 2, 4, 0)$ is a solution, then, since $y_1 = 2 > 0$, $y_2 = 2 > 0$, $y_3 = 4 > 0$, (3.26) implies that for any solution $x = (x_1, x_2, x_3, x_4)$ of the dual the first three inequalities above are actually equalities. Moreover, since $y_1 + y_2 + y_3 = 8 < 9$, in the fourth constraint, (3.25) implies $x_4 = 0$. Thus it suffices to solve

$$x_1 + 2x_2 \qquad = 2 \, ,$$
$$3x_1 + \; x_2 + x_3 = 4 \, ,$$
$$x_3 = 1 \, .$$

This yields $x = (^4/_5, \, ^3/_5, \, 1, \, 0)$ as a solution for the dual, provided $y = (2, 2, 4, 0)$ was optimal for the primal. To check we compute

$$x\,c^T = (\tfrac{4}{5}, \tfrac{3}{5}, 1, 0)\,(8, 6, 6, 9)^T = 16 = (2, 2, 4, 0)\,(2, 4, 1, 1)^T = y\,b^T \, .$$

Hence x and y are solutions by Corollary 3.1.

5. The manager's problem amounts to the linear program

Minimise $x_1 + x_2 + x_3 + x_4 + x_5 + x_6$
subject to

$$x_1 \qquad\qquad\qquad\qquad + \; x_6 \geq 2 \, ,$$
$$x_1 + x_2 \qquad\qquad\qquad\qquad \geq 10 \, ,$$
$$x_2 + x_3 \qquad\qquad\qquad \geq 14 \, ,$$
$$x_3 + x_4 \qquad\qquad \geq 8 \, ,$$
$$x_4 + x_5 \qquad \geq 10 \, ,$$
$$x_5 + x_6 \geq 3 \, ,$$

and $x_j \geq 0 \; (1 \leq j \leq 6)$

The dual is

Maximise $2y_1 + 10y_2 + 14y_3 + 8y_4 + 10y_5 + 3y_6$
subject to

$$
\begin{aligned}
y_1 + y_2 &\leq 1 , \\
y_2 + y_3 &\leq 1 , \\
y_3 + y_4 &\leq 1 , \\
y_4 + y_5 &\leq 1 , \\
y_5 + y_6 &\leq 1 , \\
y_1 \qquad\qquad\qquad + y_6 &\leq 1 ,
\end{aligned}
$$

and $y_j \geq 0 \ (1 \leq j \leq 6)$

Proceeding as in the previous two questions we easily find the solutions $\mathbf{y} = (1, 0, 1, 0, 1, 0)$ for the dual and

$$
\begin{aligned}
\mathbf{x}\,\mathbf{c}^T &= (0, 14, 0, 8, 2, 2)\,(1, 1, 1, 1, 1, 1)^T = 26 \\
&= (1, 0, 1, 0, 1, 0)\,(2, 10, 14, 8, 10, 3)^T = \mathbf{y}\,\mathbf{b}^T ,
\end{aligned}
$$

Hence $\mathbf{x}$ is indeed optimal.

6. Let $\mathbf{p} = (^1/_m, \, ^1/_m, \, \dots, \, ^1/_m)$ then by (iv)

$$
P(\mathbf{p}, \tau_j) = \frac{1}{m} \sum_{i=1}^{m} a_{ij} > q , \quad \text{for all } j , \quad 1 \leq j \leq m .
$$

Hence the value of the game is $v > q \geq 0$.

Suppose that $\mathbf{p} = (x_1{}^0, \, \dots, \, x_m{}^0)$ is an optimal strategy for player 1 with $x_j{}^0 = 0$ for some j. Then

$$
P(\mathbf{p}, \tau_j) = \sum_{i=1}^{m} x_i{}^0 a_{ij} = \sum_{i \neq j} x_i{}^0 a_{ij} \leq \sum_{i=1}^{m} x_i{}^0 q < v ,
$$

by (iii), hence $\mathbf{p}$ cannot be optimal. Thus every optimal strategy for player 1 uses every pure strategy with positive probability.

Suppose now that $\mathbf{q} = (y_1{}^0, \, \dots, \, y_m{}^0)$ is an optimal strategy for player 2. From Theorem 2.9 (ii) we have

$$
P(\sigma_i, \mathbf{q}) = \sum_{j=1}^{m} a_{ij}\, y_j{}^0 = v , \quad \text{for all } i, \quad 1 \leq i \leq m .
$$

If $y_k{}^0 = 0$ then for $i = k$ we have;

$$
P(\sigma_k, \mathbf{q}) = \sum_{j=1}^{m} a_{kj}\, y_j{}^0 = \sum_{j \neq k} a_{kj}\, y_j{}^0 \leq q < v , \quad \text{by (iii)} ,
$$

a contradiction. Thus every optimal strategy for player 2 uses every pure strategy with positive probability.

It follows that the game is completely mixed, and as a corollary every Minkowski-Leontief matrix is non-singular, by Corollary 3.5.

7. To find the complete solution of the game we use the Shapley-Snow procedure as in

Example 3.6. This gives the solution

$$p = \lambda \, (1, 0, 0) + (1 - \lambda) \, (\tfrac{1}{3}, \tfrac{2}{3}, 0) \quad (0 \le \lambda \le 1) \, ,$$

$$q = \mu \, (1, 0, 0) + (1 - \mu) \, (\tfrac{1}{2}, \tfrac{1}{2}, 0) \quad (0 \le \mu \le 1) \, ,$$

$$v = 1 \, .$$

8. $x = (0, \, {}^{16}/_{29}, \, 0, \, {}^{66}/_{29}, \, 0, \, {}^{11}/_{29})$,   Value $= - \, {}^{3}/_{29}$.

9. $x = (0, \, {}^{32}/_{27}, \, 0, \, {}^{20}/_{27})$,   Value $= {}^{52}/_{27}$.

10. $x = ({}^{11}/_{2}, \, {}^{9}/_{4}, \, 7)$,   Value $= 47$.

CHAPTER 4.

1. a) $v(\{1\}) = 1, \; v(\{2\}) = 0, \; v(\{3\}) = -1,$
        $v(\{1,2\}) = 3, \; v(\{1,3\}) = 2, \; v(\{2,3\}) = 1,$
        $v(\{1,2,3\}) = 4.$

b)   For the game on the right we
have $v(\{1\}) = v(\{2\}) = 1$ and
$v(\{1,2\}) = 2$ so (4.13) is satisfied,
but the game is not constant sum.

		{2}	
		(1,1)	(1,0)
{1}		(0,1)	(1,1)

2. By Theorem 4.10 if **x, y** are imputations in the core then

$$\sum_{i \in S} x_i \ge v(S) \, , \quad \sum_{i \in S} y_i \ge v(S) \text{ for every } S \subseteq I \, .$$

Suppose **z** is an imputation which dominates $\lambda \mathbf{x} + (1-\lambda)\mathbf{y}$, $0 \le \lambda \le 1$, through S. Then

$$z_i > \lambda \, x_i + (1 - \lambda) \, y_i \text{ for every } i \in S \, .$$

Hence

$$\sum_{i \in S} z_i > \lambda \sum_{i \in S} x_i + (1 - \lambda) \sum_{i \in S} y_i \ge v(S) \, .$$

Thus **z** is not effectively realisable by S, that is it does not satisfy (4.25), and so canno
dominate any imputation through S, a contradiction. Hence $\lambda \mathbf{x} + (1-\lambda)\mathbf{y}$, $0 \le \lambda \le 1$, i
undominated, and so the core is convex.

Theorem 4.10 specifies the core as the intersection of a finite number of closed half-spaces. Fo
any classical cooperative game one can always determine if the core is empty by applying the

phase-1 simplex algorithm of Chapter 3 which tests the feasibility of a linear program. (In general, Theorem 4.10 supplies a great number of redundant inequalities.) If the core is non-empty one can actually determine all of its vertices by similar algorithms.

3. All points **y** in the upper shaded region, not on the interior boundary lines of the region (dotted) satisfy (4.26) with respect to $S = \{1,2\}$. Similarly for $S = \{2,3\}$ and $S = \{1,3\}$. We conclude that *if* **x** is effectively realisable by one of these coalitions it will dominate all imputations in the corresponding shaded region.

Suppose now the core intersects every bounding line of the imputations set. Then the core contains each line segment

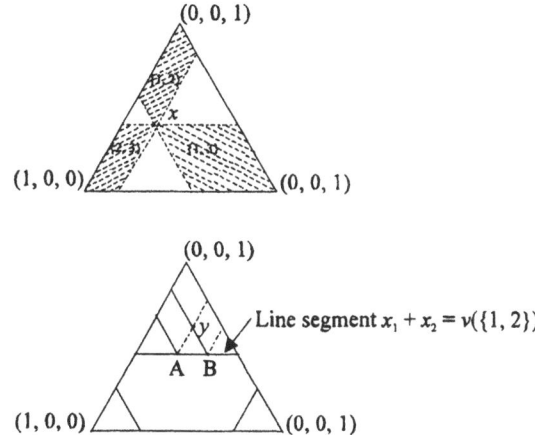

Figure S 17 Diagrams for Q4-3.

$$x_1 + x_2 = v(\{1,2\}), \quad 0 \leq x_1 \leq v(\{1,2\}),$$
$$x_1 + x_3 = v(\{1,3\}), \quad 0 \leq x_1 \leq v(\{1,3\}),$$
$$x_2 + x_3 = v(\{2,3\}), \quad 0 \leq x_2 \leq v(\{2,3\}).$$

Suppose **y** is not in the core, then by Theorem 4.10 there exists $S \subsetneq I$ such that

$$\sum_{i \in S} y_i < v(S)$$

and necessarily $S \neq \{i\}$ or I. Without loss of generality suppose $S = \{1,2\}$. Then by the first part of the question any point on the line segment AB (above) is in the core and dominates **y** through $\{1,2\}$. Algebraically, for points on the line $x_1 + x_2 = v(\{1,2\})$ we have

$$x_1 + x_2 = v(\{1,2\}) > y_1 + y_2,$$

hence $x_1 > y_1$ or $x_2 > y_2$. Suppose for example $x_1 > y_1$ and $x_2 \leq y_2$, then there exists $\theta > 0$ such that

$$x_1 - y_1 > \theta > y_2 - x_2.$$

Hence $x_1 - \theta > y_1$, $x_2 + \theta > y_2$ so that

$$(x_1 - \theta, x_2 + \theta, 1 - v(\{1,2\}))$$

is in the core and dominates **y** through $\{1,2\}$. Thus every point not in the core is dominated by some point which is in the core. Since any stable set must contain the core by Theorem 4.12, it follows that the core is the *unique* stable set, since any larger set cannot be internally stable.

4. The elements of J are

$$\gamma_1 = (0, \frac{1}{3}, \frac{1}{3}, \frac{1}{3}),$$

$$\gamma_2 = (\frac{1}{3}, 0, \frac{1}{3}, \frac{1}{3}),$$

$$\gamma_3 = (\frac{1}{3}, \frac{1}{3}, 0, \frac{1}{3}),$$

$$\gamma_4 = (\frac{1}{3}, \frac{1}{3}, \frac{1}{3}, 0),$$

and by complementarity the characteristic function for the 3-person coalitions is $v(\{1,2,3,4\})\backslash\{i\} = 1$ $(1 \le i \le 4)$.

Now γ_i has only one coordinate larger than γ_j ($j \ne i$), namely the jth one, so if γ_i dominates γ_j through S we must have $S = \{j\}$. But domination through a 1-person coalition is never possible. Hence no element of J can dominate another.

J is *not* a stable set, since no element of J dominates the imputation $y = (\frac{1}{2}, \frac{1}{2}, 0, 0)$. To see this we argue as follows.

Domination can only occur through a 2- or 3- person coalition S. Since we require $\gamma_{kl} > y_l$ for every $l \in S$, obviously $l = 3$ or 4 $(y_l = 0)$ and $k = 1$ or 2 $(\gamma_{kl} = \frac{1}{3})$. Thus the only coalition which might possibly provide a dominance is $S = \{3,4\}$, and we must look to the vectors γ_1 or γ_2. But

$$\gamma_{13} + \gamma_{14} = \gamma_{23} + \gamma_{24} = \frac{2}{3} > \frac{1}{2} = v(\{3,4\}),$$

so that neither γ_1 nor γ_2 is effectively realisable by $S = \{3,4\}$.

Hence J is not externally stable.

5. Suppose V is a stable set and $x = (x_1, x_2, ...) \in V$. Choose $j \in \mathbb{N}$ such that $x_j > 0$. Let y be an imputation with $y_j < x_j$ and

$$y_i > x_i \text{ for all } i \in \mathbb{N}\backslash\{j\} \tag{1}$$

Then $y \notin V$, since y dominates x through $\mathbb{N}\backslash\{j\}$. Since V is stable, y must be dominated by some element of V. Suppose $z \in V$ and z dominates y through S so that

$$z_i > y_i \text{ for all } i \in S \tag{2}$$

and

$$v(S) \ge \sum_{i \in S} z_i > \sum_{i \in S} y_i \ge 0.$$

Hence $v(S) > 0$, that is $v(S) = 1$, and so $\mathbb{N}\backslash S$ is finite. Since $\mathbb{N}\backslash S$ is finite, so is $\mathbb{N}\backslash(S\backslash\{j\})$ but this means

$$\sum_{i \in S\backslash\{j\}} z_i \le v(S\backslash\{j\}) = 1$$

and by (1) and (2),

$$z_i > y_i > x_i \text{ for every } i \in S\backslash\{j\} \ .$$

Hence $z \in V$ dominates $x \in V$ through $S \backslash \{j\}$, which is a contradiction against the internal stability of V. It follows that there is no stable set for this game.

6. We begin by observing that for *any* n-person game if $v(S)/|S| \le v(I)/|I|$ for every $S \subseteq I$ then

$$x_0 = \left(\frac{v(I)}{|I|}, \ \dots, \ \frac{v(I)}{|I|} \right)$$

is an imputation and is in the core. For if $x \succ_S x_0$, then $x_i > v(I)/|I|$ for every $i \in S$, and so

$$\sum_{i \in S} x_i > |S| \ \frac{v(I)}{|I|} \ge v(S)$$

which contradicts (4.25). Hence x_0 is undominated and so in the core.

Suppose, for a symmetric game, there exists $S \subset I$ such that $v(S)/|S| > v(I)/|I|$. For any imputation x consider the $|S|$ smallest components of x. Since the game is symmetric we may, without loss of generality, take S to be the coalition which corresponds to the $|S|$ smallest components of x. We then have

$$\frac{1}{|S|} \sum_{i \in S} x_i \le \frac{1}{|I|} \sum_{i \in I} x_i = \frac{v(I)}{|I|}$$

Hence

$$\sum_{i \in S} x_i \le |S| \ \frac{v(I)}{|I|} < v(S) \ ,$$

and because

$$\sum_{i \in I\backslash S} x_i = v(I) - \sum_{i \in S} x_i \ge v(S) - \sum_{i \in S} x_i$$

we can construct an imputation y with $y_i = x_i + (v(S) - \sum_{i \in S} x_i) / |S|$ for $i \in S$ such that $y \succ_S x$.

CHAPTER 5.

1. The row player's game is

$$\begin{pmatrix} -1 & 1 \\ 2 & -2 \end{pmatrix} \quad \text{(Payoffs to player 1)}$$

which has the optimal strategy $(^2/_3, \ ^1/_3)$ and value 0.

The column player's game

$$\begin{pmatrix} -1 & 1 \\ -2 & 2 \end{pmatrix} \quad (\text{Payoffs to player 2})$$

has a saddle point in the top right corner giving the value as 1. Thus the security levels are 0 and 1 respectively.

Shapley Solution The security status quo $(u^*, v^*) = (0, 1)$. To find the Shapley solution we seek the point $(\bar{u}, \bar{v})$ which maximises $(u - u^*)(v - v^*)$ on the line segment joining $(-2, 2)$ and $(1, 1)$, this is on the line $v = (-^1/_3)u + \ ^4/_3$ $(0 \le u \le 1)$. This gives $(\bar{u}, \bar{v}) = (^1/_2, \ ^7/_6)$ as the Shapley solution.

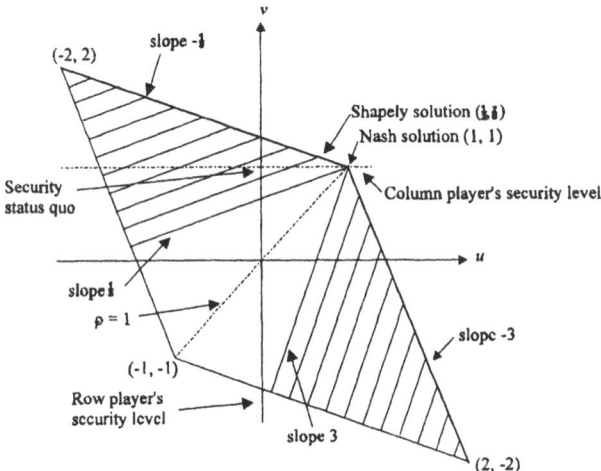

Figure S 18 Diagram for Q5-1.

Nash Solution The Pareto optimal boundary consists of the two line segments joining $(-2, 2)$ $(1, 1)$ and $(1, 1)$ $(2, -2)$ respectively. The first has slope $-^1/_3$. Taking $\rho = \ ^1/_3$ we have to solve the zero sum game $^1/_3 A - B$, that is

$$\begin{pmatrix} \dfrac{2}{3} & -\dfrac{2}{3} \\[2mm] \dfrac{8}{3} & -\dfrac{8}{3} \end{pmatrix}$$

which has a saddle point in the top right corner. This corresponds to the threat strategies (σ_1, τ_2) and leads to the Nash solution $(\bar{u}, \bar{v}) = (1, 1)$! As a matter of interest let us look at the second line segment with slope -3. Taking $\rho = 3$ we obtain the game

$$\begin{pmatrix} -2 & 2 \\ 8 & -8 \end{pmatrix}$$

which also has a saddle point corresponding to (σ_1, τ_2) giving the same Nash solution, of course.

Finally we observe that for the line through $(1, 1)$ with slope 1 the associated game $\rho A - B$ is

$$\begin{pmatrix} 0 & 0 \\ 4 & -4 \end{pmatrix}$$

which again has a saddle point at (σ_1, τ_2). But this line passes through (-1, -1), corresponding to (σ_1, τ_1). Hence τ_1 is also a best threat strategy for player 2, and so is any probability combination of τ_1 and τ_2.

The fact is that for player 2 *any* mixed strategy is equally ineffective against player 1's best threat σ_1. Player 1 is in a slightly better position and can use the threat σ_1 to enforce a marginally better payoff than he or she would obtain from the Shapley solution.

2. a) *For the row player:*

$$\begin{array}{c} \quad\quad \tau_1 \;\; \tau_2 \\ \begin{array}{c} \sigma_1 \\ \sigma_2 \end{array} \begin{pmatrix} 2 & 3 \\ 1 & 6 \end{pmatrix} \end{array} \quad (\text{Payoffs to player 1})$$

which has a saddle point at (σ_1, τ_1).

For the column player:

$$\begin{array}{c} \quad\quad \tau_1 \;\; \tau_2 \\ \begin{array}{c} \sigma_1 \\ \sigma_2 \end{array} \begin{pmatrix} 5 & 2 \\ 0 & 6 \end{pmatrix} \end{array} \quad (\text{Payoffs to player 2})$$

which has a saddle point at (σ_2, τ_1).

Hence the security status quo is $(u^*, v^*) = (2, 1)$.

Shapley solution We maximise $(u - 2)(v - 1)$ on the line $v = -u + 7$ $(2 \le u \le 6)$ to obtain $(\bar{u}, \bar{v}) = (4, 3)$.

Nash solution Since the Pareto optimal boundary is the single line segment joining (2, 5), (6, 1) having slope -1 we take $\rho = 1$ and consider the game $\rho A - B$ with matrix

$$\begin{pmatrix} -3 & 1 \\ 1 & 5 \end{pmatrix}$$

which has a saddle point at (σ_2, τ_1). Thus the threat status quo is $(u^*, v^*) = (1, 0)$. The line through (1, 0) with slope $\rho = 1$ cuts the Pareto optimal boundary at

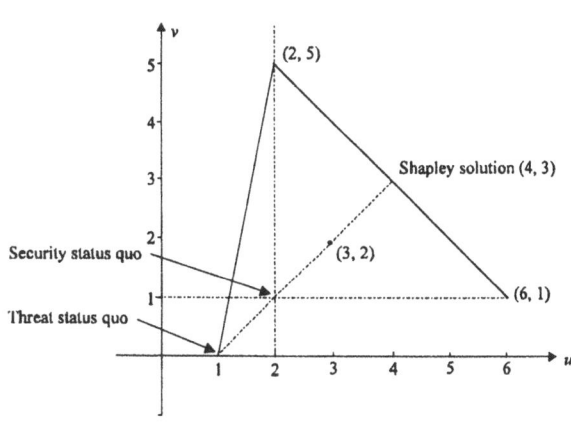

Figure S 19 Diagram for Q5-2.

(4, 3). Hence for this game the Nash solution is the same as the Shapley solution.

b) As a non-cooperative game it presents some difficulties. There are three equilibria:

$$(\sigma_1, \tau_1) \text{ giving payoffs } (2, 5)$$
$$(\sigma_2, \tau_2) \text{ giving payoffs } (6, 1)$$

and a mixed strategy equilibrium point

$$((^1/_4, {}^3/_4), (^3/_4, {}^1/_4)) \text{ giving payoffs } (^9/_4, {}^5/_4).$$

If both players go for their best equilibrium, the outcome is (σ_2, τ_1) giving payoffs (1, 0), the threat status quo, which is well below both players' security level. If player 2 uses the equilibrium strategy $(^3/_4, {}^1/_4)$ and player 1 uses $(x, 1\text{-}x)$, the payoff to player 2 is

$$P_2 = 4x + \frac{1}{4} ,$$

which means if player 1 plays $x = 0$ then player 2 gets $^1/_4$, well below that player's security level. Is this rational for player 2? Similarly if player 1 uses $(^1/_4, {}^3/_4)$ and player 2 $(y, 1\text{-}y)$, the payoff to player 1 is

$$P_1 = \frac{21}{4} - 4y .$$

Thus if player 2 plays $y = 1$, the payoff to player 1 is $^5/_4$, again below this player's security level.

On the other hand if both players 'play safe' and adopt their security level strategies the outcome is (σ_1, τ_2) giving payoffs of (3, 2), above the security level of both players. On the whole this seems to be the preferred solution, but of course it is *not* an equilibrium point.

3. *For the row player:*

$$\begin{pmatrix} 1 & -a \\ -c & 0 \end{pmatrix} \quad (\text{ Payoffs to player 1 })$$

Solving $(0 < a, c < 1)$ we obtain

Optimal strategy for player 1: $\dfrac{c}{1 + a + c}, \dfrac{1 + a}{1 + a + c}$,

Security level for player 1: $\dfrac{-ac}{1 + a + c}$.

For the column player:

$$\begin{pmatrix} 0 & -b \\ -d & 1 \end{pmatrix} \quad (\text{ Payoffs to player 2 })$$

Solving ($0 < b, d < 1$) we obtain

Optimal strategy for player 2: $\dfrac{1 + b}{1 + b + d}, \dfrac{d}{1 + b + d}$,

Security level for player 2: $\dfrac{-bd}{1 + b + d}$.

Shapley solution Taking the security status quo as

$$(u^*, v^*) = \left(\dfrac{-ac}{1 + a + c}, \dfrac{-bd}{1 + b + d} \right)$$

we maximise $(u - u^*)(v - v^*)$ on the line segment $v = -u + 1$ ($0 \le u \le 1$). This gives the Shapley solution as

$$\bar{u} = \dfrac{1}{2} + \dfrac{1}{2} \left(\dfrac{bd}{1 + b + d} - \dfrac{ac}{1 + a + c} \right),$$

$$\bar{v} = \dfrac{1}{2} + \dfrac{1}{2} \left(\dfrac{ac}{1 + a + c} - \dfrac{bd}{1 + b + d} \right).$$

We observe that

$$\dfrac{\partial \bar{u}}{\partial a} = \dfrac{-c(1 + c)}{(1 + a + c)^2} < 0, \quad \dfrac{\partial \bar{u}}{\partial b} = \dfrac{d(1 + d)}{(1 + b + d)^2} > 0,$$

$$\dfrac{\partial \bar{u}}{\partial c} = \dfrac{-a(1 + a)}{(1 + a + c)^2} < 0, \quad \dfrac{\partial \bar{u}}{\partial d} = \dfrac{b(1 + b)}{(1 + b + d)^2} > 0.$$

As threats are not relevant to the Shapley solution we interpret these relationships to mean that player 1's payoff in the arbitrated Shapley solution is a monotonic increasing function of player 1's security level and a monotonic decreasing function of player 2's security level. Similarly for player 2.

Nash solution The Pareto optimal boundary is a single line segment of slope -1. Taking $\rho = 1$ the game $\rho A - B$ is

$$\begin{pmatrix} 1 & b - a \\ d - c & -1 \end{pmatrix}$$

which because of our assumption that $0 < a, b, c, d < 1$ has a saddle point at (σ_1, τ_2). This leads to the Nash solution

$$\bar{u} = \dfrac{1}{2} + \dfrac{1}{2}(b - a),$$

$$\bar{v} = \dfrac{1}{2} + \dfrac{1}{2}(a - b).$$

Note that this depends on a and b, and the more b exceeds a the better for player 1. This is a reflection of the fact that the more b exceeds a the more effective is player 1's threat of σ_1.

4. We conclude that there is no algorithm consistsing of rational operations on the payoff entries which will compute optimal threat strategies in a finite number of steps.

INDEX

A

adjoint 128, 129
admissible 3, 4, 55, 60, 69, 70, 164
antagonistic game 222
ante 15, 41, 42, 82, 86, 239
arbitration scheme 215
Arrow 139, 140, 208
artificial variables 147, 149, 150, 153, 154
Axiom of choice 52, 53

B

badly approximable numbers 36, 45
bargaining model 215, 221, 223, 234-236
bargaining solution 215
battle of the sexes 232
Bellman 38, 45, 81, 86, 99
bimatrix game 10, 48, 60-62, 222-224, 258
binary game 35, 52
Blackwell 81, 86, 99
blotto game 92, 99
Bott 194, 208
Brouwer's fixed point theorem 59, 237, 238

C

Cassel 100, 114, 160
characteristic function 165-167, 169-171, 173, 179-182, 185, 187, 191, 197, 198, 202, 203, 205, 206, 235, 274
chess 1, 3, 8, 12, 14, 15, 31, 161, 239
cofactor 128
commodities 113, 114, 195, 197
complementarity 173, 181, 182, 274
completely mixed matrix game 133
concave 63, 64, 97, 197, 201
concave function 197, 201
constant sum game 164, 169, 175, 176, 181, 182, 186-189, 191, 193, 198, 206, 207
continuous games 64, 87, 92, 93
contract curve 195, 196
convex function 63, 93
convex hull 77, 120, 171, 212, 220

H

I

K

L

M

Z

MATHEMATICS TEACHING PRACTICE:
Guide for University and College Lecturers

JOHN H. MASON
Faculty of Mathematics & Computing, The Open University

ISBN 1-898563-79-9   408 pages   £30

Contents: Student difficulties with mathematics, statistics, modelling: techniques, concepts and logic • Lecturing: structure, employing screens, tactics, issues of interest and stimulus, repeating lectures, handouts, encouraging exploration and inventive exploration, support • Tutoring: atmosphere, debate, questions, worked examples, assent-assert, tactics, advising tutorials • Constructing tasks: purposes, aims, intentions, summary, differing tasks and purposes, mathematics forms, objectives, challenge and routine tasks, student propensities, reproducing oneself • Marking: allocating marks, student feedback • Using History: pictures, dates and potted biographies • Teaching issues, concerns, tensions, informing teaching, mathematical themes • Index of teaching tactics • Illustrations for teaching topics and projects.

REVIEW
Teaching in Higher Education: "A valuable, interesting, lively contribution in teaching mathematics and **useful also for those who teach mathematics in other disciplines**. Set out in an accessible way, it should be on the shelves of all involved. (Dr J Wilkinson Electrical Engineering Department Glasgow University

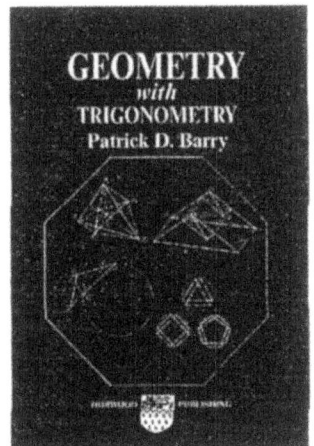

GEOMETRY with TRIGONOMETRY

PATRICK D. BARRY
Professor of Mathematics, National University of Ireland, Cork

ISBN 1-898563-69-1   235 pages   £30

Contents: Preliminaries: Basic shapes of geometry • Distance, degree measure of angles • Congruence of triangles • Parallel lines • The parallel axiom • Euclidean geometry • Cartesian coordinates • Applications • Circles and basic properties • Translations and basic properties • Isometries • Trigonometry • Sine and cosine • Addition formulae • Complex coordinates • Sensed angles • Rotations • Applications to circles • Angles between lines • Position vectors • Vector and complex number methods in geometry • Trigonometric functions in calculus • Mensuration of discs and circles.

REVIEW
The Mathematical Gazette (Mathematical Association): "Geometrical thinking is still very much alive. Dr Barry's pragmatic approach in a combination of Euclidean treatment of geometry is pitched at university degree level. Students who take trouble to become familiar with the tools introduced, will find the spirit of geometry logically unfolded."

MATHEMATICS: A Second Start, 2nd Edition

SHEILA PAGE, Mathematics Lecturer, University of Bradford
JOHN BERRY, Mathematics Education Professor, University of Plymouth
HOWARD HAMPSON, Head of Mathematics, Torquay Further Education College

ISBN 1-898563-04-7   470 pages   £20

Provides less mathematically minded students with a gentle introduction to algebra, trigonometry, calculus and statistics, and updated in a variety of disciplines, it still manages to combine clarity of presentation with liveliness of style and sympathy for students' needs. Straightforward, pragmatic and packed full of illustrative examples, exercises and self-test questions for self-study.

Contents: Using a scientific calculator • The set of real numbers; Number skills • Algebra: a basic toolkit • Algebra: More tools • Products • Factors: equations and inequalities • Manipulation of formulae • Quadratic equations • Simultaneous equations • Indices and logarithms • Functions and graphs • Linear graphs and their use in experimental work • Calculus - the mathematics of change • Applications of differentiation • Integration • Introduction to trigonometry • Calculus of trigonometry • The law of natural growth • Some applications of logarithms • A first look at statistics • Probability • Probability distributions

FUNDAMENTALS OF UNIVERSITY MATHEMATICS 2nd Edition

COLIN McGREGOR, JOHN NIMMO and WILSON STOTHERS, Glasgow University

ISBN 1-898563-10-1   560 pages   £27.50

Contents: Preliminaries • Functions and inverse functions • Polynomials & rational functions • Induction • The binomial theorem • Trigonometry • Complex numbers • Limits and continuity • Differentiation fundamentals • Differentiation applications • Curve sketching matrices and linear equations • Vectors and three dimensional geometry • Products of vectors • Integration fundamentals • Logarithms and exponentials • Integration methods and applications • Ordinary differential equations • Sequences and series • Numerical methods • Appendices.

REVIEW

Mathematics Today (Institute of Mathematics & Applications): "A sound beginning to mathematics degrees, and a reference book for anyone who uses maths regularly. If you seek a first year university text, certainly look at this one. I wish that I had a copy as an undergraduate." (Dr. Rochead, Defence Science & Technological Laboratory, Chertsey)

The Mathematical Gazette (Mathematical Association): "This book sits firmly on the university side of the difficult and tricky borderline between school and university mathematics. Definitions, theorems and corollaries, figures and notes carefully numbered for reference. Theorems are proved clearly in modern format and helpful, plentiful figures. If you are looking for a first year university text you should carefully look at this, a unifier of mathematical ideas at this level. I found it most valuable, and saw mathematics at this level more clearly as a single subject and less as a disparate collection of topics."

NETWORKS AND GRAPHS:
Techniques and Computational Methods

DAVID K. SMITH,
School of Mathematical Sciences, University of Exeter

ISBN 1-898563-91-8   200 pages   £25

Dr Smith here presents essential mathematical and computational ideas of network optimisation for senior undergraduate and postgraduate students in mathematics, computer science and operational research. He shows how algorithms can be used for finding optimal paths and flows, identifying trees in networks, and optimal matching. Later chapters discuss postman and salesperson tours, and demonstrate how many network problems are related to the "minimal-cost feasible-flow" problem. Techniques are presented both informally and with mathematical rigour, and aspects of computation, especially of complexity, have been included. Numerous examples and diagrams illustrate the techniques and applications. Problem exercises with tutorial hints.

Contents: Ideas and definitions • Spanning trees • Shortest path problems • Flow problems • How to store details of a network in a computer • Minimal-cost, feasible-flow problems • Matchings • Postman problems • Salesperson problems • Future research

MATHEMATICAL MODELLING:
A Way of Life ICTMA 11

S.J. LAMON, Marquette University, Milwaukee
W.A. PARKER, University of Ulster
S.K. HOUSTON, University of Ulster

ISBN 1-904275-03-6   *ca* 400 pages   £45

This book encourages teachers to help students to model a variety of real phenomena appropriate to mathematical backgrounds and interests. Many students have difficulty studying mathematical modelling because of their beliefs about what it means to do mathematics. Without prior experience in building, interpreting and applying mathematical models, many may never come to view and regard modelling as a "way of life."

Contents: Mathematical modelling in the elementary school • Modelling with middle and secondary students • Post-secondary modelling • Research perspectives • Reflections on the 20th anniversary of ICTMA

Lightning Source UK Ltd.
Milton Keynes UK
UKOW06f1211191115

262898UK00033B/40/P